Physical Constants

Description	Symbol	Value
Elementary charge	q	1.60218×10^{-19} C
Electron volt	eV	1.60218×10^{-19} J
Boltzmann's constant	k	1.38066×10^{-23} J/K
Free electron mass	m_0	9.1095×10^{-31} kg
Permittivity in vacuum	ε_0	8.85418×10^{-12} F/m
Planck's constant	h	6.62617×10^{-34} J·s
Reduced Planck's constant ($h/2\pi$)	$\hbar$	1.05458×10^{-34} J·s
Speed of light in vacuum	c	2.99792×10^{8} m/s
Thermal voltage at T = 300 K	kT/q	0.0259 V
Wavelength of 1-eV photon	λ	1.23977 μm

Unit Conversion

Quantity	Symbol	Value/Dimension
Meter	m	1 m = 10^2 cm
Millimeter	mm	1 mm = 10^{-1} cm = 10^{-3} m
Micrometer, micron	μm	1 μm = 10^4 Å = 10^3 nm = 10^{-4} cm
Nanometer	nm	1 nm = 10 Å = 10^{-3} μm = 10^{-7} cm
Angstrom	Å	1 Å = 10^{-4} μm = 10^{-8} cm = 10^{-10} m
Electron volt	eV	1 eV = 1.60218×10^{-19} J
Electric charge (Coulomb)	C	A·s
Current (Ampere)	A	C/s
Frequency (Hertz)	Hz	1/s
Energy (Joule)	J	N·m
Power (Watt)	W	J/s
Potential (Volt)	V	J/C
Conductance (Siemens)	S	A/V
Resistance (Ohm)	Ω	V/A
Capacitance (Farad)	F	C/V

Modern Semiconductor Devices for Integrated Circuits

Modern Semiconductor Devices for Integrated Circuits

- **CHENMING CALVIN HU**
  University of California, Berkeley

Prentice Hall

Upper Saddle River Boston Columbus San Francisco New York
Indianapolis London Toronto Sydney Singapore Tokyo Montreal
Dubai Madrid Hong Kong Mexico City Munich Paris Amsterdam Cape Town

VICE PRESIDENT AND EDITORIAL DIRECTOR, ECS: Marcia J. Horton
ASSOCIATE EDITOR: Alice Dworkin
EDITORIAL ASSISTANT: William Opaluch
DIRECTOR OF TEAM-BASED PROJECT MANAGEMENT: Vince O'Brien
SENIOR MANAGING EDITOR: Scott Disanno
PRODUCTION LIAISON: Jane Bonnell
PRODUCTION EDITOR: Debjani Gill, Laserwords
SENIOR OPERATIONS SPECIALIST: Alan Fischer
OPERATIONS SPECIALIST: Lisa McDowell
MARKETING MANAGER: Tim Galligan
MARKETING ASSISTANT: Mack Patterson
ART DIRECTOR: Kenny Beck
COVER DESIGNER: Jason Hu
INTERIOR DESIGNER: Laura Ierardi
ART EDITOR: Greg Dulles
MEDIA EDITOR: Daniel Sandin
MEDIA PROJECT MANAGER: Danielle Leone
COMPOSITION/FULL-SERVICE PROJECT MANAGEMENT: Laserwords Private Limited, Chennai, India

---

Copyright © 2010 Pearson Higher Education. Upper Saddle River, New Jersey 07458.
All rights reserved. Manufactured in the United States of America. This publication is protected by Copyright and permissions should be obtained from the publisher prior to any prohibited reproduction, storage in a retrieval system, or transmission in any form or by any means, electronic, mechanical, photocopying, recording, or likewise. To obtain permission(s) to use materials from this work, please submit a written request to Pearson Higher Education, Permissions Department, 1 Lake Street, Upper Saddle River, NJ 07458.

The author and publisher of this book have used their best efforts in preparing this book. These efforts include the development, research, and testing of the theories and programs to determine their effectiveness. The author and publisher make no warranty of any kind, expressed or implied, with regard to these programs or the documentation contained in this book. The author and publisher shall not be liable in any event for incidental or consequential damages in connection with, or arising out of, the furnishing, performance, or use of these programs.

Library of Congress Cataloging-in-Publication Data

Hu, Chenming.
  Modern semiconductor devices for integrated circuits / Chenming Calvin Hu.
      p. cm.
  Includes bibliographical references and index.
  ISBN 0-13-608525-3
  1. Semiconductors. 2. Integrated circuits. I. Title.
  TK7871.85.H746 2010
  621.3815'31--dc22
                           2008055140

Prentice Hall
is an imprint of

www.pearsonhighered.com

25 17

ISBN-13: 978-0-13-608525-6
ISBN-10: 0-13-608525-3

*Dedicated to the semiconductor technologists
of the past, present, and future.
Their ingenuity makes the modern world possible.*

Contents

Preface xiii

About the Author xv

• 1 •
Electrons and Holes in Semiconductors 1

1.1 • Silicon Crystal Structure 1
1.2 • Bond Model of Electrons and Holes 4
1.3 • Energy Band Model 8
1.4 • Semiconductors, Insulators, and Conductors 11
1.5 • Electrons and Holes 12
1.6 • Density of States 15
1.7 • Thermal Equilibrium and the Fermi Function 16
1.8 • Electron and Hole Concentrations 19
1.9 • General Theory of n and p 25
1.10 • Carrier Concentrations at Extremely High and Low Temperatures 28
1.11 • Chapter Summary 29
PROBLEMS 30
REFERENCES 33
GENERAL REFERENCES 34

• 2 •
Motion and Recombination of Electrons and Holes 35

2.1 • Thermal Motion 35
2.2 • Drift 38
2.3 • Diffusion Current 46
2.4 • Relation Between the Energy Diagram and V, $\mathscr{E}$ 47
2.5 • Einstein Relationship Between D and μ 48
2.6 • Electron–Hole Recombination 50
2.7 • Thermal Generation 52

2.8 • Quasi-Equilibrium and Quasi-Fermi Levels   52
2.9 • Chapter Summary   54
PROBLEMS   56
REFERENCES   58
GENERAL REFERENCES   58

3

Device Fabrication Technology   59

3.1 • Introduction to Device Fabrication   60
3.2 • Oxidation of Silicon   61
3.3 • Lithography   64
3.4 • Pattern Transfer—Etching   68
3.5 • Doping   70
3.6 • Dopant Diffusion   73
3.7 • Thin-Film Deposition   75
3.8 • Interconnect—The Back-End Process   80
3.9 • Testing, Assembly, and Qualification   82
3.10 • Chapter Summary—A Device Fabrication Example   83
PROBLEMS   85
REFERENCES   87
GENERAL REFERENCES   88

4

PN and Metal–Semiconductor Junctions   89

Part I: PN Junction   89
4.1 • Building Blocks of the PN Junction Theory   90
4.2 • Depletion-Layer Model   94
4.3 • Reverse-Biased PN Junction   97
4.4 • Capacitance-Voltage Characteristics   98
4.5 • Junction Breakdown   100
4.6 • Carrier Injection Under Forward Bias—Quasi-Equilibrium Boundary Condition   105
4.7 • Current Continuity Equation   107
4.8 • Excess Carriers in Forward-Biased PN Junction   109
4.9 • PN Diode IV Characteristics   112
4.10 • Charge Storage   115
4.11 • Small-Signal Model of the Diode   116

Part II: Application to Optoelectronic Devices   117
4.12 • Solar Cells   117
4.13 • Light-Emitting Diodes and Solid-State Lighting   124

4.14 • Diode Lasers 128
4.15 • Photodiodes 133

Part III: Metal–Semiconductor Junction 133

4.16 • Schottky Barriers 133
4.17 • Thermionic Emission Theory 137
4.18 • Schottky Diodes 138
4.19 • Applications of Schottky Diodes 140
4.20 • Quantum Mechanical Tunneling 141
4.21 • Ohmic Contacts 142
4.22 • Chapter Summary 145
PROBLEMS 148
REFERENCES 156
GENERAL REFERENCES 156

• 5 •
MOS Capacitor 157

5.1 • Flat-Band Condition and Flat-Band Voltage 158
5.2 • Surface Accumulation 160
5.3 • Surface Depletion 161
5.4 • Threshold Condition and Threshold Voltage 162
5.5 • Strong Inversion Beyond Threshold 164
5.6 • MOS C–V Characteristics 168
5.7 • Oxide Charge—A Modification to V_{fb} and V_t 172
5.8 • Poly-Si Gate Depletion—Effective Increase in T_{ox} 174
5.9 • Inversion and Accumulation Charge-Layer Thicknesses —Quantum Mechanical Effect 176
5.10 • CCD Imager and CMOS Imager 179
5.11 • Chapter Summary 184
PROBLEMS 186
REFERENCES 193
GENERAL REFERENCES 193

• 6 •
MOS Transistor 195

6.1 • Introduction to the MOSFET 195
6.2 • Complementary MOS (CMOS) Technology 198
6.3 • Surface Mobilities and High-Mobility FETs 200
6.4 • MOSFET V_t, Body Effect, and Steep Retrograde Doping 207
6.5 • Q_{INV} in MOSFET 209
6.6 • Basic MOSFET IV Model 210

- 6.7 • CMOS Inverter—A Circuit Example 214
- 6.8 • Velocity Saturation 219
- 6.9 • MOSFET IV Model with Velocity Saturation 220
- 6.10 • Parasitic Source-Drain Resistance 225
- 6.11 • Extraction of the Series Resistance and the Effective Channel Length 226
- 6.12 • Velocity Overshoot and Source Velocity Limit 228
- 6.13 • Output Conductance 229
- 6.14 • High-Frequency Performance 230
- 6.15 • MOSFET Noises 232
- 6.16 • SRAM, DRAM, Nonvolatile (Flash) Memory Devices 238
- 6.17 • Chapter Summary 245

PROBLEMS 247
REFERENCES 256
GENERAL REFERENCES 257

• 7 •
MOSFETs in ICs—Scaling, Leakage, and Other Topics 259

- 7.1 • Technology Scaling—For Cost, Speed, and Power Consumption 259
- 7.2 • Subthreshold Current—"Off" Is Not Totally "Off" 263
- 7.3 • V_t Roll-Off—Short-Channel MOSFETs Leak More 266
- 7.4 • Reducing Gate-Insulator Electrical Thickness and Tunneling Leakage 270
- 7.5 • How to Reduce W_{dep} 272
- 7.6 • Shallow Junction and Metal Source/Drain MOSFET 274
- 7.7 • Trade-Off Between I_{on} and I_{off} and Design for Manufacturing 276
- 7.8 • Ultra-Thin-Body SOI and Multigate MOSFETs 277
- 7.9 • Output Conductance 282
- 7.10 • Device and Process Simulation 283
- 7.11 • MOSFET Compact Model for Circuit Simulation 284
- 7.12 • Chapter Summary 285

PROBLEMS 286
REFERENCES 288
GENERAL REFERENCES 289

• 8 •
Bipolar Transistor 291

- 8.1 • Introduction to the BJT 291
- 8.2 • Collector Current 293
- 8.3 • Base Current 297
- 8.4 • Current Gain 298
- 8.5 • Base-Width Modulation by Collector Voltage 302

8.6 • Ebers–Moll Model   304
8.7 • Transit Time and Charge Storage   306
8.8 • Small-Signal Model   310
8.9 • Cutoff Frequency   312
8.10 • Charge Control Model   314
8.11 • Model for Large-Signal Circuit Simulation   316
8.12 • Chapter Summary   318
PROBLEMS   319
REFERENCES   323
GENERAL REFERENCES   323

Appendix I
Derivation of the Density of States   325

Appendix II
Derivation of the Fermi–Dirac Distribution Function   329

Appendix III
Self-Consistencies of Minority Carrier Assumptions   333

Answers to Selected Problems   337

Index   341

Preface

Modern Semiconductor Devices for Integrated Circuits is primarily intended for use by undergraduate students, although it is also suitable for graduate students and practicing engineers and scientists.

The manuscript of this book has been used very successfully at the University of California, Berkeley, as the text of a one-semester course. It can also serve as the text of a course taught over two quarters. It has been well received by students with interests in fields such as semiconductor technology, IC design, MEMS, optical devices, nanotechnology, and materials science.

Readers should have had an introduction to elementary differential equations, modern physics, and electronics. From that background, this book develops a deep understanding of modern device theory and practice to help prepare students for further studies or professional careers. It presents more information on modern transistors and their impact on circuit design than typical texts in this field. In so doing, it aims to provide a strong foundation for understanding future issues of devices and design.

Modern Semiconductor Devices for Integrated Circuits emphasizes the commonality among devices by avoiding the usual compartmental organization along the lines of electronic, optical, microwave devices, and so on. The strong focus on a few basic devices, PN junction, metal–semiconductor contact, bipolar transistor, and especially MOSFET achieves the desired depth of treatment. These devices provide the theoretical background and chapter homes for introducing other important devices such as solar cell, LED, diode laser, CCD, CMOS imager, HEMT, and memory devices. The goal is to achieve depth and breadth in a more concise, integrated, and hopefully interesting way.

An Instructor's Manual, art PowerPoints, and lecture PowerPoint slides are available to professors only. The lecture slides have been developed through several years of classroom use.

I would like to thank many people who have helped to bring this book to fruition. Bingliang Yang, Hyuck Choo, and Vivian Lin helped to type the text and prepare the figures and the solutions manual. Jemin Park, Kanghoon Jeon, Chung-Hsun Lin, and especially Babak Heydari contributed substantive materials. Many students including Anupama Bowonder, Pratik Patel, and Darsen Lu helped to proofread the manuscript. Grateful acknowledgment is made to the panel of reviewers: John Beresford, Brown University; Tim Dallas, Texas Tech University; Todd J. Kaiser, Montana State University; Long Que, Louisiana Tech University; Pritpal Singh, Villanova University; Nina Telang, University of Texas at Austin; Pouya Valizadeh, Concordia University; and Joshua M. O. Zide, University of Delaware.

Thanks are due to the staff at Pearson Prentice Hall, especially to Jane Bonnell, who guided the production expertly. I give special thanks to my wife, Margaret, and my sons—Jason, who designed the beautiful book cover, and Raymond—for their understanding and support when I extracted myself from their lives for long hours spent writing this book.

Chenming Calvin Hu
University of California, Berkeley

About the Author

Chenming Calvin Hu holds the TSMC Distinguished Professor Chair of Microelectronics at the University of California, Berkeley. He is a member of the U.S. Academy of Engineering and a foreign member of the Chinese Academy of Sciences. From 2001 to 2004, he was the Chief Technology Officer of TSMC, the world's largest IC foundry company. A fellow of the Institute of Electrical and Electronics Engineers (IEEE), he has been honored with the Jack Morton Award in 1997 for his research on transistor reliability, the Solid State Circuits Award in 2002 for codeveloping the first international standard transistor model for circuit simulation, and the Jun-ichi Nishizawa Medal in 2009 for exceptional contributions to device physics and scaling. He has supervised over 60 Ph.D. student theses, published 800 technical articles, and received more than 100 U.S. patents. His other honors include the Sigma Xi Moni Ferst Award, R&D 100 Award, and UC Berkeley's highest award for teaching—the Berkeley Distinguished Teaching Award.

Modern Semiconductor Devices for Integrated Circuits

1

Electrons and Holes in Semiconductors

CHAPTER OBJECTIVES

This chapter provides the basic concepts and terminology for understanding semiconductors. Of particular importance are the concepts of energy band, the two kinds of electrical charge carriers called electrons and holes, and how the carrier concentrations can be controlled with the addition of dopants. Another group of valuable facts and tools is the Fermi distribution function and the concept of the Fermi level. The electron and hole concentrations are closely linked to the Fermi level. The materials introduced in this chapter will be used repeatedly as each new device topic is introduced in the subsequent chapters. When studying this chapter, please pay attention to (1) concepts, (2) terminology, (3) typical values for Si, and (4) all boxed equations such as Eq. (1.7.1).

The title and many of the ideas of this chapter come from a pioneering book, *Electrons and Holes in Semiconductors* by William Shockley [1], published in 1950, two years after the invention of the transistor. In 1956, Shockley shared the Nobel Prize in physics for the invention of the transistor with Brattain and Bardeen (Fig. 1–1).

The materials to be presented in this and the next chapter have been found over the years to be useful and necessary for gaining a deep understanding of a large variety of semiconductor devices. Mastery of the terms, concepts, and models presented here will prepare you for understanding not only the many semiconductor devices that are in existence today but also many more that will be invented in the future. It will also enable you to communicate knowledgeably with others working in the field of semiconductor devices.

1.1 ● SILICON CRYSTAL STRUCTURE ●

A crystalline solid consists of atoms arranged in a repetitive structure. The periodic structure can be determined by means of X-ray diffraction and electron microscopy. The large cubic unit shown in Fig. 1–2 is the **unit cell** of the silicon

● Inventors of the Transistor ●

Born on three different continents (Brattain in Amoy, China; Bardeen in Madison, Wisconsin, USA; and Shockley in London, England), they all grew up in the United States and invented the transistor in 1947–1948 at Bell Telephone Laboratories. Brattain was an experimentalist while Bardeen and Shockley contributed more to the concepts and theories. Their reflections on that historic event:

"... after fourteen years of work, I was beginning to give up ..."
—Walter H. Brattain (1902–1987)

"Experiments that led to the invention of the point-contact transistor by Walter Brattain and me were done in November and December, 1947, followed closely by the invention of the junction transistor by Shockley."
—John Bardeen (1908–1991)

"All of us who were involved had no doubt that we had opened a door to a new important technology."
—William B. Shockley (1910–1988)

FIGURE 1–1 Transistor inventors John Bardeen, William Shockley, and Walter Brattain (left to right) at Bell Telephone Laboratories. (Courtesy of Corbis/Bettmann.)

crystal. Each sphere represents a silicon atom. This unit cell is repeated in all three directions many times to form a silicon crystal. The length of the unit cell, e.g., 5.43 Å in Fig. 1–2, is called the **lattice constant**.

The most important information from Fig. 1–2 is the simple fact that *each and every silicon atom has four other silicon atoms as its nearest neighbor atoms.* This fact is illustrated in Fig. 1–2 with the darkened cluster of a center atom having four neighboring atoms. This cluster is called the **primitive cell**. Silicon is a group IV element in the periodic table and has four valence electrons. These four electrons are shared with the nearest neighbors so that eight covalent electrons are associated

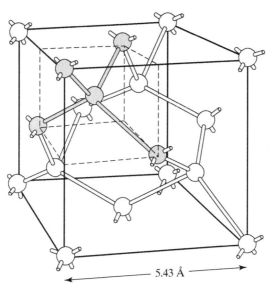

FIGURE 1–2 The unit cell of the silicon crystal. Each sphere is a Si atom. Each Si atom has four nearest neighbors as illustrated in the small cube with darkened atoms. (Adapted from Shockley [1].) For an interactive model of the unit cell, see http://jas.eng.buffalo.edu/

with each atom. The structure shown in Fig. 1–2 is known as the **diamond structure** because it is also the unit cell of the diamond crystal with each sphere representing a carbon atom. Germanium, the semiconductor with which the first transistor was made, also has the diamond crystal structure.

Figure 1–3 introduces a useful system of denoting the orientation of the silicon crystal. The cube in Fig. 1–3a represents the Si unit cell shown in Fig. 1–2 and each darkened surface is a crystal plane. The (100) crystal plane in the leftmost drawing in Fig. 1–3a, for example, is simply the plane in Fig. 1–2 closest to the reader. It intersects the x axis at 1 lattice constant and the y and z axes at infinity. One might refer to this plane as the 1 ∞ ∞ plane. However, it is standard practice to refer to it as the (1/1 1/∞ 1/∞), or the (100), plane. In general, the (abc) plane intersects the x, y, and z axes at $1/a$, $1/b$, and $1/c$ lattice constants. For example, the (011) plane in the middle drawing in Fig. 1–3a intersects the x axis at infinity and the y and z axes at 1 lattice constant. The numerals in the parentheses are called the **Miller indices**. The related symbol [abc] indicates the direction in the crystal normal to the (abc) plane. For example, when an electron travels in the [100] direction, it travels perpendicular to the (100) plane, i.e., along the x axis.

Figure 1–3b shows that the silicon wafers are usually cut along the (100) plane to obtain uniformity and good device performance. A flat or a notch is cut along the (011) plane in order to precisely and consistently orient the wafer as desired during device fabrication. Different surface orientations have different properties such as the rate of oxidation and the electronic quality of the oxide/semiconductor interface. Both the surface orientation and the direction of current flow along the surface affect the speed performance of a surface-base device such as metal-oxide-semiconductor field-effect transistor (MOSFET, see Section 6.3.1). The most important semiconductor materials used in microelectronics are crystalline. However, most everyday solids are not single crystals as explained in the sidebar in Section 3.7.

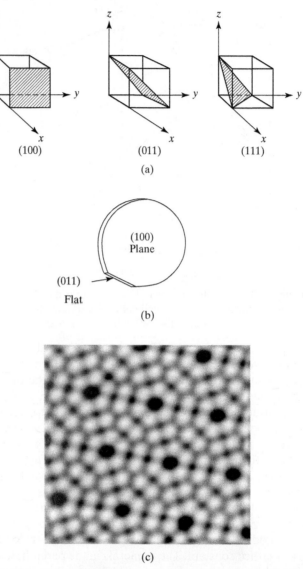

FIGURE 1–3 (a) A system for describing the crystal planes. Each cube represents the unit cell in Fig. 1–2. (b) Silicon wafers are usually cut along the (100) plane. This sample has a (011) flat to identify wafer orientation during device fabrication. (c) Scanning tunneling microscope view of the individual atoms of silicon (111) plane.

1.2 • BOND MODEL OF ELECTRONS AND HOLES •

Each silicon atom is surrounded by four nearest neighbors as illustrated by the shaded cluster in Fig. 1–2. We can represent the silicon crystal structure with the two-dimensional drawing shown in Fig. 1–4. An Si atom is connected to each neighbor with two dots representing the two shared electrons in the covalent bond. Figure 1–4 suggests that there are no free electrons to conduct electric current. This is strictly true

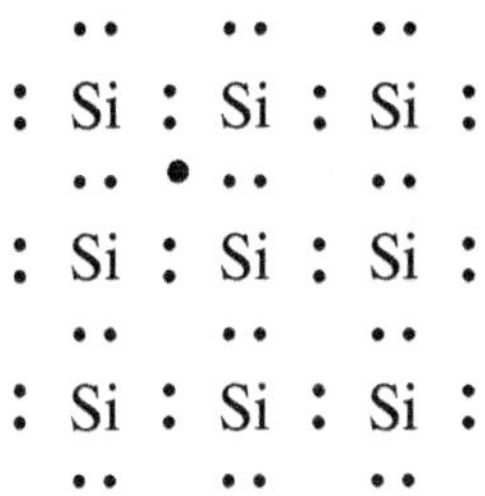

```
        ..    ..    ..
      : Si : Si : Si :
        ..  • ..   ..
      : Si : Si : Si :
        ..    ..    ..
      : Si : Si : Si :
        ..    ..    ..
```

FIGURE 1–4 The silicon crystal structure in a two-dimensional representation.

only at the absolute zero temperature. At any other temperature, thermal energy will cause a small fraction of the covalent electrons to break loose and become **conduction electrons** as illustrated in Fig. 1–5a. Conduction electrons can move around in a crystal and therefore can carry electrical currents. For this reason, the conduction electrons are of more interest to the operation of devices than valence electrons.

An interesting thing happens when an electron breaks loose and becomes free. It leaves behind a void, or a **hole** indicated by the open circle in Fig. 1–5a. The hole can readily accept a new electron as shown in Fig. 1–5b. This provides another means for electrons to move about and conduct currents. An alternative way to think of this process is that the hole moves to a new location. It is much easier to think of this second means of current conduction as the motion of a positive hole than the motion of negative electrons moving in the opposite direction just as it is much easier to think about the motion of a bubble in liquid than the liquid movement that creates the moving bubble.

In semiconductors, current conduction by holes is as important as electron conduction in general. *It is important to become familiar with thinking of the holes as mobile particles carrying positive charge, just as real as conduction electrons are mobile particles carrying negative charge.* It takes about 1.1 eV of energy to free a covalent electron to create a conduction electron and a hole. This energy can be determined, for example, from a photoconductivity experiment. When light shines on a Si sample, its conductivity increases because of the generation of mobile electrons and holes. The minimum photon energy required to induce photoconductivity is 1.1 eV.

The densities of thermally generated electrons and holes in semiconductors are generally very small at room temperature given that the thermal energy, kT, is 26 meV at room temperature. A much larger number of conduction electrons can be introduced if desired by introducing suitable impurity atoms—a process called **doping**

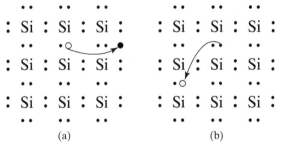

FIGURE 1–5 (a) When a covalent electron breaks loose, it becomes mobile and can conduct electrical current. It also creates a void or a hole represented by the open circle. The hole can also move about as indicated by the arrow in (b) and thus conduct electrical current.

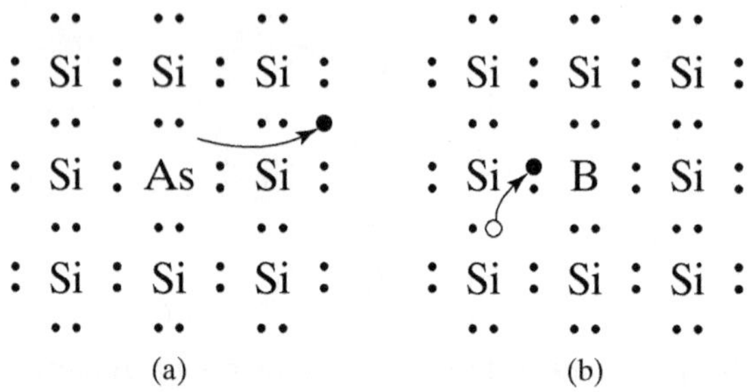

```
    ··    ··    ··              ··   ··   ··
  : Si : Si : Si :            : Si : Si : Si :
    ··    ··   ··↗•             ··   ··   ··
  : Si : As : Si :            : Si↗ B : Si :
    ··    ··    ··              ·○   ··   ··
  : Si : Si : Si :            : Si : Si : Si :
    ··    ··    ··              ··   ··   ··
         (a)                          (b)
```

FIGURE 1–6 Doping of a semiconductor is illustrated with the bond model. (a) As is a donor. (b) B is an acceptor.

the semiconductor. For example, group V elements such as As shown in Fig. 1–6a bring five valence electrons with each atom. While four electrons are shared with the neighboring Si atoms, the fifth electron may escape to become a mobile electron, leaving behind a positive As ion. Such impurities are called **donors** for they *donate* electrons. Notice that in this case, no hole is created in conjunction with the creation of a conduction electron. Semiconductors containing many mobile electrons and few holes are called **N-type semiconductors** because electrons carry negative (N) charge. As and P are the most commonly used donors in Si.

Similarly, when boron, a group III impurity, is introduced into Si as shown in Fig. 1–6b, each B atom can accept an extra electron to satisfy the covalent bonds, thus creating a hole. Such dopants are called **acceptors**, for they *accept* electrons. Semiconductors doped with acceptors have many holes and few mobile electrons, and are called **P type** because holes carry positive (P) charge. Boron is the most commonly used acceptor in Si. In and Al are occasionally used.

The energy required to ionize a donor atom (i.e., to free the extra electron and leave a positive ion behind) may be estimated by modifying the theory of the ionization energy of a hydrogen atom,

$$E_{\text{ion}} = \frac{m_0 q^4}{8\varepsilon_0^2 h^2} = 13.6 \text{ eV} \qquad (1.2.1)$$

where m_0 is the free electron mass, ε_0 is the permittivity of free space, and h is Planck's constant. The modification involves replacing ε_0 with $12\varepsilon_0$ (where 12 is the relative permittivity of silicon) and replacing m_0 with an electron effective mass, m_n, which is a few times smaller than m_0 as explained later. The result is about 50 meV. Because donors have such small ionization energies, they are usually fully ionized at room temperature. For example, 10^{17}cm^{-3} of donor atoms would lead to 10^{17}cm^{-3} of conduction electrons. The same conclusion also applies to the acceptors.

● GaAs, III–V Compound Semiconductors and Their Dopants ●

GaAs and similar **compound semiconductors**, such as InP and GaN, are dominant in optoelectronic devices such as light-emitting diodes and semiconductor lasers (see Sections 4.13 and 4.14). GaAs also plays a role in high-frequency electronics (see Sections 6.3.2 and 6.3.3). Its crystal structure is shown in Fig. 1–7 and Fig. 1–8. The similarity to the Si crystal is obvious. The shaded spheres represent As atoms and the light spheres represent Ga atoms. Each Ga atom has four As neighbors and each As atom has four Ga neighbors. The lattice constant is 5.65 Å. Ga is a group III element and As is a group V element. GaAs is known as a **III–V compound semiconductor**, as are GaP and AlAs, which also have the same crystal structure as illustrated in Fig. 1–7.

It is probably obvious that group VI elements such as S and Se can replace the group V As and serve as donors in GaAs. Similarly, group II elements such as Zn can replace Ga and serve as acceptors.

But, are group IV elements such as Si and Ge donors or acceptors in GaAs? The answer is that they can be either donors or acceptors, depending on whether they substitute for Ga atoms (which have three valence electrons) or As atoms (which have five valence electrons). Such impurities are called **amphoteric dopants**. It turns out that Si is a donor and Ge is an acceptor in GaAs because it is energetically more favorable for the small Si atoms to substitute for the small Ga atoms and for the larger Ge to substitute for the larger As.

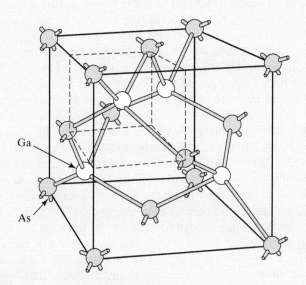

FIGURE 1–7 The GaAs crystal structure.

$$\begin{array}{ccc} \vdots\text{Ga}\vdots & \text{As}\vdots & \text{Ga}\vdots \\ \vdots\text{As}\vdots & \text{Ga}\vdots & \text{As}\vdots \\ \vdots\text{Ga}\vdots & \text{As}\vdots & \text{Ga}\vdots \end{array}$$

FIGURE 1–8 Bond model of GaAs.

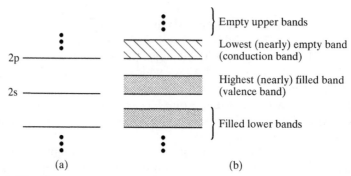

FIGURE 1–9 The discrete energy states of a Si atom (a) are replaced by the energy bands in a Si crystal (b).

1.3 • ENERGY BAND MODEL •

While the bond model described in the previous section is conceptually simple, it is not complete enough for understanding semiconductor devices. The most useful model involves the concept of **energy bands**.

Recall that electrons in an atom occupy discrete energy levels as shown in Fig. 1–9a. If two atoms are in close proximity, each energy level will split into two due to the Pauli exclusion principle that states that each quantum state can be occupied by no more than one electron in an electron system such as an atom molecule, or crystal. When many atoms are brought into close proximity as in a crystal, the discrete energy levels are replaced with *bands* of energy states separated by gaps between the bands as shown in Fig. 1–9b. One may think of an energy band as a semicontinuum of a very large number of energy states.

Naturally, the electrons tend to fill up the low energy bands first. The lower the energy, the more completely a band is filled. In a semiconductor, most of the energy bands will be basically totally filled (completely filled at absolute zero), while the higher energy bands are basically totally empty. Between the (basically) totally filled and totally empty bands lie two bands that are only nearly filled and nearly empty as shown in Fig. 1–9b. They are of utmost interest to us. The top nearly filled band is called the **valence band** and the lowest nearly empty band is called the **conduction band**. The gap between them is called the **band gap**. The electrons in a totally filled band do not have a net velocity and do not conduct current, just as the water in a totally filled bottle does not slosh about. Similarly, a totally empty band cannot contribute to current conduction. These are the reasons the valence band and the conduction band are the only energy bands that contribute to current flows in a semiconductor.

1.3.1 Energy Band Diagram

Figure 1–10 is the **energy band diagram** of a semiconductor, a small portion of Fig. 1–9. It shows the top edge of the valence band, denoted by E_v, and the bottom edge of the conduction band, denoted by E_c. The difference between E_c and E_v is the **band-gap energy** or **energy gap**, E_g. Clearly, $E_g = E_c - E_v$. For silicon, the energy gap is about 1.1 eV. The electrons in the valence band are those associated with the

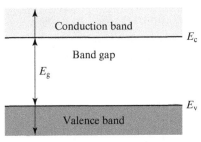

FIGURE 1–10 The energy band diagram of a semiconductor.

• Photoconductor as Light Detector •

When light is absorbed by a semiconductor sample and electron–hole pairs are created as shown in Fig. 1–11, the number of electrons and holes (and therefore the conductivity of the semiconductor) increase in proportion to the light intensity. By putting two electrodes on the semiconductor and applying a voltage between the electrodes, one can measure the change in the semiconductor conductance and thus detect changes in light intensity. This simple yet practical photodetector is called a **photoconductor**.

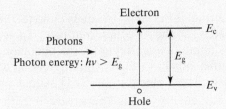

FIGURE 1–11 E_g can be determined from the minimum energy ($h\nu$) of photons that are absorbed by the semiconductor.

covalent bonds in the bond model discussed in the previous section, and the electrons in the conduction band are the conduction or mobile electrons. The band-gap energies of some semiconductors are listed in Table 1–1 to illustrate their wide range. The band-gap energy has strong influence on the characteristics and performance of optoelectronic devices (see Section 4.12.4 and Table 4–1). By mixing multiple semiconductors, the band-gap energy can be precisely tuned to desired values. This is widely practiced for optical semiconductor devices (see Section 4.13).

The band-gap energy can be determined by measuring the absorption of light by the semiconductor as a function of the photon energy, $h\nu$. The light is strongly absorbed only when $h\nu$ is larger than E_g. The absorbed photon energy is consumed to create an electron–hole pair as shown in Fig. 1–12. As $h\nu$ is reduced below E_g, the specimen becomes transparent to the light. E_g can be determined by observing this critical $h\nu$. Values of E_g listed in Table 1–1 are basically obtained in this way.

TABLE 1–1 • Band-gap energies of selected semiconductors.

Semiconductor	InSb	Ge	Si	GaAs	GaP	ZnSe	Diamond
E_g (eV)	0.18	0.67	1.12	1.42	2.25	2.7	6.0

EXAMPLE 1–1 Measuring the Band-Gap Energy

If a semiconductor is transparent to light with a wavelength longer than 0.87 μm, what is its band-gap energy?

SOLUTION:

Photon energy of light with 0.87 μm wavelength is, with c being the speed of light

$$h\nu = h\frac{c}{\lambda} = \frac{6.63 \times 10^{-34}\,(\text{J}\cdot\text{s}) \times 3 \times 10^{8}\,\text{m/s}}{0.87\,\mu\text{m}} = \frac{1.99 \times 10^{-19}\,(\text{J}\cdot\mu\text{m})}{0.87\,\mu\text{m}}$$

$$= \frac{1.99 \times 10^{-19}\,(\text{eV}\cdot\mu\text{m})}{1.6 \times 10^{-19} \times 0.87\,\mu\text{m}} = \frac{1.24\,(\text{eV}\cdot\mu\text{m})}{0.87\,\mu\text{m}} = 1.42\,\text{eV}$$

Therefore, the band gap of the semiconductor is 1.42 eV. The semiconductor is perhaps GaAs (see Table 1–1).

USEFUL RELATIONSHIP: $h\nu\,(\text{eV}) = \dfrac{1.24}{\lambda\,(\mu\text{m})}$

The visible spectrum is between 0.5 and 0.7 μm. (Silicon and GaAs have band gaps corresponding to the $h\nu$ of infrared light. Therefore they absorb visible light strongly and are opaque.) Some semiconductors such as indium and tin oxides have sufficiently large E_g's to be transparent to the visible light and be used as the transparent electrode in LCD (liquid crystal display) flat panel displays.

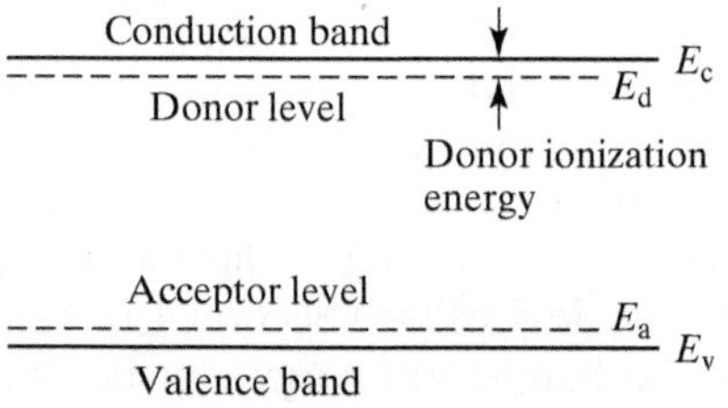

FIGURE 1–12 Energy levels of donors and acceptors.

1.3.2 Donors and Acceptors in the Band Model

The concept of donors and acceptors is expressed in the energy band model in the following manner. Although less important than E_c and E_v, two other energy levels are present in the energy band diagram, the donor energy level E_d and the acceptor energy level E_a (Fig. 1–12). Recall that it takes the donor ionization energy (about 50 meV) to free the extra electron from the donor atom into a conduction electron. Therefore, the donor electron, before the electron is freed, must occupy a state at about 50 meV below E_c. That is to say, $E_c - E_d$ is the donor ionization energy. Similarly, $E_a - E_v$ is the acceptor ionization energy (i.e., the

TABLE 1–2 • Ionization energy of selected donors and acceptors in silicon.

	Donors			Acceptors		
Dopant	Sb	P	As	B	Al	In
Ionization energy, $E_c - E_d$ or $E_a - E_v$ (meV)	39	44	54	45	57	160

energy it takes for an acceptor atom to receive an extra electron from the valence band, creating a hole there). Some donor and acceptor ionization energies in silicon are listed in Table 1–2 for reference. As, P, Sb, and B are the most commonly used dopants for silicon. Acceptor and donor levels with small ionization energies, such as these four, are called **shallow levels**. Deep levels can be created with other impurities such as copper and gold, and they affect silicon properties in very different ways (see Section 2.6).

1.4 • SEMICONDUCTORS, INSULATORS, AND CONDUCTORS •

Based on the energy band model, we can now understand the differences among semiconductors, insulators, and conductors. A semiconductor has a nearly filled valence band and a nearly empty conduction band separated by a band gap as illustrated in Fig. 1–13a. The band diagram of an insulator is similar to that of a semiconductor except for a larger E_g, which separates a completely filled band and a completely empty band (see Fig. 1–13b). Totally filled bands and totally empty bands do not contribute to current conduction, just as there can be no motion of liquid in totally filled jars and totally empty jars. A conductor has a quite different energy band diagram. As depicted in Fig. 1–13c, a conductor has a partially filled band. This is the conduction band of the conductor and it holds the conduction electrons. The abundance of the conduction electrons makes the resistivity of a typical conductor much lower than that of semiconductors and insulators.

Why do some materials have a partially filled band and are therefore conductors? Each energy band can hold two electrons per atom.[1] This is why

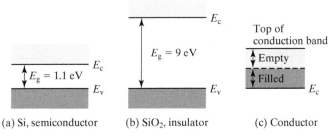

(a) Si, semiconductor (b) SiO$_2$, insulator (c) Conductor

FIGURE 1–13 Energy band diagrams for a semiconductor (a), an insulator (b), and a conductor (c).

---

[1] This is a simplified picture. Actually, each band can hold two electrons per primitive cell, which may contain several atoms. A primitive cell is the smallest repeating structure that makes up a crystal. The darkened part of Fig. 1–2 is the primitive cell of the Si crystal.

elemental solids with odd atomic numbers (and therefore odd numbers of electrons) such as Au, Al, and Ag are conductors. Elements with even atomic numbers such as Zn and Pb can still be conductors because a filled band and an empty band may overlap in energy, thus leaving the combined band partially filled. These elements are known as **semimetals**.

An insulator has a filled valence band and an empty conduction band that are separated by a larger E_g. How large an E_g is large enough for the material to be classified as an insulator? There is no clear boundary, although 4 eV would be an acceptable answer. However, even diamond, with $E_g \sim 6$ eV (often cited as a textbook example of an insulator) exhibits semiconductor characteristics. It can be doped N type and P type, and electronic devices such as rectifiers and transistors have been made with diamond.

One may say that semiconductors differ from insulators in that semiconductors can be made N type or P type with low resistivities through impurity doping. This characteristic of the semiconductors is very important for device applications.

1.5 • ELECTRONS AND HOLES •

Although the term **electrons** can be used in conjunction with the valence band as in "nearly all the energy states in the valence band are filled with electrons," we should assume that the term usually means **conduction-band** electrons. **Holes** are the electron voids in the valence band. Electrons and holes carry negative and positive charge ($\pm q$) respectively. As shown in Fig. 1–14, a higher position in the energy band diagram represents a higher electron energy. The minimum conduction electron energy is E_c. Any energy above E_c is the electron kinetic energy. Electrons may gain energy by getting accelerated in an electric field and may lose energy through collisions with imperfections in the crystal.

A lower location in the energy diagram represents a higher hole energy as shown in Fig. 1–14. It requires energy to move a hole "downward" because that is equivalent to moving an electron upward. E_v is the minimum hole energy. *We may think of holes as bubbles in liquid, floating up in the energy band. Similarly, one may think of electrons as water drops that tend to fall to the lowest energy states in the energy band.*

1.5.1 Effective Mass

When an electric field, $\mathscr{E}$, is applied, an electron or a hole will accelerate according to

$$\text{Acceleration} = \frac{-q\mathscr{E}}{m_n} \quad \text{electrons} \tag{1.5.1}$$

$$\text{Acceleration} = \frac{q\mathscr{E}}{m_p} \quad \text{holes} \tag{1.5.2}$$

In order to describe the motion of electrons and holes with the laws of motion of the classical particles, we must assign **effective masses** (m_n and m_p) to them. The electron and hole effective masses of a few semiconductors are listed in Table 1–3.

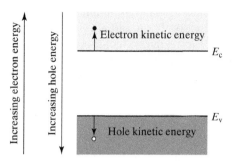

FIGURE 1–14 Both electrons and holes tend to seek their lowest energy positions. Electrons tend to fall in the energy band diagram. Holes float up like bubbles in water.

TABLE 1–3 • Electron and hole effective masses, m_n and m_p, normalized to the free electron mass.

	Si	Ge	GaAs	InAs	AlAs
m_n/m_0	0.26	0.12	0.068	0.023	2.0
m_p/m_0	0.39	0.30	0.50	0.30	0.3

The electrons and holes in a crystal interact with a periodic coulombic field in the crystal. They surf over the periodic potential of the crystal, and therefore m_n and m_p are not the same as the free electron mass.

A complete description of the electrons in a crystal must be based on their wave characteristics, not just the particle characteristics. The electron wave function is the solution of the three-dimensional Schrödinger wave equation [2].

$$-\frac{\hbar^2}{2m_0}\nabla^2\psi + V(r)\psi = E\psi \qquad (1.5.3)$$

where $\hbar = h/2\pi$ is the reduced Planck constant, m_0 is the free electron mass, $V(r)$ is the potential energy field that the crystal presents to the electron in the three-dimensional space, and E is the energy of the electron. The solution is of the form $\exp(\pm k \cdot r)$, which represents an electron wave k, called the **wave vector**, is equal at 2π/electron wavelength and is a function of E. In other words, for each k there is a corresponding E (see Fig. 4–27 for a schematic E–k diagram). It can further be shown [2] that, assuming the E–k relationship has spherical symmetry, an electric field, $\mathscr{E}$, would accelerate an electron wave packet with

$$\text{Acceleration} = -\frac{q\mathscr{E}}{\hbar^2}\frac{d^2E}{dk^2} \qquad (1.5.4)$$

In order to interpret the acceleration in the form of F/m, it is convenient to introduce the concept of the effective mass

$$\text{Effective mass} \equiv \frac{\hbar^2}{d^2E/dk^2} \qquad (1.5.5)$$

Each semiconductor material has a unique E–k relationship (due to the unique $V(r)$) for its conduction band and another unique E–k relationship for its valence band. Therefore, each semiconductor material has its unique m_n and m_p.

The values listed in Table 1–3 are experimentally measured values. These values agree well with the effective masses obtained by solving the Schrödinger wave equation with computers.

1.5.2 How to Measure the Effective Mass[2]

If you wonder how one may measure the effective mass of electrons or holes in a semiconductor, let us study a powerful technique called **cyclotron resonance**.

Consider an electron in an N-type semiconductor located in a magnetic field, **B**, as shown in Fig. 1–15. A moving electron will trace out a circular path in a plane normal to **B**. (In addition, there may be electron motion parallel to **B**. We may ignore this velocity component for the present analysis). The magnetic field exerts a Lorentzian force of $qv\mathbf{B}$, where v is the electron velocity and **B** is the magnetic flux density. Equating this force to the centripetal force corresponding to the circular motion with radius r, we obtain

$$\frac{m_n v^2}{r} = qv\mathbf{B} \tag{1.5.6}$$

$$v = \frac{q\mathbf{B}r}{m_n} \tag{1.5.7}$$

The frequency of the circular motion is

$$f_{cr} = \frac{v}{2\pi r} = \frac{q\mathbf{B}}{2\pi m_n} \tag{1.5.8}$$

This is the cyclotron resonance frequency. Notice that the resonance frequency is independent of r and v. Now, if a circularly polarized electric field of the same frequency f_{cr} (typically in the gigahertz range) is applied to this semiconductor, the

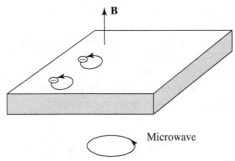

FIGURE 1–15 The motion of electrons in an N-type semiconductor in the presence of a magnetic field, **B**, and a microwave with rotating electric field (the direction of rotation is indicated by the arrow).

---

[2] This section may be omitted in an accelerated course.

electrons will strongly absorb the microwave energy. They do so by accelerating to a higher velocity and tracing circles of increasing radius [see Eq. (1.5.7)] without changing their frequency of circular motion [see Eq. (1.5.8)], losing the energies through collisions, and starting the acceleration process again. Obviously, the absorption would be much weaker if the frequency of the applied field is not equal to f_{cr}, i.e., when the applied field is out of sync with the electron motion.

By varying the frequency of the electric field or varying **B** until a peak in absorption is observed, one can calculate m_n using Eq. (1.5.8). One can also perform the same measurement on a P-type semiconductor to measure the effective mass of holes.

1.6 • DENSITY OF STATES •

It is useful to think of an energy band as a collection of discrete energy states. Figure 1–16a emphasizes this picture. In quantum mechanics terms, each state represents a unique spin (up and down) and unique solution to the Schrodinger's wave equation for the periodic electric potential function of the semiconductor [3]. Each state can hold either one electron or none. If we count the number of states in a small range of energy, ΔE, in the conduction band, we can find the **density of states**:

$$D_c(E) \equiv \frac{\text{number of states in } \Delta E}{\Delta E \times \text{volume}} \qquad (1.6.1)$$

This conduction-band density of states is a function of E (i.e., where ΔE is located). Similarly, there is a valence-band density of states, $D_v(E)$. D_c and D_v, graphically illustrated in Fig. 1–16b, can be shown to be proportional to $\sqrt{E - E_c}$ and $\sqrt{E_v - E}$ at least for a range of E. The derivation is presented in Appendix I, "Derivation of the Density of States."

$$D_c(E) = \frac{8\pi m_n \sqrt{2m_n(E - E_c)}}{h^3}, \quad E \geq E_c \qquad (1.6.2a)$$

$$D_v(E) = \frac{8\pi m_p \sqrt{2m_p(E_v - E)}}{h^3}, \quad E \leq E_v \qquad (1.6.2b)$$

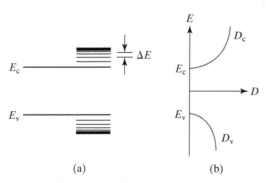

FIGURE 1–16 (a) Energy band as a collection of discrete energy states. (b) D is the density of the energy states.

$D_c(E)$ and $D_v(E)$ have the dimensions of number per cubic centimeter per electronvolt.

It follows from Eq. (1.6.1) that the product $D_c(E)\,dE$ and $D_v(E)\,dE$ are the numbers of energy states located in the energy range between E and $E + dE$ per cubic centimeter of the semiconductor volume. We will use this concept in Section 1.8.1.

1.7 • THERMAL EQUILIBRIUM AND THE FERMI FUNCTION •

We have mentioned in Section 1.6 that most of the conduction-band electrons will be found near E_c, where the electron energy is the lowest. Most holes will "float" toward E_v, where the hole energy is the lowest. In this section, we will examine the distribution of electrons and holes in greater detail.

1.7.1 An Analogy for Thermal Equilibrium

Let us perform the following mental experiment. Spread a thin layer of sand on the bottom of a shallow dish sitting on a table as shown in Fig. 1–17. The sand particles represent the electrons in the conduction band. A machine shakes the table and therefore the dish up and down. The vibration of the dish represents the thermal agitation experienced by the atoms, electrons and holes at any temperature above absolute zero. The sand will move and shift until a more or less level surface is created. This is the equilibrium condition. The **equilibrium** condition is the lowest energy configuration in the *presence of the thermal agitation*. If a small sand dune exists as a protrusion above the flat surface, it would not be the lowest energy condition because the gravitational energy of the system can be reduced by flattening the dune. This is the equivalent of electrons preferring to occupy the lower energy states of the conduction band.

Notice, however, that the agitation prevents the particles from taking only the lowest possible energy positions. This fact can be dramatized by vibrating the dish more vigorously. Now, some sand particles jump up into the air and fall back. The system is at the lowest possible energy consistent with the presence of the "thermal" agitation. This is **thermal equilibrium**.

Because the sand particles bounce up and down, some higher energy states, in the air, are occupied, too. There is a lower probability (smaller fraction of time) for the sand particles to be higher in the air, i.e., to occupy higher energy states. Similarly, electrons and holes in semiconductors receive and exchange energy from

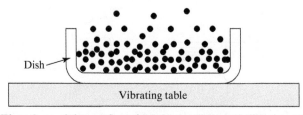

FIGURE 1–17 Elevations of the sand particles in the dish on a vibrating table represent the energies of the electrons in the conduction band under the agitation of thermal energy.

or with the crystal and one another and *every energy state in the conduction and valence bands has a certain probability of being occupied by an electron.* This probability is quantified in the next section.

1.7.2 Fermi Function—The Probability of an Energy State Being Occupied by an Electron

One can derive an expression to describe the probability of finding an electron at a certain energy. A statistical thermodynamic analysis without regard to the specifics of how particles bounce off the atoms or one another and only assuming that the number of particles and the total system energy are held constant yield the following result. (The derivation is presented in Appendix II, "Derivation of the Fermi–Dirac Distribution Function.)

$$\boxed{f(E) = \frac{1}{1 + e^{(E-E_F)/kT}}} \qquad (1.7.1)$$

An equation that is highlighted with a box, such as Eq. (1.7.1), is particularly important and often cited. Equation (1.7.1) is the **Fermi function**, or the **Fermi–Dirac distribution function**, or the **Fermi–Dirac statistics**. E_F is called the **Fermi energy** or the **Fermi level**. $f(E)$ is the probability of a state at energy E being occupied by an electron. Figure 1–18 depicts the Fermi function. At large E (i.e., $E - E_F \gg kT$) the probability of a state being occupied decreases exponentially with increasing E. In this energy region, Eq. (1.7.1) can be approximated with

$$\boxed{f(E) \approx e^{-(E-E_F)/kT}} \qquad (1.7.2)$$

Equation (1.7.2) is known as the **Boltzmann approximation**. In the low energy region (i.e., $E - E_F \ll -kT$), the occupation probability approaches 1. In other words, the low energy states tend to be fully occupied. Here Eq. (1.7.1) can be approximated as

$$f(E) \approx 1 - e^{-(E_F - E)/kT} \qquad (1.7.3)$$

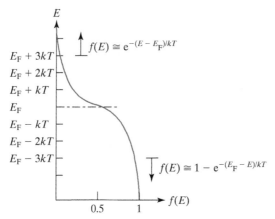

FIGURE 1–18 The Fermi function diagram. For an interactive illustration of the Fermi function, see http://jas.eng.buffalo.edu/education/semicon/fermi/functionAndStates/functionAndStates.html

In this energy region, the probability of a state *not* being occupied, i.e., being *occupied by a hole* is

$$1 - f(E) \approx e^{-(E_F - E)/kT} \qquad (1.7.4)$$

From Eq. (1.7.1) and Fig. 1–18, the probability of occupation at E_F is 1/2. The probability approaches unity if E is much lower than E_F, and approaches zero at E much higher than E_F. *A very important fact to remember about E_F is that there is only one Fermi level in a system at thermal equilibrium.* This fact will be used often in the rest of this book. Later, you will see that the value of E_F depends on the number of electrons or holes present in the system.

● What Determines E_F ? ●

This example is designed to show that the Fermi level, E_F, is determined by the available electrons and states in the system. Figure 1–19 shows the energy states of an electron system at room temperature. Each energy state can hold either one electron or none, i.e., be either occupied or empty. It is known that there is only one electron in this system. Since there is only one electron in the system, the sum of the probabilities that each state is occupied must be equal to 1. By trial and error, convince yourself that only one specific E_F, somewhere above E_1 and below E_2, can satisfy this condition. Do not calculate this value. *Hint*: Ask yourself how many electrons will be in the system if E_F is, say, above E_3 and how many if E_F is below E_1.

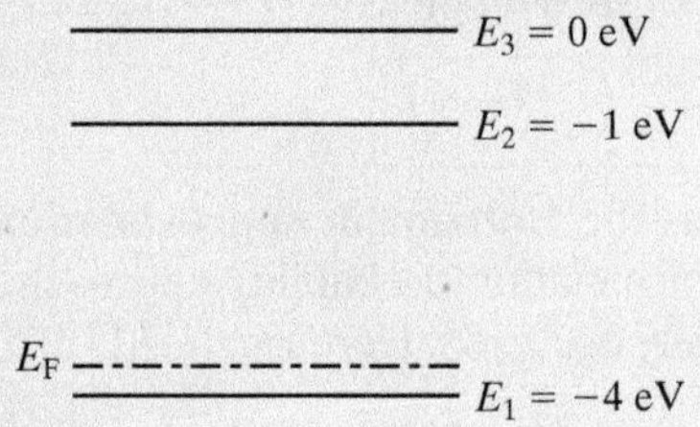

$$E_3 = 0 \text{ eV}$$
$$E_2 = -1 \text{ eV}$$
$$E_F \text{ -----}$$
$$E_1 = -4 \text{ eV}$$

FIGURE 1–19 A simple electron system at room temperature for illustration of what determines the Fermi energy, E_F.

EXAMPLE 1–2 Oxygen Concentration versus Altitude

We all know that there is less oxygen in the air at higher altitudes. What is the ratio of the oxygen concentration at 10 km above sea level, N_h, to the concentration at sea level, N_0, assuming a constant temperature of 0°C?

SOLUTION:

There are fewer oxygen molecules at higher altitudes because the gravitational potential energy of an oxygen molecule at the higher altitude, E_h, is larger than at sea level, E_0. According to Eq. (1.7.2)

$$\frac{N_h}{N_0} = \frac{e^{-E_h/kT}}{e^{-E_0/kT}} = e^{-(E_h - E_0)/kT}$$

$E_0 - E_h$ is the potential energy difference, i.e., the energy needed to lift an oxygen molecule from sea level to 10 km.

$$E_h - E_0 = \text{altitude} \times \text{weight of } O_2 \text{ molecule} \times \text{acceleration of gravity}$$
$$= 10^4 m \times O_2 \text{ molecular weight} \times \text{atomic mass unit} \times 9.8 \ m \cdot s^{-2}$$
$$= 10^4 m \times 32 \times 1.66 \times 10^{-27} kg \times 9.8 \ m \cdot s^{-2}$$
$$= 5.2 \times 10^{-21} J$$

$$\therefore \frac{N_h}{N_0} = e^{-5.2 \times 10^{-21} J / 1.38 \times 10^{-10} J \cdot K^{-1} \times 273 K}$$
$$= e^{-1.38} = 0.25$$

So, the oxygen concentration at 10 km is 25% of the sea level concentration.

This example and the sand-in-a-dish analogy are presented to demystify the concept of equilibrium, and to emphasize that each electron energy state has a probability of being occupied that is governed by the Fermi function.

Additional question: See Problem 1.4 for a follow-up question.

1.8 • ELECTRON AND HOLE CONCENTRATIONS •

We have stated that if a semiconductor is doped with 10^{16} donors per cubic centimeter, the electron concentration will be $10^{16} cm^{-3}$. But, what would the hole concentration be? What are the carrier concentrations in undoped semiconductors? These questions will be answered after the important relationships between the Fermi level and the carrier concentrations are derived in this section.

1.8.1 Derivation of n and p from D(E) and f(E)

First, we will derive the concentration of electrons in the conduction band, known as the **electron concentration**. Since $D_c(E) \ dE$ is the number of energy states between E and $E + dE$ for each cubic centimeter, the product $f(E)D_c(E) \ dE$ is then the number of electrons between E and $E + dE$ per cubic centimeter of the semiconductor. Therefore, the number of electrons per cubic centimeter in the entire conduction band is

$$n = \int_{E_c}^{\text{Top of conduction band}} f(E) D_c(E) \ dE \quad (1.8.1)$$

Graphically, this integration gives the shaded area in Fig. 1–20, which yields the density of electrons. We now substitute Eqs. (1.6.2a) and (1.7.2) into Eq. (1.8.1) and set the upper limit of integration at infinity. Resetting the upper limit is acceptable because of the rapid fall of $f(E)$ with increasing E as shown in Fig. 1–20. This allows us to obtain a closed form expression for n

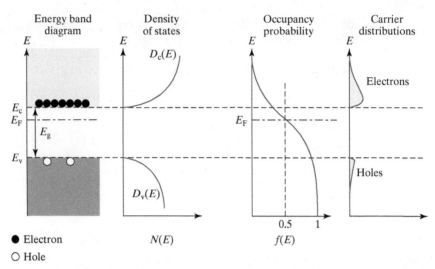

FIGURE 1–20 Schematic band diagram, density of states, Fermi–Dirac distribution, and carrier distributions versus energy.

$$n = \frac{8\pi m_n \sqrt{2m_n}}{h^3} \int_{E_c}^{\infty} \sqrt{E - E_c}\, e^{-(E - E_F)/kT} dE \tag{1.8.2}$$

$$= \frac{8\pi m_n \sqrt{2m_n}}{h^3} e^{-(E_c - E_F)/kT} \int_0^{\infty} \sqrt{E - E_c}\, e^{-(E - E_c)/kT} d(E - E_c) \tag{1.8.3}$$

Introducing a new variable

$$x = (E - E_c)/kT$$

reveals that the integral in Eq. (1.8.3) is of a form known as a **gamma function** and is equal to $\sqrt{\pi}/2$, i.e.,

$$\int_0^{\infty} \sqrt{x}\, e^{-x} dx = \sqrt{\pi}/2 \tag{1.8.4}$$

Applying Eq. (1.8.4) to Eq. (1.8.3) leads to the following two equations:

$$\boxed{n = N_c\, e^{-(E_c - E_F)/kT}} \tag{1.8.5}$$

$$N_c \equiv 2\left[\frac{2\pi m_n kT}{h^2}\right]^{3/2} \tag{1.8.6}$$

N_c is called the **effective density of states**. Equation (1.8.5) is an important equation and should be memorized. It is easy to remember this equation if we understand why N_c is called the **effective** density of states. It is as if all the energy states in the

TABLE 1–4 • Values of N_c and N_v for Ge, Si, and GaAs at 300 K.

	Ge	Si	GaAs
N_c (cm^{-3})	1.04×10^{19}	2.8×10^{19}	4.7×10^{17}
N_v (cm^{-3})	6.0×10^{18}	1.04×10^{19}	7.0×10^{18}

conduction band were **effectively** squeezed into a single energy level, E_c, which can hold N_c electrons (per cubic centimeter). As a result, the electron concentration in Eq. (1.8.5) is simply the product of N_c and the probability that an energy state at E_c is occupied.

An expression for the hole concentration can be derived in the same way. The probability of an energy state being occupied by a hole is the probability of it *not* being occupied by an electron, i.e., $1 - f(E)$. Therefore,

$$p = \int_{\text{Valence band bottom}}^{E_v} D_v(E)(1 - f(E)) \, dE \qquad (1.8.7)$$

Substituting Eqs. (1.6.2b) and (1.7.4) into Eq. (1.8.7) yields

$$\boxed{p = N_v \, e^{-(E_F - E_v)/kT}} \qquad (1.8.8)$$

$$N_v \equiv 2\left[\frac{2\pi m_p kT}{h^2}\right]^{3/2} \qquad (1.8.9)$$

N_v is the **effective density of states of the valence band.** (The full name of N_c is the effective density of states of the conduction band.) The values of N_c and N_v, both about 10^{19} cm^{-3}, differ only because m_n and m_p are different. N_c and N_v vary somewhat from one semiconductor to another because of the variation in the effective masses, too. N_c and N_v for Ge, Si, and GaAs are listed in Table 1–4.[3]

1.8.2 Fermi Level and the Carrier Concentrations

We will use Eqs. (1.8.5) and (1.8.8) time and again. Right now, they can help to remove the mystery of the Fermi level by linking E_F to the electron and hole concentrations.

Figure 1–21 shows the location of the Fermi level as a function of the carrier concentration. Note that the solid lines stop when E_F is about 20 meV (~kT) from

---

[3] The effective mass in Table 1–3 is called the *conductivity effective mass* and is an average over quantum mechanical wave vectors appropriate for describing carrier motions [3]. The effective mass in Eqs. (1.6.2a) and (1.6.2b), (1.8.6), and (1.8.9) is called the *density-of-states effective mass* and has a somewhat different value (because it is the result of a different way of averaging that is appropriate for describing the density of states).

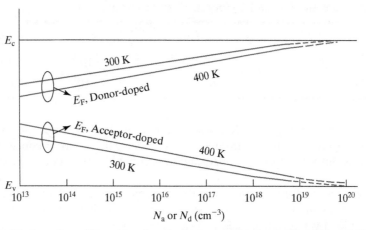

FIGURE 1–21 Location of Fermi level vs. dopant concentration in Si at 300 and 400 K.

EXAMPLE 1-3 Finding the Fermi Level in Si

Where is E_F located in the energy band of silicon, at 300K with $n = 10^{17} \text{cm}^{-3}$? And for $p = 10^{14} \text{cm}^{-3}$?

SOLUTION: From Eq. (1.8.5)

$$E_c - E_F = kT \cdot \ln(N_c/n)$$
$$= 0.026 \ln(2.8 \times 10^{19}/10^{17})$$
$$= 0.146 \text{ eV}$$

Therefore, E_F is located at 146 meV below E_c, as shown in Fig. 1–22a.
For $p = 10^{14} \text{cm}^{-3}$, from Eq. (1.8.8),

$$E_F - E_v = kT \cdot \ln(N_v/p)$$
$$= 0.026 \ln(1.04 \times 10^{19}/10^{14})$$
$$= 0.31 \text{ eV}$$

Therefore E_F is located at 0.31 eV above E_v.

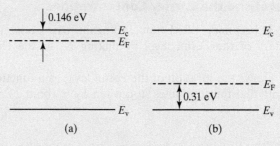

FIGURE 1–22 Location of E_F when $n = 10^{17} \text{cm}^{-3}$ (a), and $p = 10^{14} \text{cm}^{-3}$ (b).

E_c or E_v. Beyond this range, the use of the Boltzmann approximation in the derivation of Eqs. (1.8.5) and (1.8.8) is not quantitatively valid. Those equations are not accurate when the semiconductor is heavily doped ($>10^{19}$cm^{-3}) or **degenerate**. Please remember that n decreases as E_F moves farther below E_c, and vice versa; p decreases as E_F moves farther above E_v.

1.8.3 The *np* Product and the Intrinsic Carrier Concentration

Since E_F cannot be close to both E_c and E_v, n and p cannot both be large numbers at the same time. When Eqs. (1.8.5) and (1.8.8) are multiplied together, we obtain

$$np = N_c N_v e^{-(E_c - E_v)/kT} = N_c N_v e^{-E_g/kT} \quad (1.8.10)$$

Equation (1.8.10) states that the *np* product is a constant for a given semiconductor and T, independent of the dopant concentrations. It is an important relationship and is usually expressed in the following form:

$$np = n_i^2 \quad (1.8.11)$$

$$n_i = \sqrt{N_c N_v} e^{-E_g/2kT} \quad (1.8.12)$$

According to Eq. (1.8.11), there are always some electrons and holes present—whether dopants are present or not. If there are no dopants present, the semiconductor is said to be **intrinsic**. In an intrinsic semiconductor, the nonzero n and p are the results of thermal excitation, which moves some electrons from the valence band into the conduction band. Since such movements create electrons and holes in pairs, $n = p$ in intrinsic semiconductors. This fact and Eq. (1.8.11) immediately suggest that, in intrinsic semiconductors,

$$n = p = n_i \quad (1.8.13)$$

Therefore, n_i is called the **intrinsic carrier concentration**. n_i is a strong function of E_g and T according to Eq. (1.8.12), but is independent of the dopant concentration. n_i at room temperature is roughly 10^{10}cm^{-3} for Si and 10^7cm^{-3} for GaAs, which has a larger band gap than Si. For silicon, the *np* product is therefore 10^{20}cm^{-6} regardless of the conductivity type (P type or N type) and the dopant concentrations.

One may explain why the *np* product is a constant this way: the electron–hole recombination rate is proportional to the *np* product. When $np = n_i^2$, the recombination rate happens to be equal to the rate of thermal generation of electron–hole pairs. This is the same mass action principle that keeps the product of the concentrations of [H$^+$] and [OH$^-$] constant in aqueous solutions, whether strongly acidic, strongly alkaline, or neutral.

EXAMPLE 1-4 Carrier Concentrations

QUESTION: What is the hole concentration in an N-type semiconductor with 10^{15}cm^{-3} of donors?

SOLUTION: For each ionized donor, an electron is created. Therefore $n = 10^{15} \text{cm}^{-3}$.

$$p = \frac{n_i^2}{n} \approx \frac{10^{20} \text{ cm}^{-3}}{10^{15} \text{ cm}^{-3}} = 10^5 \text{cm}^{-3}$$

With a modest temperature increase of 60°C, n remains the same at 10^{15}cm^{-3}, while p increases by about a factor of 2300 because n_i^2 increases according to Eq. (1.8.12).

QUESTION: What is n if $p = 10^{17} \text{cm}^{-3}$ in a P-type silicon wafer?

SOLUTION: $n = \dfrac{n_i^2}{p} = \dfrac{10^{20}}{10^{17}} = 10^3 \text{cm}^{-3}$

The electron and hole concentrations in a semiconductor are usually very different. In an N-type semiconductor, the abundant electrons are called the **majority carriers** and the almost nonexistent holes are the **minority carriers**. In P-type semiconductors, the holes are the majority carriers and the electrons are the minority carriers.

EXAMPLE 1-5 The Intrinsic Fermi Level

In an intrinsic semiconductor, $n = p$. Therefore $E_c - E_F \approx E_F - E_v$ and the Fermi level is nearly at the middle of the band gap, i.e., $E_F \approx E_c - E_g/2$. This level is called the **intrinsic Fermi level**, E_i. Here we derive a more exact expression for E_i. Rewriting Eq. (1.8.12) for $\ln n_i$, yields

$$\ln n_i = \ln \sqrt{N_c N_v} - E_g/2kT$$

Writing Eq. (1.8.5) for the intrinsic condition where $n = n_i$,

$$n_i = N_c e^{-(E_c - E_i)/kT} \tag{1.8.14}$$

$$\therefore E_i = E_c - kT \ln \frac{N_c}{n_i} = E_c + kT \ln n_i - kT \ln N_c = E_c - \frac{E_g}{2} - kT \ln \sqrt{\frac{N_c}{N_v}} \tag{1.8.15}$$

We see that E_i would be at the midgap, $E_c - E_g/2$, if $N_c = N_v$. For silicon, E_i is very close to the midgap and the small last term in Eq. (1.8.15) is only of academic interest.

1.9 • GENERAL THEORY OF n AND p

The shallow donor and acceptor levels (E_d and E_a) in Fig. 1–12 are energy states and their occupancy by electrons is governed by the Fermi function. Since E_d is usually a few kT above E_F, the donor level is nearly empty of electrons. We say that nearly all the donor atoms are ionized (have lost the extra electrons). Similarly, all the acceptor atoms are ionized. *For simplicity, we can assume that all shallow donors and acceptors are ionized.*

EXAMPLE 1–6 Complete Ionization of the Dopant Atoms

In a silicon sample doped with $10^{17} cm^{-3}$ of phosphorus atoms, what fraction of the donors are not ionized (i.e., what fraction are occupied by the "extra" electrons)?

SOLUTION:

First assume that all the donors are ionized and each donor donates an electron to the conduction band.

$$n = N_d = 10^{17} cm^{-3}$$

From Fig. 1–20, Example 1–3, E_F is located at 146 meV below E_c. The donor level E_d is located at 45 meV below E_c for phosphorus (see Table 1–2 and Figure 1–23).

The probability that a donor is not ionized, i.e., the probability that it is occupied by the "extra" electron, according to Eq. (1.7.1), is

$$\text{Probability of non-ionization} \approx \frac{1}{1 + \frac{1}{2} e^{(E_d - E_F)/kT}}$$

$$\frac{1}{1 + \frac{1}{2} e^{((146-45)meV)/26meV}} = 3.9\%$$

(The factor 1/2 in the denominators stems from the complication that a donor atom can hold an electron with upspin or downspin. This increases the probability that a donor state is occupied by an electron.)

Therefore, it is reasonable to assume complete ionization, i.e., $n = N_d$.

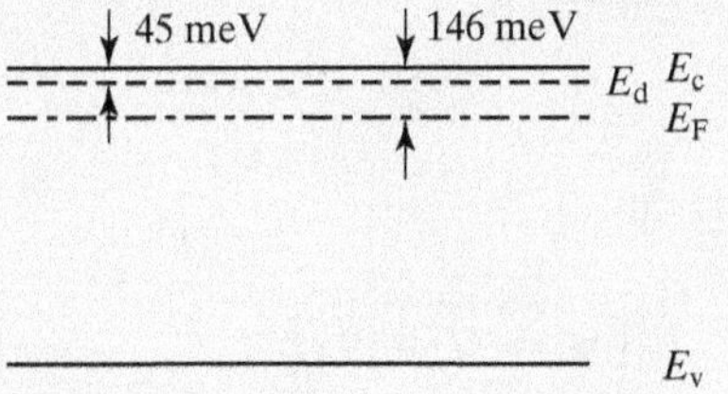

FIGURE 1–23 Location of E_F and E_d. Not to scale.

DISCUSSION: You may have noticed that as N_d increases, E_F rises toward E_d and the probability of nonionization can become quite large. In reality, the impurity level broadens into an **impurity band** that merges with the conduction band in heavily doped semiconductor (i.e., when donors or acceptors are close to one another). This happens for the same reason energy levels broaden into bands when atoms are brought close to one another to form a crystal (see Fig. 1–9). The electrons in the impurity band are also in the conduction band. *Therefore, the assumption of $n = N_d$ (or complete ionization) is reasonable even at very high doping densities.* The same holds true in P-type materials.

There are four types of charged species in a semiconductor: electrons, holes, positive donor ions, and negative acceptor ions. Their densities are represented by the symbols n, p, N_d, and N_a. In general, all samples are free of net charge. **Charge neutrality** requires that the densities of the negative particles and positive particles are equal:

$$n + N_a = p + N_d \tag{1.9.1}$$

Equations (1.8.11) and (1.9.1) can be solved for n and p:

$$n = \frac{N_d - N_a}{2} + \left[\left(\frac{N_d - N_a}{2}\right)^2 + n_i^2\right]^{1/2} \tag{1.9.2a}$$

$$p = \frac{N_a - N_d}{2} + \left[\left(\frac{N_a - N_d}{2}\right)^2 + n_i^2\right]^{1/2} \tag{1.9.2b}$$

Although it is interesting to know that n and p can be calculated for arbitrary values of N_a and N_d, the complicated Eq. (1.9.2) is rarely used. Instead, one of the following two cases is almost always valid:

1. $N_d - N_a \gg n_i$ (i.e., N type),

$$\boxed{n = N_d - N_a} \tag{1.9.3a}$$

$$\boxed{p = n_i^2/n} \tag{1.9.3b}$$

If, furthermore, $N_d \gg N_a$, then

$$n = N_d \quad \text{and} \quad p = n_i^2/N_d \tag{1.9.4}$$

2. $N_a - N_d \gg n_i$ (i.e., P type),

$$\boxed{p = N_a - N_d} \tag{1.9.5a}$$

$$\boxed{n = n_i^2/p} \tag{1.9.5b}$$

If, furthermore, $N_a \gg N_d$, then

$$p = N_a \quad \text{and} \quad n = n_i^2/N_a \tag{1.9.6}$$

We have intuitively assumed Eqs. (1.9.4) and (1.9.6) to be true in the previous sections. It is worthwhile to remember that Eqs. (1.9.3) and (1.9.5) as the more exact expressions. We see that an acceptor can effectively negate the effect of a donor in Eq. (1.9.3a) and vice versa in Eq. (1.9.5a). This fact is known as **dopant compensation**. One can even start with P-type Si and convert a portion of it into N-type simply by adding enough donors. This is one of the techniques employed to make complex devices.

EXAMPLE 1–7 Dopant Compensation

What are n and p in a Si sample with $N_d = 6 \times 10^{16} \text{cm}^{-3}$ and $N_a = 2 \times 10^{16} \text{cm}^{-3}$? With additional $6 \times 10^{16} \text{cm}^{-3}$ of acceptors?

SOLUTION: As shown in Fig. 1–24a:

$$n = N_d - N_a = 4 \times 10^{16} \text{cm}^{-3}$$

$$p = n_i^2/n = 10^{20}/4 \times 10^{16} = 2.5 \times 10^3 \text{cm}^{-3}$$

With the additional acceptors, $N_a = 2 \times 10^{16} + 6 \times 10^{16} = 8 \times 10^{16} \text{cm}^{-3}$, holes become the majority,

$$p = N_a - N_d = 8 \times 10^{16} - 6 \times 10^{16} = 2 \times 10^{16} \text{cm}^{-3}$$

$$n = n_i^2/p = 10^{20}/(2 \times 10^{16}) = 5 \times 10^3 \text{cm}^{-3}$$

The addition of acceptors has converted the Si to P-type as shown in Fig. 1–24b.

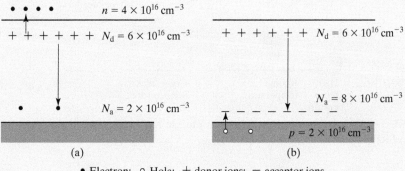

• Electron; ○ Hole; + donor ions; − acceptor ions

FIGURE 1–24 Graphical illustration of dopant compensation.

1.10 • CARRIER CONCENTRATIONS AT EXTREMELY HIGH AND LOW TEMPERATURES[4] •

At very high temperatures, n_i is large [see Eq. (1.8.12)], and it is possible to have $n_i \gg |N_d - N_a|$. In that case, Eq. (1.9.2) becomes

$$n = p = n_i \qquad (1.10.1)$$

In other words, the semiconductor becomes "intrinsic" at very high temperatures.

At the other extreme of very low temperature, E_F may rise above E_d, and most of the donor (or acceptor, in the case of P-type material) atoms can remain nonionized. The fifth electrons stay with the donor. This phenomenon is called **freeze-out**. In this case, if the doping is not heavy enough to form an impurity band (see Section 1.9), the dopants are not totally ionized. The carrier concentration may be significantly less than the dopant concentration. The exact analysis is complicated, but the result is [4]

$$n = \left[\frac{N_c N_d}{2}\right]^{1/2} e^{-(E_c - E_d)/2kT} \qquad (1.10.2)$$

Freeze-out is a concern when semiconductor devices are operated at, for example, the liquid–nitrogen temperature (77 K) in order to achieve low noise and high speed.

Figure 1–25 summarizes the temperature dependence of majority carrier concentrations. The slope of the curve in the intrinsic regime is $E_g/2k$, and the slope in the freeze-out portion is $(E_c - E_d)/2k$ (according to Eq. (1.10.2)). These facts may be used to determine E_g and $E_c - E_d$.

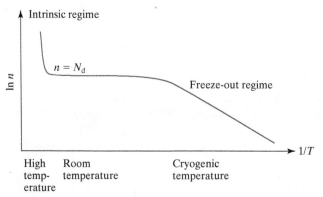

FIGURE 1–25 Variation of carrier concentration in an N-type semiconductor over a wide range of temperature.

---

[4] This section may be omitted in an accelerated course.

● **Infrared Detector Based on Freeze-Out** ●

Often it is desirable to detect or image the black-body radiation emitted by warm objects, e.g., to detect tumors (which restrict blood flow and produce cold spots), to identify inadequately insulated building windows, to detect people and vehicles at night, etc. This requires a photodetector that responds to photon energies around 0.1 eV. For this purpose, one can use a semiconductor photoconductor with E_g less than 0.1 eV, such as HgPbTe operating in the mode shown in Fig. 1–11. Alternatively, one can use a more common semiconductor such as doped Si operating in the freeze-out mode shown in Fig. 1–26. In Fig. 1–26, conduction electrons are created when the infrared photons provide the energy to ionize the donor atoms, which are otherwise frozen-out. The result is a lowering of the detector's electrical resistance, i.e., photoconductivity.

At long enough wavelength or low enough photon energy $h\nu$, light will no longer be absorbed by the specimen shown in Fig. 1–26. That critical $h\nu$ corresponds to $E_c - E_d$. This is a method of measuring the dopant ionization energy, $E_c - E_d$.

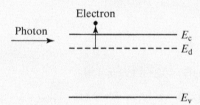

FIGURE 1–26 Infrared photons can ionize the frozen-out donors and produce conduction electrons.

1.11 ● CHAPTER SUMMARY ●

In a silicon crystal, each Si atom forms covalent bonds with its four neighbors. In an intrinsic Si crystal, there are few mobile electrons and holes. Their concentrations are equal to n_i (~10^{10} cm^{-3} for Si), the intrinsic carrier concentration. The bond model explains why group V atoms can serve as donors and introduce electrons, and group III atoms can serve as acceptors and introduce holes. Electrons and holes should be thought of as particles of equal importance but carrying negative and positive charge, respectively, and having effective masses, m_n and m_p, which are a fraction of the free electron mass.

The band model is needed for quantitative analysis of semiconductors and devices. The valence band and conduction band are separated by an energy gap. This band gap is 1.12 eV for Si. Thermal agitation gives each energy state a certain probability of being occupied by an electron. That probability is expressed by the Fermi function, which reduces to simple exponential functions of $(E - E_F)/kT$ for E more than a few kT's above the Fermi level, E_F. E_F is related to the density of electrons and holes in the following manner:

$$n = N_c \, e^{-(E_c - E_F)/kT} \tag{1.8.5}$$

$$p = N_v e^{-(E_F - E_v)/kT} \tag{1.8.8}$$

The effective densities of states, N_c and N_v, are around 10^{19}cm^{-3}. The majority carrier concentrations are

$$n = N_d - N_a \quad \text{for N-type semiconductor} \tag{1.9.3a}$$

$$p = N_a - N_d \quad \text{for P-type semiconductor} \tag{1.9.5a}$$

where N_d and N_a are the concentrations of donors and acceptors, which are usually assumed to be completely ionized. The minority carrier concentrations can be found from

$$np = n_i^2 \tag{1.8.11}$$

n_i is the intrinsic carrier concentration, about 10^{10}cm^{-3} for Si at 300 K. It is a function of E_g and T.

You are now ready to study how electrons and holes move and produce current in the next chapter.

● PROBLEMS ●

● Visualization of the Silicon Crystal ●

1.1 **(a)** How many silicon atoms are there in each unit cell?

(b) How many silicon atoms are there in one cubic centimeter?

(c) Knowing that the length of a side of the unit cell (the silicon **lattice constant**) is 5.43 Å, Si atomic weight is 28.1, and the Avogdaro's number is 6.02×10^{23} atoms/mole, find the silicon density in g/cm^3.

● Fermi Function ●

1.2 **(a)** Under equilibrium condition, what is the probability of an electron state being occupied if it is located at the Fermi level?

(b) If E_F is positioned at E_c, determine the probability of finding electrons in states at $E_c + kT$. (A numerical answer is required.)

(c) The probability of a state being filled at $E_c + kT$ is equal to the probability of a state being empty at $E_c + 3\,kT$. Where is the Fermi level located?

1.3 **(a)** What is the probability of an electron state being filled if it is located at the Fermi level?

(b) If the probability that a state being filled at the conduction band edge (E_c) is precisely equal to the probability that a state is empty at the valence band edge (E_v), where is the Fermi level located?

(c) The Maxwell–Boltzmann distribution is often used to approximate the Fermi–Dirac distribution function. On the same set of axes, sketch both distributions as a function of $(E - E_F)/kT$. Consider only positive values of $E - E_F$. For what range of $(E - E_F)/kT$ is the Maxwell–Boltzmann approximation accurate to within 10%?

1.4 Refer to the oxygen concentration example in Sec. 1.7.2.
  (a) Given that nitrogen is lighter in weight than oxygen, is N_2 concentration at 10 km more or less that 25% of the sea level N_2 concentration?
  (b) What is the ratio of N_2 concentration to O_2 at 10 km? At sea level, the ratio is 4 to 1.

1.5 Show that the probability of an energy state being occupied ΔE above the Fermi level is the same as the probability of a state being empty ΔE below the Fermi level.

$$f(E_F + \Delta E) = 1 - f(E_F - \Delta E)$$

1.6 (a) Sketch the Fermi–Dirac distribution $f(E)$ at room temperature (300 K) and at a lower temperature such as 150 K. (Qualitative hand drawing.)
  (b) The state distribution in a system is given in Fig. 1–27, where each circle represents two electron states (one is spin-up; one is spin-down). Each electron state can be occupied by one electron. There is no state below E_{min}. The Fermi level at 0 K is given in Fig. 1–27. How many electrons are there in the system?

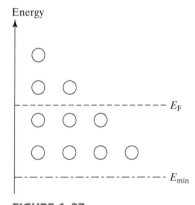

FIGURE 1–27

● **Energy: Density of States** ●

1.7 The carrier distributions in the conduction and valence bands were noted to peak at energies close to the band edges. (Refer to carrier distribution in Fig. 1–20.) Using Boltzmann approximation, show that the energy at which the carrier distribution peaks is $E_c + kT/2$ and $E_v - kT/2$ for the conduction and valence bands, respectively.

1.8 For a certain semiconductor, the densities of states in the conduction and valance bands are constants A and B, respectively. Assume non-degeneracy, i.e., E_F is not close to E_c or E_v.
  (a) Derive expressions for electron and hole concentrations.
  (b) If A = 2B, determine the location of the intrinsic Fermi energy (E_i) at 300 K with respect to the mid-bandgap of the semiconductor.
  Hint: These relationships may be useful:

$$\int_0^\infty x^{n-1} e^{-x} dx = \Gamma(n) \quad \text{(Gamma function)}$$

$$\Gamma(2) = \Gamma(1) = 1, \Gamma(3) = 2, \Gamma(4) = 6$$

$$\Gamma(1/2) = \sqrt{\pi}, \Gamma(3/2) = 1/2\sqrt{\pi}, \Gamma(5/2) = 1/3\sqrt{\pi}.$$

1.9 For a certain semiconductor, the densities of states in the conduction and valence bands are: $D_c(E) = A \cdot (E - E_c) \cdot u(E - E_c)$ and $D_v(E) = B \cdot (E_v - E) \cdot u(E_v - E)$, respectively. $u(x)$, the unit step function, is defined as $u(x) = 0$ if $x < 0$ and $u(x) = 1$ if $x > 0$. Assume nondegeneracy, i.e. not too highly doped. You may find this fact useful:

$$\int_0^\infty x e^{-x} dx = 1$$

(a) Derive expressions for electron and hole concentrations as functions of the Fermi energy, E_F.

(b) If $A = 2B$, compute the intrinsic Fermi energy at 300 K.

1.10 The Maxwell–Boltzmann distribution function $f(E) = e^{-(E - E_f)/kT}$ is often used as an approximation to the Fermi–Dirac function. Use this approximation and the densities of the states in the conduction band $D_c(E) = A(E - E_c)^{1/2}$ to find:

(a) The energy at which one finds the most electrons ($1/cm^3 \cdot eV$).

(b) The conduction-band electron concentration (explain any approximation made).

(c) The ratio of the peak electron concentration at the energy of (a) to the electron concentration at $E = E_c + 40\,kT$ (about 1eV above E_c at 300 K). Does this result justify one of the approximations in part(b)?

(d) The average kinetic energy, $E - E_c$ of the electrons.
Hint: These relationships may be useful:

$$\int_0^\infty x^{n-1} e^{-x} dx = \Gamma(n) \quad \text{(Gamma function)}$$

$\Gamma(2) = \Gamma(1) = 1, \Gamma(3) = 2, \Gamma(4) = 6$

$\Gamma(1/2) = \sqrt{\pi}, \Gamma(3/2) = 1/2\sqrt{\pi}, \Gamma(5/2) = 3/4\sqrt{\pi}$.

● **Electron and Hole Concentrations** ●

1.11 (a) The electron concentration in a piece of Si at 300 K is 10^5 cm^{-3}. What is the hole concentration?

(b) A semiconductor is doped with impurity concentrations N_d and N_a such that $N_d - N_a \gg n_i$ and all the impurities are ionized. Determine n and p.

(c) In a silicon sample at $T = 300$ K, the Fermi level is located at 0.26 eV ($10\,kT$) above the intrinsic Fermi level. What are the hole and electron concentrations?

(d) What are the hole and electron concentration at $T = 800$ K for the sample in part (c), and where approximately is E_F? Comment on your results.

● **Nearly Intrinsic Semiconductor** ●

1.12 For a germanium sample at room temperature, it is known that $n_i = 10^{13}$ cm^{-3}, $n = 2p$, and $N_a = 0$. Determine n and N_d.

1.13 Boron atoms are added to a Si film resulting in an impurity density of 4×10^{16} cm^{-3}.

(a) What is the conductivity type (N-type or P-type) of this film?

(b) What are the equilibrium electron and hole densities at 300 K and 600 K?

(c) Why does the mobile carrier concentration increase at high temperatures?

(d) Where is the Fermi level located if $T = 600$ K?

● **Incomplete Ionization of Dopants and Freeze-Out** ●

1.14 Suppose you have samples of Si, Ge, and Ge, and GaAs at T = 300 K, all with the same doping level of $N_d^+ - N_a^- = 3 \times 10^{15}$ /cm^3. Assuming all dopants are ionized, for which material is p most sensitive to temperature (the sensitivity of p is define by $\delta p/\delta T$)? What is your conclusion regarding the relation between E_g and temperature sensitivity of minority carrier concentration? Repeat the problem using $(\delta p/\delta T)/p$ as the definition of sensitivity.

1.15 An N-type sample of silicon has uniform density ($N_d = 10^{19}$/cm^{-3}) of arsenic, and a P-type silicon sample has a uniform density ($N_a = 10^{15}$/cm^{-3}) of boron. For each sample, determine the following:

(a) The temperature at which the intrinsic concentration n_i exceeds the impurity density by factor of 10.

(b) The equilibrium minority-carrier concentrations at 300 K. Assume full ionization of impurities.

(c) The Fermi level relative to the valence–band edge E_v in each material at 300 K.

(d) The electron and hole concentrations and the Fermi level if both types of impurities are present in the same sample.

1.16 A silicon sample is doped with $N_d = 10^{17}$cm^{-3} of As atoms.

(a) What are the electron and hole concentrations and the Fermi level position (relative to E_c or E_v) at 300 K? (Assume full ionization of impurities.)

(b) Check the full ionization assumption using the calculated Fermi level, (i.e., find the probability of donor states being occupied by electrons and therefore not ionized.) Assume that the donor level lies 50 meV below the conduction band, (i.e., $E_c - E_D = 50$ meV.)

(c) Repeat (a) and (b) for $N_d = 10^{19}$cm^{-3}. (Discussion: when the doping concentration is high, donor (or acceptor) **band** is formed and that allows all dopant atoms to contribute to conduction such that "full ionization" is a good approximation after all).

(d) Repeat (a) and (b) for $N_d = 10^{17}$cm^{-3} but T = 30K. (This situation is called dopant freeze-out.)

1.17 Given N-type silicon sample with uniform donor doping of (a) $N_d = 10^{18}$/cm^3, (b) $N_d = 10^{19}$/cm^3, and (c) $N_d = 10^6$/cm^3, calculate the Fermi levels at room temperature assuming full ionization for all cases. Check whether the above assumption of full ionization of each case is correct with the calculated Fermi level. When this is not correct, what is the relative position of E_F and E_D? Assume that

$$E_c - E_D = 0.05 \text{ eV}.$$

● **REFERENCES** ●

1. Shockley, W. *Electrons and Holes in Semiconductors*. Princeton, NJ: Van Nostrand, 1950.

2. Shur, M. *Physics of Semiconductor Devices*. Englewood Cliffs, NJ: Prentice-Hall, Inc., 1990.

3. Neamen, D. *Semiconductor Physics and Devices*, 3rd ed. New York: McGraw-Hill, 2003.

4. Smith, R.A. *Semiconductors*, 2nd ed. London: Cambridge University Press, 1979.

5. Streetman, B.G. *Solid State Electronic Devices*, 6th ed. Upper Saddle River, NJ: Prentice-Hall, Inc., 2006.

6. Pierret, R. F. *Modular Series on Solid State Devices*, Vol. I. Reading, MA: Addison-Wesley Publishing Co., 1983.

7. Sze, S. M. *Physics of Semiconductor Devices*, 2nd ed. New York: John Wiley & Sons, 1981.
8. Taur, Y., and T. H. Ning. *Fundamentals of Modern VLSI Devices*. Cambridge, UK: Cambridge University Press, 1998.

● **GENERAL REFERENCES** ●

1. Neamen, D. *Semiconductor Physics and Devices,* 3rd ed. New York: McGraw-Hill, 2003.
2. Streetman, B.G. *Solid State Electronic Devices*, 6th ed. Upper Saddle River, NJ: Prentice-Hall, Inc., 2006.

2

Motion and Recombination of Electrons and Holes

CHAPTER OBJECTIVES

The first chapter builds the necessary model for understanding semiconductors at equilibrium. This chapter will consider how the electrons and holes respond to an electric field and to a gradient in the carrier concentration. It is the response of charge carriers to these disturbances that gives life to the myriad of semiconductor devices. This chapter also introduces recombination and its opposite, generation. They are nature's ways of restoring the carrier concentrations to the equilibrium value by annihilating and creating electron–hole pairs.

2.1 • THERMAL MOTION •

Even without an applied electric field, carriers are not at rest but possess finite kinetic energies. The average **kinetic energy** of electrons, $E - E_c$, can be calculated in the following manner:

$$\text{Average electron kinetic energy} = \frac{\text{total kinetic energy}}{\text{number of electrons}}$$
$$= \frac{\int f(E)D(E)(E - E_c)dE}{\int f(E)D(E)dE} \quad (2.1.1)$$

The integration in Eq. (2.1.1) is to be carried out over the conduction band, and the same approximations used in the derivation of Eq. (1.8.5) can be used here. The result is

$$\text{Average kinetic energy} = \frac{3}{2}kT \quad (2.1.2)$$

It can be shown that Eq. (2.1.2) is true for both electrons and holes.[1] The kinetic energy in Eq. (2.1.2) may be used to estimate the **thermal velocity**, v_{th}, of electrons and holes by equating the energy with $m_n v_{th}^2/2$ or $m_p v_{th}^2/2$.

$$v_{th} = \sqrt{\frac{3kT}{m}} \qquad (2.1.3)$$

EXAMPLE 2-1   Thermal Velocity

What are the approximate thermal velocities of electrons and holes in silicon at room temperature?

SOLUTION: Assume $T = 300$ K and recall $m_n = 0.26\, m_0$.

$$\text{Kinetic energy} = \frac{1}{2}m_n v_{th}^2 = \frac{3}{2}kT$$

$$v_{th} = \sqrt{\frac{3kT}{m}} \qquad (2.1.3)$$

$$= (3 \times 1.38 \times 10^{-23} \text{J/K} \times (300 \text{ K}/0.26 \times 9.1 \times 10^{-31} \text{kg}))^{1/2}$$

$$= 2.3 \times 10^5 \text{m/s} = 2.3 \times 10^7 \text{cm/s}$$

Note that $1\, J = 1\, \text{kg·m}^2/\text{s}^2$. Using $m_p = 0.39\, m_0$ instead of m_n, one would find the hole thermal velocity to be 2.2×10^7 cm/s. So, the typical thermal velocity of electrons and holes is 2.5×10^7 cm/s, which is about 1000 times slower than the speed of light and 100 times faster than the sonic speed.

Electrons and holes move at the thermal velocity but not in a simple straight-line fashion. Their directions of motion change frequently due to **collisions** or **scattering** with imperfections in the crystal, more about which will be said in Section 2.2. The carriers move in a zigzag fashion as shown in Fig. 2–1. The **mean free time** between collisions is typically 10^{-13}s or 0.1 ps (picosecond), and the distance between collisions is a few tens of nanometers or a few hundred angstroms. The *net* thermal velocity (averaged over time or over a large number of carriers at any given time) is zero. Thus, thermal motion does not create a steady electric current, but it does introduce a thermal noise.

---

[1] In fact, Eq. (2.1.2) is applicable to many kinds of particles and is known as the **equal-partition principle** because the kinetic energy of motion is equally partitioned among the three dimensions $(x, y, z) - kT/2$ for each direction.

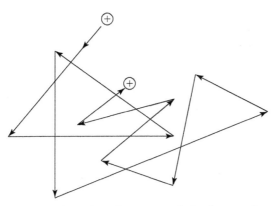

FIGURE 2–1 The thermal motion of an electron or a hole changes direction frequently by scattering off imperfections in the semiconductor crystal.

● **Hot-Point Probe, Thermoelectric Generator and Cooler** ●

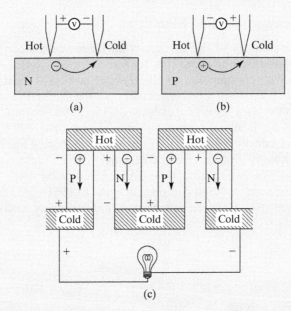

FIGURE 2–2 (a, b) Hot-point probe test can determine the doping type of a sample. (c) A thermoelectric generator converts heat into electric power.

Thermal motion can create a net current flow when there is a temperature difference. Figure 2–2a shows an N-type semiconductor sample. A cold (room-temperature) metal probe is placed on the sample close to a hot probe, perhaps a soldering iron. The electrons around the hot probe have higher thermal velocity and therefore on average move toward the cold side at a higher rate than the electrons on the cold side move to the hot side. The imbalance causes the electrons to accumulate on the cold side and build up a negative voltage, which is detected with a voltmeter.

Figure 2–2b shows that a positive voltage would be registered on the cold side if the sample is P-type. This is a practical and simple technique of testing the doping type of a semiconductor sample. It is called the **hot-point probe test**.

With engineering optimization, not only voltage but also significant electric power can be created with a temperature difference across P- and N-type semiconductor elements. Figure 2–2c is a schematic drawing of a **thermoelectric generator**. It powered some early space satellites using the radioactive decay of radioactive materials as the heat source. If, instead of extracting power from the device, current is fed into it, one set of the junctions become cooler than the other. The device is then a **thermoelectric cooler**. It can be used to cool a hot IC in a circuit board or beverages in a battery-powered portable cooler.

2.2 • DRIFT •

Drift is the motion of charge carriers caused by an electric field. Clearly, drift is usually at play when voltages are applied to a semiconductor device.

2.2.1 Electron and Hole Mobilities

The average velocity of the carriers is no longer zero when an electric field $\mathcal{E}$ is applied to the semiconductor. This nonzero velocity is called the **drift velocity**. The drift velocity is superimposed on the thermal motion as illustrated in Fig. 2–3. The drift velocity is so much more important than the thermal velocity in semiconductor devices that the term velocity usually means the *drift* velocity. A faster carrier velocity is desirable, for it allows a semiconductor device or circuit to operate at a higher speed. We can develop a model for the drift velocity using Fig. 2–3. Consider the case for holes. Assume that the mean free time between collisions is τ_{mp} and that the carrier loses its entire drift momentum, $m_p v$, after each collision.[2] The drift momentum gained between collisions is equal to the force, $q\mathcal{E}$, times the mean free time. Equating the loss to the gain, we can find the steady state drift velocity, v.

$$m_p v = q\mathcal{E}\tau_{mp} \qquad (2.2.1)$$

$$v = \frac{q\mathcal{E}\tau_{mp}}{m_p} \qquad (2.2.2)$$

Equation (2.2.2) is usually written as

$$\boxed{\begin{aligned} v &= \mu_p \mathcal{E} \\ \mu_p &= \frac{q\tau_{mp}}{m_p} \end{aligned}} \qquad \begin{aligned} &(2.2.3a)\\ &(2.2.4a) \end{aligned}$$

---

[2] Actually, it may take more than one collision for the carrier to lose its drift momentum. Another name for τ_{mp}, the *mean time for momentum relaxation*, is therefore more accurate. Although we will study a simplified analysis, be assured that a full analysis does lead to the same results as presented here.

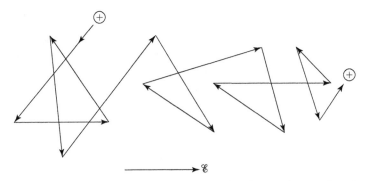

FIGURE 2–3 An electric field creates a drift velocity that is superimposed on the thermal velocity.

Equation (2.2.3a) simply says that the drift velocity is proportional to $\mathcal{E}$. The proportionality constant μ_p is the **hole mobility**, a metric of how mobile the holes are. Similarly, electron drift velocity and **electron mobility** are

$$v = -\mu_n \mathcal{E} \tag{2.2.3b}$$

$$\mu_n = \frac{q\tau_{mn}}{m_n} \tag{2.2.4b}$$

The negative sign in Eq. (2.2.3b) means that the electrons drift in a direction opposite to the field $\mathcal{E}$. They do so because the electron is negatively charged. We should memorize these statements rather than the negative sign.

Carrier mobility has the same dimension as $v/\mathcal{E}$, i.e., cm^2/V·s. Table 2–1 shows some mobility values. Notice that GaAs has a much higher μ_n than Si (due to a smaller m_n). Thus, higher-speed transistors can be made with GaAs, which are typically used in communications equipment. InAs has an even higher μ_n, but the technology of fabricating InAs devices has not yet been fully developed.

TABLE 2–1 • Electron and hole mobilities at room temperature of selected lightly doped semiconductors.

	Si	Ge	GaAs	InAs
μ_n (cm^2/V·s)	1400	3900	8500	30,000
μ_p (cm^2/V·s)	470	1900	400	500

> **EXAMPLE 2–2 Drift Velocity, Mean Free Time, and Mean Free Path**
>
> Given $\mu_p = 470$ cm^2/V·s for Si, what is the hole drift velocity at $\mathscr{E} = 10^3$ V/cm? What is τ_{mp} and what is the average distance traveled between collisions, i.e., the **mean free path**?
>
> **SOLUTION:** $v = \mu_p \mathscr{E} = 470$ cm^2/V·s $\times 10^3$ V/cm $= 4.7 \times 10^5$ cm/s
>
> This is much lower than the thermal velocity, $\sim 2.1 \times 10^7$ cm/s.
>
> From Eq. (2.2.4b),
>
> $$\tau_{mp} = \mu_p m_p/q = 470 \text{ cm}^2 \times 0.39 \times 9.1 \times 10^{-31} \text{kg}/1.6 \times 10^{-19} \text{C}$$
>
> $$= 0.047 \text{ m}^2 \times 2.2 \times 10^{-12} \text{ kg/C} = 1 \times 10^{-13} \text{s} = 0.1 \text{ ps}$$
>
> $$\text{Mean free path} = \tau_{mp} v_{th} \sim 1 \times 10^{-13} \text{s} \times 2.2 \times 10^7 \text{cm/s}$$
>
> $$= 2.2 \times 10^{-6} \text{cm} = 220 \text{ Å} = 22 \text{ nm}$$

2.2.2 Mechanisms of Carrier Scattering

We will now present a more detailed description of carrier collisions and show that τ_{mn} and τ_{mp} in Eq. (2.2.4) can vary significantly with temperature and the doping concentration.

What are the imperfections in the crystal that cause carrier collisions or scattering? There are two main causes: **phonon scattering** and **ionized impurity scattering**. Phonons are the particle representation of the vibration of the atoms in the crystal—the same sort of vibration that carries sound, hence the term phonons. Crystal vibration distorts the periodic crystal structure and thus scatters the electron waves. Instead of electron waves and vibration waves, it is more intuitive to think of electron particles scattering off phonon particles. The mobility due to phonon scattering alone, $\mu_{phonon} = q\tau_{ph}/m$, is proportional to τ_{ph}, the mean free time of phonon scattering. But, what determines the phonon scattering mean free time? Let us use the pinball machine for analogy. In a pinball machine, the mean time of collisions between the ball and the pins is inversely proportional to the density of the pins and the speed of the ball. Similarly, the mean free time of phonon scattering is inversely proportional to the phonon density and the electron speed, which is basically the thermal velocity. In addition, the phonon density is known to be proportional to the absolute temperature, T.

$$\mu_{phonon} \propto \tau_{ph}$$

$$\propto \frac{1}{\text{phonon density} \times \text{carrier thermal velocity}} \propto \frac{1}{T \cdot T^{1/2}} \propto T^{-3/2} \quad (2.2.5)$$

So, the phonon scattering mobility decreases when temperature rises. What about the impurity scattering mobility? The dopant ions are fixed charge in the semiconductor crystal. They can make electrons and holes change the direction of motion through the coulombic force. An electron can be scattered by either a donor (positive) ion or an acceptor (negative) ion as shown in Fig. 2–4. The same is true for a hole.

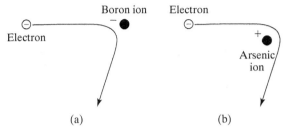

FIGURE 2–4 An electron can be scattered by an acceptor ion (a) and a donor ion (b) in a strikingly similar manner, even though the ions carry opposite types of charge. The same is true for a hole (not shown).

In Fig. 2–4a, the repulsive coulombic force between the electron and the negative ion deflects the motion of the electron. Figure 2–4b shows that an attractive coulombic force can induce the same effect on the electron trajectory.[3] The mobility due to impurity scattering is therefore inversely proportional to the sum of the donor and acceptor ion concentrations. It is also proportional to $T^{3/2}$. Why is the mobility higher, i.e., the scattering weaker, at a higher T? At a higher temperature, the electron in Fig. 2–4 has a higher thermal velocity and flies by the ion in a shorter time, and its direction of motion is thus less affected by the ion. A sports analogy: a ball carrier that charges by a blocker at high speed gives the blocker less of a chance to stop him.

$$\mu_{impurity} \propto \frac{T^{3/2}}{N_a + N_d} \tag{2.2.6}$$

When there is more than one scattering mechanism, the total scattering rate $(1/\tau)$ and therefore the total mobility are determined by the sum of the inverses.

$$\frac{1}{\tau} = \frac{1}{\tau_{phonon}} + \frac{1}{\tau_{impurity}} \tag{2.2.7a}$$

$$\frac{1}{\mu} = \frac{1}{\mu_{phonon}} + \frac{1}{\mu_{impurity}} \tag{2.2.7b}$$

Figure 2–5 shows that the silicon hole mobility is about one-third of the electron mobility. Part of this difference in mobility can be explained by the difference in the effective mass (see Eq. (2.2.4)). The rest is attributable to the difference in the scattering mean free time. The mobilities may be expressed as (with N_a and N_d in per cubic centimeter) [1, 2].

$$\mu_p(cm^2/V \cdot s) = \frac{420}{1 + [(N_a + N_d)/1.6 \times 10^{17}]^{0.7}} + 50 \tag{2.2.8}$$

---

[3] This is how a space probe uses the attractive gravitational force of a planet to change its course in a "slingshot" manner.

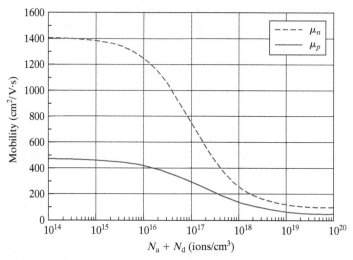

FIGURE 2–5 The electron and hole mobilities of silicon at 300 K. At low dopant concentration, the electron mobility is dominated by phonon scattering; at high dopant concentration, it is dominated by impurity ion scattering. (After [3].)

$$\mu_n(\text{cm}^2/\text{V} \cdot \text{s}) = \frac{1318}{1 + [(N_a + N_d)/1 \times 10^{17}]^{0.85}} + 92 \quad (2.2.9)$$

You may notice that the inverse proportionality to dopant density (Eq. (2.2.6)) is not followed in Eqs. (2.2.8) and (2.2.9) at the limit of very large N_a or N_d. The reason is **free-carrier screening.** When the carrier concentration is large, the carriers can distribute themselves to partially screen out the coulombic field of the dopant ions.

QUESTION • *The electron mobility of an N-type silicon sample at room temperature is measured to be 600 $\text{cm}^2/\text{V·s}$. Independent measurement shows that the electron concentration is $n = 5 \times 10^{16}$ cm^{-3}. According to Fig. 2–5, μ_n should be significantly larger than 600 $\text{cm}^2/\text{V·s}$ if $N_d = 5 \times 10^{16}$ cm^{-3}. What do you think may be responsible for the discrepancy? Be as quantitative as you can. Hint: Consider possible dopant compensation (Section 1.9).*

Figure 2–6 shows a schematic plot of the temperature dependence of μ_n. At small dopant concentrations, μ decreases with increasing T, indicative of the dominance of phonon scattering (Eq. (2.2.5)). At very high dopant concentration and low temperature, where impurity scattering is expected to dominate, μ indeed increases with increasing T (Eq. (2.2.6)). The mobility data in Figs. 2–5 and 2–6 (and Fig. 2–8) do not agree perfectly. This goes to show that it is not easy to measure mobility accurately and that we should presume the existence of uncertainties in experimental data in general.

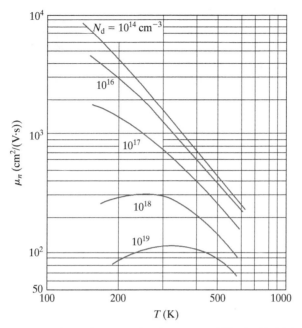

FIGURE 2–6 Temperature dependence of the electron mobility in Si. (After [4], reprinted by permission of John Wiley & Sons, Inc.)

● **Velocity Saturation and Ballistic Transport** ●

In small devices, the electric field can easily reach 10^5 V/cm. If the electron mobility is 10^3 cm²/V·s, the drift velocity, according to $v = \mathscr{E}\mu$, should be 10^8 cm/s. In reality, electron and hole velocities saturate at around 10^7 cm/s and do not increase beyond that, no matter how large $\mathscr{E}$ is (as shown in Figs. 6–20 and 6–21). The culprit is **optical phonon scattering**. Optical phonons are high-energy phonons that interact strongly with the electrons and holes. When the kinetic energy of a carrier exceeds the optical phonon energy, E_{opt}, it generates an optical phonon and loses the kinetic energy. Therefore, the velocity does not rise above **saturation velocity**, v_{sat}.

Mobility and even velocity saturation (see Section 2.2.2) are concepts that describe the carrier motion averaged over many scattering events. These concepts become fuzzy when we deal with devices whose sizes are smaller than the mean free path. The motion of carriers in a nearly scattering-free environment is called **ballistic transport**. Section 6.12 presents an example of this situation.

$$\frac{1}{2}mv_{\text{sat}}^2 \approx E_{\text{opt}}$$

$$v_{\text{sat}} \approx \sqrt{2E_{\text{opt}}/m}$$

E_{opt} is about 40 meV, which puts v_{sat} at around 10^7 cm/s. **Velocity saturation** has a deleterious effect on device speed as shown in Chapter 6.

2.2.3 Drift Current and Conductivity

Let us turn our attention to the current that flows in a semiconductor as a result of carrier drift. The current density, J, is the charge per second crossing a unit area plane normal to the direction of current flow. In the P-type semiconductor bar of unit area shown in Fig. 2–7, the hole current density is

$$J_{p,\text{drift}} = qpv \tag{2.2.10}$$

For example, if $p = 10^{15}\,\text{cm}^{-3}$ and $v = 10^4\,\text{cm/s}$, then $J_{p,\text{drift}} = 1.6 \times 10^{-19}\text{C} \times 10^{15}\,\text{cm}^{-3} \times 10^4\,\text{cm/s} = 1.6\,\text{C/s·cm}^2 = 1.6\,\text{A/cm}^2$.

Employing Eq. (2.2.3a), Eq. (2.2.10) can be written as

$$J_{p,\text{drift}} = qp\mu_p \mathscr{E} \tag{2.2.11}$$

Similarly, the electron current density can be expressed as

$$J_{n,\text{drift}} = -qnv = qn\mu_n \mathscr{E} \tag{2.2.12}$$

The total drift current density is the sum of the electron and the hole components:

$$J_{\text{drift}} = J_{n,\text{drift}} + J_{p,\text{drift}} = (qn\mu_n + qp\mu_p)\mathscr{E} \tag{2.2.13}$$

The quantity in the parentheses is the **conductivity**, σ, of the semiconductor

$$\sigma = qn\mu_n + qp\mu_p \tag{2.2.14}$$

Usually only one of the components in Eq. (2.2.14) is significant because of the large ratio between the majority and minority carrier densities. The resistivity, ρ, is the reciprocal of the conductivity. The standard units of σ and ρ are A/V·cm (or S/cm, S being siemens) and Ω·cm, respectively. ρ is shown as a function of the dopant density in Fig. 2–8.

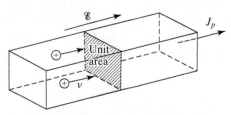

FIGURE 2–7 A P-type semiconductor bar of unit area is used to demonstrate the concept of current density.

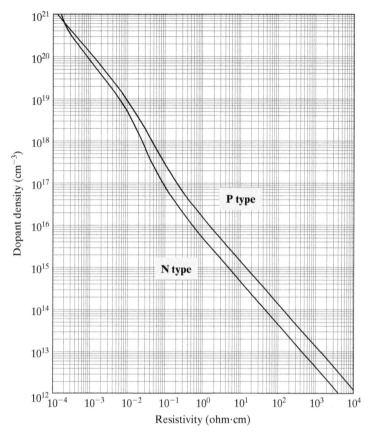

FIGURE 2–8 Conversion between resistivity and dopant density of silicon at room temperature. (After [3].)

EXAMPLE 2-3 Temperature Dependence of Resistance

a. What is the resistivity, ρ, of silicon doped with 10^{17}cm^{-3} of arsenic?

b. What is the resistance, R, of a piece of this silicon material 1 μm long and 0.1 μm² in cross-sectional area?

c. By what factor will R increase (or decrease) from $T = 300$ K to T = 400 K?

d. What As concentration should one choose if she wishes to minimize the change in (c)?

SOLUTION:

a. Using the N-type curve in Fig. 2–8, we find that $\rho = 0.084$ Ω·cm. You can also answer this question using Fig. 2–5 and Eq. (2.2.14).

b. $R = \rho \times \text{length/area} = 0.084$ Ω·cm × 1 μm / 0.1 μm²
   $= 0.084$ Ω·cm × 10^{-4}cm /10^{-9}cm^{-2} = 8.4×10^3 Ω.

c. The temperature dependent factor in σ (Eq. (2.2.14)) (and therefore in ρ) is μ_n. Figure 2–6 (10^{17}cm^{-3} curve) shows μ_n to decrease from 770 at 300 K to 400 at 400 K. We conclude that R **increases** by

$$\frac{770}{400} = 1.93$$

d. The 10^{19}cm^{-3} curve indicates nearly equal mobilities at 300 and 400 K. Therefore, that N_d would be a good choice. 1.1×10^{19} cm^{-3} would be an even better choice.

2.3 • DIFFUSION CURRENT •

In addition to the drift current, there is a second component of current called the **diffusion current**. Diffusion current is generally not an important consideration in metals because of their high conductivities. The low conductivity and the ease of creating nonuniform carrier densities make diffusion an important process in semiconductors.

Diffusion is the result of particles undergoing thermal motion as depicted in Fig. 2–1. It is the familiar process by which particles move from a point of higher particle density toward a point of lower density, as shown in Fig. 2–9. The aroma of a cup of coffee travels across a room by the diffusion of flavor molecules through the air. A drop of soy sauce spreads in a bowl of clear soup by diffusion, too.

It is known that the rate of particle movement by diffusion is proportional to the concentration gradient. If the electron concentration is not uniform, there will be an electron diffusion current, which is proportional to the gradient of the electron concentration.

$$J_{n,\text{diffusion}} \propto \frac{dn}{dx} \tag{2.3.1}$$

$$J_{n,\text{diffusion}} = qD_n\frac{dn}{dx} \tag{2.3.2}$$

We have introduced the proportional constant qD_n. q is the elementary charge $(+1.6 \times 10^{-19}$ C), and D_n is called the electron **diffusion constant**. The larger the D_n

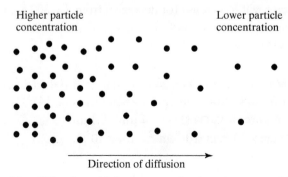

FIGURE 2–9 Particles diffuse from high-concentration locations toward low-concentration locations.

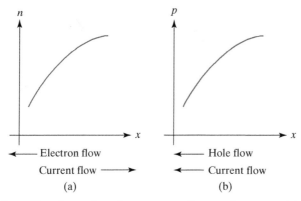

FIGURE 2–10 A positive slope of carrier concentration produces a positive electron diffusion current (a), but a negative hole diffusion current (b).

is, the faster the electrons diffuse. In Section 2.4, we will see what determines D_n. For holes,

$$J_{p,\text{diffusion}} = -qD_p \frac{dp}{dx} \qquad (2.3.3)$$

Equation (2.3.3) has a negative sign, while Eq. (2.3.2) has a positive sign. Instead of memorizing the signs, memorize Fig. 2–10. In Fig. 2–10, (a) shows a positive dn/dx (n increases as x increases) and (b) shows a positive dp/dx. In (a), electrons diffuse to the left (toward the lower concentration point). Because electrons carry negative charge, the diffusion current flows to the *right*. In (b), holes diffuse to the left, too. Because holes are positively charged, the hole current flows to the *left*, i.e., the current is negative.

In general, both drift and diffusion may contribute to the current. Therefore,

$$J_n = J_{n,\text{drift}} + J_{n,\text{diffusion}} = qn\mu_n \mathscr{E} + qD_n \frac{dn}{dx} \qquad (2.3.4)$$

$$J_p = J_{p,\text{drift}} + J_{p,\text{diffusion}} = qp\mu_p \mathscr{E} - qD_p \frac{dp}{dx} \qquad (2.3.5)$$

$$J = J_n + J_p \qquad (2.3.6)$$

2.4 • RELATION BETWEEN THE ENERGY DIAGRAM AND V, $\mathscr{E}$ •

When a voltage is applied across a piece of semiconductor as shown in Fig. 2–11a, it alters the band diagram. By definition, a positive voltage raises the potential energy of a positive charge and lowers the energy of a negative charge. It therefore lowers the energy diagrams since the energy diagram plots the energy of an electron (a negative charge). Figure 2–11c shows that the energy diagram is lower (at the left) where the voltage is higher. The band diagram is higher where the voltage is lower. (E_c and E_v are always separated by a constant, E_g.) *The point to remember is that E_c and E_v vary in the opposite direction from the voltage. E_c and E_v are higher where the voltage is lower.* That is to say

$$E_c(x) = \text{constant} - qV(x) \qquad (2.4.1)$$

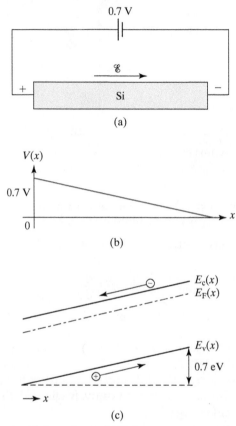

FIGURE 2–11 Energy band diagram of a semiconductor under an applied voltage. 0.7 eV is an arbitrary value.

The q takes care of the difference between the units, eV and V. The "constant" takes care of the unspecified and inconsequential zero references for E_c and V. The "constant" drops out when one considers the electric field.

$$\mathcal{E}(x) \equiv -\frac{dV}{dx} = \frac{1}{q}\frac{dE_c}{dx} = \frac{1}{q}\frac{dE_v}{dx} \qquad (2.4.2)$$

In other words, the slope of E_c and E_v indicates the electric field. The direction of $\mathcal{E}$ in Fig. 2–11c is consistent with Eq. (2.4.2). Figure 2–11c suggests the following analogies: *the electrons roll downhill like stones in the energy band diagram and the holes float up like bubbles.*

2.5 • EINSTEIN RELATIONSHIP BETWEEN D AND μ •

Consider a bar of semiconductor, whose band diagram is shown in Fig. 2–12. The semiconductor is at equilibrium, and therefore the Fermi level E_F is constant (Section 1.7). The left side is more heavily doped than the right side, and so E_c is

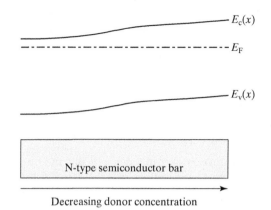

FIGURE 2–12 A piece of N-type semiconductor in which the dopant density decreases toward the right.

closer to E_F on the left side. Because E_c is not a constant, there is an electric field equal to $(1/q)\,dE_c/dx$, according to Eq. (2.4.2). This field is internally created and is as real as a field created by an external voltage. Because the semiconductor is at equilibrium, there cannot be any J_n (or J_p). From Eq. (2.3.4),

$$J_n = 0 = qn\mu_n \mathscr{E} + qD_n \frac{dn}{dx} \qquad (2.5.1)$$

Recall Eq. (1.8.5)

$$n = N_c e^{-(E_c - E_F)/kT} \qquad (1.8.5)$$

$$\frac{dn}{dx} = \frac{-N_c}{kT} e^{-(E_c - E_F)/kT} \frac{dE_c}{dx} \qquad (2.5.2)$$

$$= \frac{-n}{kT} \frac{dE_c}{dx} \qquad \text{(Eq. (1.8.5) is used)} \qquad (2.5.3)$$

$$= \frac{-n}{kT} q\mathscr{E} \qquad \text{(Eq. (2.4.2) is used)} \qquad (2.5.4)$$

dn/dx in Eq. (2.5.1) will now be substituted with Eq. (2.5.4)

$$0 = qn\mu_n \mathscr{E} - qn\frac{qD_n}{kT}\mathscr{E} \qquad (2.5.5)$$

$$\therefore \boxed{D_n = \frac{kT}{q}\mu_n} \qquad (2.5.6a)$$

The electron drift and diffusion currents will perfectly cancel each other out for an arbitrary doping profile, if and only if Eq. (2.5.6a) is satisfied. This remarkably simple relationship between the diffusion constant, D_n, and mobility was derived by Albert Einstein. A close relationship between μ and D becomes plausible when one realizes that all scattering mechanisms (e.g. phonon and impurity scatterings) that impede electron drift would also impede electron diffusion. Equation (2.5.6a) and its counterpart for holes

$$D_p = \frac{kT}{q}\mu_p \qquad (2.5.6b)$$

are known as the **Einstein relationship**.

EXAMPLE 2–4 Diffusion Constant

What is the hole diffusion constant in a piece of silicon doped with $3 \times 10^{15} \text{cm}^{-3}$ of donors and $7 \times 10^{15} \text{cm}^{-3}$ of acceptors at 300 K? at 400 K?

SOLUTION: Figure 2–5 shows that, for $N_a + N_d = 3 \times 10^{15} + 7 \times 10^{15} = 1 \times 10^{16} \text{cm}^{-3}$, μ_p is about 410 cm²/V·s at 300 K. $D_p = (kT/q)\mu_p = 26$ mV × 410cm²/V·s = 11cm²/s. Do you remember $kT/q = 26$ mV at room temperature?

For the 400 K case, we turn to Fig. 2–6, which shows for $N = 10^{16} \text{cm}^{-3}$ and $T = 127°C$, $\mu_p = 220$ cm²/V·s.

$D_p = \mu_p(kT/q) = 220\text{cm}^2/\text{V·s} \times 26 \text{ mV} \times (400\text{ K}/300\text{ K}) = 7.6\text{cm}^2/\text{V·s}$.

2.6 • ELECTRON–HOLE RECOMBINATION •

The electron and hole concentrations introduced in Chapter 1 (for example, $n = N_d$ and $p = n_i^2/N_d$ for an N-type sample) are the **equilibrium carrier concentrations** and they will be denoted with n_0 and p_0 from now on. The electron and hole concentrations can be different from n_0 and p_0, for example, when light shines on the sample and generates electrons and holes (as shown in Fig. 1–12). The differences are known as the **excess carrier concentrations** denoted by n' and p'.

$$n \equiv n_0 + n' \qquad (2.6.1a)$$
$$p \equiv p_0 + p' \qquad (2.6.1b)$$

If n' and p' are created by light, n' and p' are equal because the electrons and holes are created in pairs. If n' and p' are introduced by other means, they will still be equal to each other because of **charge neutrality** (Eq. (1.9.1)). Since charge neutrality is satisfied at equilibrium when $n' = p' = 0$, any time a non-zero n' is present, an equal p' must be present to maintain the charge neutrality, and vice versa. Otherwise, the net charge will attract or repel the abundant majority carriers until neutrality is restored. This conclusion for a charge neutral sample can be written as

$$n' \equiv p' \qquad (2.6.2)$$

If the light is suddenly turned off, n' and p' will decay with time until they become zero and n and p return to their equilibrium values, n_0 and p_0. The process of decay is **recombination**, whereby an electron and a hole recombine and annihilate each other. The time constant of the decay is called the **recombination time** or **carrier lifetime**, τ.

$$\frac{dn'}{dt} = -\frac{n'}{\tau} = -\frac{p'}{\tau} \quad (2.6.3)$$

The recombination rate (per cubic centimeter per second) is proportional to n' and p'.

$$\text{Recombination rate} = \frac{n'}{\tau} = \frac{p'}{\tau} \quad (2.6.4)$$

τ has the dimension of time and is typically around 1 μs in Si. It may range from 1 ns to 1 ms, depending on the density of trace metal impurities such as Au and Pt, which form traps in the band gap with several energy levels deep in the band gap. These **deep traps** can capture electrons or holes to facilitate recombination (as shown in Fig. 2–13) and thus shorten the recombination time (they are also called **recombination centers**). Too small a τ is bad for device leakage current, and hence extreme cleanliness is maintained in the semiconductor fabrication plants partly to avoid these metallic contaminants. The other recombination process shown in Fig. 2–13, **direct recombination**, or **radiative recombination**, is very inefficient and unimportant in silicon because the electrons and holes at the edges of the band gap do not have the same wave vectors (see Section 1.5.1 and Figure 4–27b). These types of semiconductors are called **indirect gap semiconductors**. The radiative recombination process is very efficient in **direct gap semiconductors** such as GaAs because the electrons and holes have the same wave vectors (see Figure 4–27a) and is responsible for light emission in light-emitting diodes and lasers, which will be presented in Chapter 4.

Echoes in a canyon eventually die out. So do ripples produced by a cast stone. Nature provides ways to restore equilibrium. Recombination is nature's way of restoring n' and p' to zero.

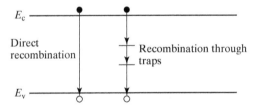

FIGURE 2–13 An electron–hole pair recombines when an electron drops from the conduction band into the valence band. In silicon, direct recombination is unimportant and the lifetime is highly variable and determined by the density of recombination centers.

EXAMPLE 2-5 Photoconductors

A bar of Si is doped with boron at 10^{15} cm^{-3}. It is exposed to light such that electron–hole pairs are generated throughout the volume of the bar at the rate of 10^{20}/s·cm^3. The recombination lifetime is 10 μs. What are (a) p_0, (b) n_0, (c) p', (d) n', (e) p, (f) n, and (g) the np product? (h) If the light is suddenly turned off at t = 0, find $n'(t)$ for t > 0.

SOLUTION:

a. $p_0 = N_a = 10^{15}$ cm^{-3} is the equilibrium hole concentration.

b. $n_0 = n_i^2/p_0 \approx 10^5$ cm^{-3} is the equilibrium electron concentration.

c. In steady state, the rate of generation is equal to the rate of recombination (Eq. (2.6.4)).

$$10^{20}/\text{s·cm}^3 = p'/\tau$$

$$\therefore p' = 10^{20}/\text{s·cm}^3 \cdot t = 10^{20}/\text{s·cm}^3 \cdot 10^{-5}\text{s} = 10^{15}\text{cm}^{-3}$$

d. $n' = p' = 10^{15}\text{cm}^{-3}$ (Eq. (2.6.2))

e. $p = p_0 + p' = 10^{15}\text{cm}^{-3} + 10^{15}\text{cm}^{-3} = 2 \times 10^{15}\text{cm}^{-3}$

f. $n = n_0 + n' = 10^5\text{cm}^{-3} + 10^{15}\text{cm}^{-3} \approx 10^{15}\text{cm}^{-3}$. **The non-equilibrium minority carrier concentration is often much much larger than the small equilibrium concentration.**

g. $np \approx 2 \times 10^{15}\text{cm}^{-3} \times 10^{15}\text{cm}^{-3} = 2 \times 10^{30}\text{cm}^{-6} \gg n_i^2 = 10^{20}\text{cm}^{-6}$. **The np product can be very different from n_i^2.**

h. The solution to Eq. (2.6.4) is

$$n'(t) = n'(0)e^{-t/\tau} = 10^{15}\text{cm}^{-3}e^{-t/\tau}$$

Therefore, n' decays exponentially toward its equilibrium value of zero. The characteristic time of the exponential decay is the carrier lifetime, τ.

2.7 • THERMAL GENERATION •

The reverse process of recombination is called **thermal generation**. At any nonzero temperature, electron–hole pairs are constantly being generated *and* lost (by recombination). If $n' = p' = 0$, the rate of recombination equals the rate of generation and the net rate of change is zero. If n' is positive, there is a net recombination rate as shown in Eq. (2.6.4). If n' is negative, i.e., there are fewer electrons than the equilibrium concentration, nature sees to it that there is a net rate of thermal generation rather than recombination. Equation (2.6.3) confirms this fact by predicting a positive dn'/dt. Later we will see that thermal generation is responsible for the leakage current in diodes.

When the np product is equal to n_i^2, the rate of thermal generation is equal to the rate of recombination. Under this condition, n and p are said to be at thermal equilibrium. When $np > n_i^2$, there is net recombination; when $np < n_i^2$, there is net generation. The terms recombination and generation rates generally refer to the *net* rates of recombination and generation.

2.8 • QUASI-EQUILIBRIUM AND QUASI-FERMI LEVELS •

Whenever $np \neq n_i^2$, the semiconductor is not at equilibrium. More precisely, the electrons and holes are not at equilibrium with each other. Nonetheless, we would like to preserve and use, as much as possible, the following equilibrium relationships, which we have found to be very useful.

2.8 • Quasi-Equilibrium and Quasi-Fermi Levels

$$n = N_c e^{-(E_c - E_F)/kT} \qquad (1.8.5)$$

$$p = N_v e^{-(E_F - E_v)/kT} \qquad (1.8.8)$$

The problem is that the above equations, when multiplied together, lead to $np = n_i^2$. We saw in Example 2-5(g) that the presence of excess carriers can easily make the np product much larger than n_i^2. This problem can be addressed by introducing two **quasi-Fermi levels**, E_{Fn} and E_{Fp}, such that

$$n = N_c e^{-(E_c - E_{Fn})/kT} \qquad (2.8.1)$$

$$p = N_v e^{-(E_{Fp} - E_v)/kT} \qquad (2.8.2)$$

E_{Fn} and E_{Fp} are the electron and hole quasi-Fermi levels. When electrons and holes are at equilibrium, i.e., when $np = n_i^2$, E_{Fn} and E_{Fp} coincide and this is known as E_F. Otherwise, $E_{Fn} \neq E_{Fp}$. Equations (2.8.1) and (2.8.2) indicate that even when electrons and holes, as two groups, are not at equilibrium with each other, the electrons (and holes) can still be at equilibrium among themselves. Electrons and holes, as two groups of particles, can get out of equilibrium easily because they are only loosely coupled by the recombination/generation mechanism, which is a slow process (has a long time constant around 1 μs). In contrast, the electrons (or holes) are strongly coupled among themselves by exchanging positions and energy through thermal motion at high speed and by scattering with 0.1 ps mean free time. The usefulness of this **quasi-equilibrium** concept will become clear in later applications.

EXAMPLE 2–6 Quasi-Fermi Levels and Low-Level Injection

Consider an Si sample with $N_d = 10^{17} \text{cm}^{-3}$.

a. Find the location of E_F.

b. Find the location of E_{Fn} and E_{Fp} when excess carriers are introduced such that $n' = p' = 10^{15} \text{cm}^{-3}$.

Notice that n' and p' are much less than the majority carrier concentration. This condition is commonly assumed and is called **low-level injection**. The opposite condition, **high-level injection**, is often encountered in bipolar transistors (Section 8.4.4).

SOLUTION:

a. Using Eq. (1.8.5)

$$n = N_d = 10^{17} \text{cm}^{-3} = N_c e^{-(E_c - E_F)/kT}$$

$$E_c - E_F = kT \ln \frac{N_c}{10^{17} \text{cm}^{-3}} = 26 \text{ meV} \cdot \ln \frac{2.8 \times 10^{19} \text{cm}^{-3}}{10^{17} \text{cm}^{-3}} = 0.15 \text{ eV}$$

E_F is below E_c by 0.15 eV.

b. $n = n_0 + n' = N_d + n' = 1.01 \times 10^{17}\,\text{cm}^{-3}$
Using Eq. (2.8.1),

$$1.01 \times 10^{17}\,\text{cm}^{-3} = N_c e^{-(E_c - E_{Fn})/kT}$$

$$E_c - E_{Fn} = kT \ln \frac{N_c}{1.01 \times 10^{17}\,\text{cm}^{-3}} = 26\,\text{meV} \cdot \ln \frac{2.8 \times 10^{19}\,\text{cm}^{-3}}{1.01 \times 10^{17}\,\text{cm}^{-3}} = 0.15\,\text{eV}$$

E_{Fn} is basically unchanged from the E_F in (a) as illustrated in Fig. 2–14.

$$p = p_0 + p' = \frac{n_i^2}{N_d} + p' = 10^3\,\text{cm}^{-3} + 10^{15}\,\text{cm}^{-3} = 10^{15}\,\text{cm}^{-3}$$

Using Eq. (2.8.2)

$$10^{15}\,\text{cm}^{-3} = N_v e^{-(E_{Fp} - E_v)/kT}$$

$$E_{Fp} - E_v = kT \ln \frac{N_v}{10^{15}\,\text{cm}^{-3}} = 26\,\text{meV} \cdot \ln \frac{1.04 \times 10^{19}\,\text{cm}^{-3}}{10^{15}\,\text{cm}^{-3}} = 0.24\,\text{eV}$$

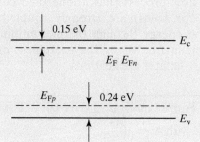

FIGURE 2–14 Location of E_F, E_{Fn}, and E_{Fp}.

2.9 ● CHAPTER SUMMARY ●

In the presence of an electric field, charge carriers gain a **drift velocity** and produce a **drift current density** in proportion to $\mathscr{E}$;

$$v_p = \mu_p \mathscr{E} \qquad (2.2.3a)$$

$$v_n = -\mu_n \mathscr{E} \qquad (2.2.3b)$$

$$J_{p,\text{drift}} = qp\mu_p \mathscr{E} \qquad (2.2.11)$$

$$J_{n,\text{drift}} = qn\mu_n \mathscr{E} \qquad (2.2.12)$$

μ_p and μ_n are called the **hole and electron mobility**. They are determined by how frequently the carriers collide with phonons or dopant ions and lose their drift momentum. Mobilities are functions of temperature and the total dopant concentration. Measured mobility data are routinely presented in figures.

The second important transport mechanism is **diffusion**. **Diffusion current density** is proportional to the gradient of the carrier concentration.

$$J_{n,\text{diffusion}} = qD_n \frac{dn}{dx} \qquad (2.3.2)$$

$$J_{p,\text{diffusion}} = -qD_p \frac{dp}{dx} \qquad (2.3.3)$$

D_n and D_p are the electron and hole diffusion constants. Both drift and diffusion are perturbations to the same thermal motion, and both are slowed down by the same collisions that are responsible for the zigzag paths of the thermal motion. As a result, D and μ are related by the **Einstein relationship**:

$$D_n = \frac{kT}{q}\mu_n \qquad (2.5.6a)$$

$$D_p = \frac{kT}{q}\mu_p \qquad (2.5.6b)$$

The sum of J_{drift} and $J_{\text{diffusion}}$ is the total current density.

The minority carrier concentration, e.g., p in an N-type semiconductor, can easily be increased from its **equilibrium concentration** p_0 by orders of magnitude with light or by another means to be presented in Chapter 4, so that

$$p \equiv p_0 + p' \gg p_0 \qquad (2.6.1b)$$

p' is the **excess hole concentration**. In a charge neutral region,

$$n' \equiv p' \qquad (2.6.2)$$

Charge nonneutrality will generate an electric field that causes the majority carriers to redistribute until neutrality is achieved.

The electron–hole **recombination rate** is proportional to n' (=p'):

$$\text{Recombination rate} = \frac{n'}{\tau} = \frac{p'}{\tau}$$

τ is the recombination lifetime and ranges from nanoseconds to milliseconds for Si, depending on the density of trace metal impurities that form deep traps.

When excess minority carriers are present, the pn product can be orders of magnitude larger than n_i^2. Clearly, electrons and holes as two groups of particles are *not* at equilibrium with each other. Within each group, however, the carriers are still at equilibrium among themselves and share one common (quasi) Fermi level at different locations. This situation is called **quasi-equilibrium** and the following relationships are useful:

$$n = N_c e^{-(E_c - E_{Fn})/kT} \qquad (2.8.1)$$

$$p = N_v e^{-(E_{Fp} - E_v)/kT} \qquad (2.8.2)$$

E_{Fn} and E_{Fp} are the quasi-Fermi levels of electrons and holes.

• PROBLEMS •

• Mobility •

2.1 (a) For an electron mobility of 500 cm²/V·s, calculate the time between collisions. (Take $m_n = m_0$ in these calculations.)

(b) For an electric field 100 V/cm, calculate the distance an electron travels by drift between collisions.

2.2 An electron is moving in a piece of very lightly doped silicon under an applied field such that its drift velocity is one-tenth of its thermal velocity. Calculate the average number of collisions it will experience in traversing by drift a region 1 μm long. What is the voltage across this region?

2.3 The electron mobility is determined by collisions that come in two flavors:

(1) scattering due to phonons (lattice vibrations) and

(2) scattering due to ionized impurities.

The mobilities from phonon interactions alone, μ_1, and from ionized impurities alone μ_2, depend on the electron effective mass m_n, ionized impurity density N_i, and temperature as follows:

$$\mu_1 = \frac{2.2 \times 10^{-70}}{m_n^{5/2} T^{3/2}} kg^{5/2} K^{3/2} \, cm^2/(V \cdot s)$$

$$\mu_2 = \frac{7.45 T^{3/2}}{m_n^{1/2} N_i} kg^{1/2} /(cm \cdot K^{3/2} \cdot V \cdot s)$$

Consider a uniformly doped N-type semiconductor with $N_d = 10^{17} cm^{-3}$ and $m_n = 0.27 m_0$.

(a) Make a plot of $\log(\mu_1)$ and $\log(\mu_2)$ versus temperature from 100 to 700 K.

(b) What is the total electron mobility at 300 K?

(c) Calculate the electron drift current density if the sample is biased as shown in Fig. 2–15 ($T = 300$ K):

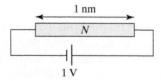

FIGURE 2–15

• Drift •

2.4 Phosphorus donor atoms at a concentration of $10^{16} cm^{-3}$ are distributed uniformly throughout a silicon sample.

(a) What is the sample resistivity at 300 K?

(b) If $10^{17} cm^{-3}$ of boron is included in addition to the phosphorus, what is the resulting resistivity and conductivity type (N-type or P-type material)?

(c) Sketch the energy band diagram for part (a) and for part (b) and show the position of the Fermi level.

2.5 An N-type silicon sample has a uniform density $N_d = 10^{17} \text{cm}^{-3}$ of arsenic, and a P-type sample has $N_a = 10^{15} \text{cm}^{-3}$. A third sample has both impurities present at the same time.
  (a) Find the equilibrium minority carrier concentrations at 300 K in each sample.
  (b) Find the conductivity of each sample at 300 K.
  (c) Find the Fermi level in each material at 300 K with respect to either the conduction band edge (E_c) or the valence band edge (E_v).

2.6 (a) A silicon sample maintained at $T = 300$ K is uniformly doped with $N_d = 10^{16} \text{cm}^{-3}$ donors. Calculate the resistivity of the sample.
  (b) The silicon sample of part (a) is "compensated" by adding $N_a = 10^{16} \text{cm}^{-3}$ acceptors. Calculate the resistivity of the compensated sample. (Exercise caution in choosing the mobility values to be employed in this part of the problem.)
  (c) Compute the resistivity of intrinsic ($N_a = 0$, $N_d = 0$) silicon at $T = 300$ K. Compare it with the result of part (b) and comment.

2.7 A sample of N-type silicon is at the room temperature. When an electric field with a strength of 1000 V/cm is applied to the sample, the hole velocity is measured and found to be 2×10^5 cm/sec.
  (a) Estimate the thermal equilibrium electron and hole densities, indicating which is the minority carrier.
  (b) Find the position of E_F with respect to E_c and E_v.
  (c) The sample is used to make an integrated circuit resistor. The width and height of the sample are 10 µm and 1.5 µm, respectively, and the length of the sample is 20 µm. Calculate the resistance of the sample.

● **Diffusion** ●

2.8 A general relationship for the current density carried by electrons of density n is $J = qnv$, where q is the electronic charge and v is the electron velocity.
  (a) Find the velocity of electrons, $v(x)$, that are moving only by diffusion if they have a density distribution of $n(x) = n_0 \exp(-x/\lambda)$. The electric field is zero.
  (b) What would be the electric field, $\mathcal{E}(x)$, that would lead to an electron drift velocity equal to that of the diffusion velocity in part (a)?
  (c) At 300 K, what value of λ would make the field in part (b) to be 1000 V/cm?

2.9 Figure 2–16 is a part of the energy band diagram of a P-type semiconductor bar under equilibrium conditions (i.e., E_F is constant). The valence band edge is sloped because doping is nonuniform along the bar. Assume that E_v rises with a slope of Δ/L.

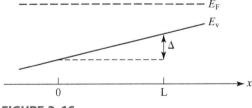

FIGURE 2–16

  (a) Write an expression for the electric field inside this semiconductor bar.
  (b) Within the Boltzmann approximation, what is the electron concentration $n(x)$ along the bar? Assume that $n(x = 0)$ is n_0. Express your answer in terms of n_0, Δ, and L.

(c) Given that the semiconductor bar is under equilibrium, the total electron and hole currents are individually zero. Use this fact and your answers to parts (a) and (b) to derive the Einstein relation ($D_n/\mu_n = kT/q$) relating electron mobility and diffusion constant.

● REFERENCES ●

1. Baccarani, G., and P. Ostoja. "Electron Mobility Empirically Related to the Phosphorus Concentration in Silicon." *Solid State Electronics,* 18, 1975, 579.
2. Antoniadis, D. A., A. G. Gonzalez, and R. W. Dutton. "Boron in Near-Intrinsic <100> and <111> Silicon Under Inert and Oxidizing Ambients—Diffusion and Segregation," *J. Electrochem. Soc.: Solid-State Science and Technology,* 5, 1978, 813.
3. Beadle, W. E., J. C. C. Tsai, and R. D. Plummer. *Quick Reference Manual for Silicon Integrated Circuit Technology.* New York: John Wiley & Sons, 1985. 2–22, 2–23, 2–27.
4. Sze, S. M. *Semiconductor Devices.* New York: Wiley, 1985, 33.

● GENERAL REFERENCES ●

1. Neamen, D. A. *Semiconductor Physics and Devices,* 3rd ed. New York: McGraw-Hill, 2003.
2. Pierret, R. F. *Semiconductor Device Fundamentals.* Reading, MA: Addison-Wesley, 1996.
3. Sze, S. M. *Semiconductor Devices.* New York: Wiley, 1985.

3

Device Fabrication Technology[1]

CHAPTER OBJECTIVES

While the previous chapters explain the properties of semiconductors, this chapter will explain how devices are made out of the semiconductors. It introduces the basic techniques of defining physical patterns by lithography and etching, changing the doping concentration by ion implantation and diffusion, and depositing thin films over the semiconductor's substrate. One section describes the techniques of fabricating the important metal interconnection structures. It is useful to remember the names of the key techniques and their acronyms, as they are often used in technical discussions.

With rapid miniaturization and efficient high-volume processing, over 10^{19} transistors (or a billion for every person in the world) are produced every year. Massive integration of transistors has made complex circuits in the form of **integrated circuits** (ICs) inexpensive and a wide range of electronic applications practical and affordable. Semiconductor devices are responsible for the arrival of the "computer age" or the "second industrial revolution." At the heart of the information and communication technologies, ICs of all descriptions also find applications in consumer electronics, automobiles, medical equipment, and industrial electronics. As a result, semiconductor devices are making contributions to every segment of the global economy and every branch of human endeavors.

Many large semiconductor companies both design and fabricate ICs. They are called **integrated semiconductor companies**. An even larger number of companies only design the circuits. They are called **fabless** design companies. They leave the fabrication to silicon **foundries**, which specialize in manufacturing. So an IC company may or may not fabricate the chips that they design.

---

[1] Readers who are more interested in devices than fabrication technology may proceed to Chapter 4 after reading the introduction and Section 3.1 of this chapter. Some subsequent chapters will refer back to specific parts of Chapter 3 and afford the reader the opportunity to pick up the needed information on fabrication technology.

> ● **VLSI! ULSI! GSI!** ●
>
> The complexity or density of integration of ICs is sometimes described by the names LSI (large-scale integration, 10^4 transistors on a chip), **VLSI** (very large-scale integration, 10^6 transistors on a chip), **ULSI** (ultra-large-scale integration), and **GSI** (giga-scale integration). In actuality, all these terms are used to describe circuits and technologies of wide ranges of size and complexity and simply mean "large IC."

3.1 ● INTRODUCTION TO DEVICE FABRICATION ●

A handful of companies produce most of the silicon wafers (Fig. 1–3b) used in the world. Hundreds of silicon device fabrication lines purchase these wafers as their starting material. A large **wafer fab** can process 40,000 silicon wafers into circuits each month.

The simple example of the device fabrication process shown in Fig. 3–1 includes (a) formation of an SiO_2 layer, (b) its selective removal, (c) introduction of dopant atoms into the wafer surface, and (d) dopant diffusion into silicon.

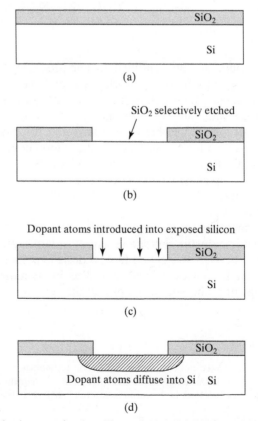

FIGURE 3–1 Some basic steps in the silicon device fabrication process: (a) oxidation of silicon; (b) selective oxide removal; (c) introduction of dopant atoms; and (d) diffusion of dopant atoms into silicon.

Combination of these and other fabrication steps can produce complex devices and circuits. This step-by-step and layer-upon-layer method of making circuits on a wafer substrate is called **planar technology**.

A major advantage of the planar process is that each fabrication step is applied to the entire silicon wafer. Therefore, it is possible to not only make and interconnect many devices with high precision to build a complex IC, but also fabricate many IC chips on one wafer at the same time. A large IC, for example, a central processor unit or **CPU**, may be 1–2 cm on a side, and a wafer (perhaps 30 cm in diameter) can produce hundreds of these chips. There is a clear economic advantage to reduce the area of each IC, i.e., to reduce the size of devices and metal interconnects because the result is more chip per wafer and lower cost per chip.

Since 1960, the world has made a huge investment in the planar microfabrication technology. Variations of this technology are also used to manufacture flat-panel displays, micro-electro-mechanical systems (MEMS), and even DNA chips for DNA screening. The rest of this chapter provides an introduction to the modern device processing technology. Perhaps the most remarkable advances have occurred in the fields of lithography (Section 3.3) and interconnect technology (Section 3.8). These are also the two areas that soak up the largest parts of the IC fabrication cost.

3.2 ● OXIDATION OF SILICON ●

In ICs, silicon dioxide is used for several purposes, ranging from serving as a mask against dopant introduction into silicon to serving as the most critical component in the metal-oxide-semiconductor transistor, the subject of Chapters 5–7.

SiO_2 layers of precisely controlled thickness are produced during IC fabrication by reacting Si with either oxygen gas or water vapor at an elevated temperature. In either case the oxidizing species diffuses through the existing oxide and reacts at the Si–SiO_2 interface to form more SiO_2. The relevant overall reactions are

$$Si + O_2 \rightarrow SiO_2 \qquad (3.2.1a)$$

$$Si + 2H_2O \rightarrow SiO_2 + 2H_2 \qquad (3.2.1b)$$

Growth of SiO_2 using oxygen and water vapor is referred to as **dry** and **wet oxidation**, respectively. Dry oxidation is used to form thin oxide films. Wet oxidation, on the other hand, proceeds at a faster rate and is therefore preferred in forming the thicker oxides. Water vapor diffuses through SiO_2 faster than oxygen.

Figures 3–2a and b show a **horizontal furnace**. Oxidation may also be carried out in a **vertical furnace** as shown in Fig. 3–2c. A simplified sketch of the furnace is presented in Fig. 3–3. Oxidation temperatures of 700– 1,200 °C are produced in the furnace by electrical resistance heating coils. The tube at the center of the furnace is usually made of clear fused quartz, although SiC and polycrystalline Si tubes are also used. The Si wafers to be oxidized are loaded onto a quartz boat and pushed into the center of the furnace. During dry oxidation, the oxygen gas is fed into the

FIGURE 3–2 Examples of furnace systems that may be used for oxidation and other processes. (a) is a horizontal furnace and (b) is a close-up photo showing sillicon wafers waiting to be pushed into the furnace. (© Steed Technology, Inc. Used by permission.) (c) shows a newer vertical furnace. (Copyright © ASM International N.V. Used by permission.) The vertical furnaces occupy less floor space.

tube. Wet oxidation is performed by bubbling a carrier gas (Ar or N_2) through water in a heated flask (see Fig. 3–3) or by burning O_2 and H_2 to form H_2O at the input to the tube. Generally, in a production system, processes such as wafer loading, insertion into the furnace, ramping of the furnace temperature, and gas control are all automated. The thickness of the oxide grown depends on the furnace temperature, the oxidation time, the ambient gas, and the Si surface orientation. Representative dry and wet oxidation growth curves are shown in Fig. 3–4. Wafers

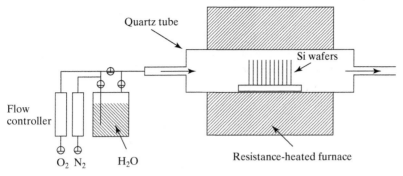

FIGURE 3–3 Schematic drawing of an oxidation system.

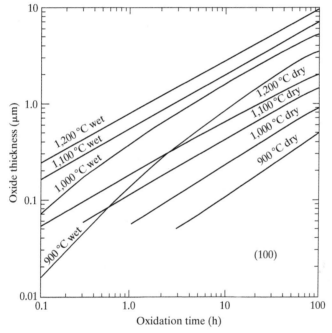

FIGURE 3–4 The SiO_2 thickness formed on (100) silicon surfaces as a function of time. (From [2]. Reprinted by permission of Pearson Education, Inc., Upper Saddle River, NJ.)

used in IC productions are predominantly cut in the (100) plane because the interface trap density (see Section 5.7) is low due to the low density of unsaturated bonds in this plane relative to the other planes. Also, the electron surface mobility (see Section 6.3.1) is high.

> **EXAMPLE 3–1 Two-Step Oxidation**
>
> a. How long does it take to grow 0.1 µm of dry oxide at 1,000°C?
> b. After step (a), how long will it take to grow an additional 0.2 µm of oxide at 900 °C in a wet ambient so that the total oxide thickness is 0.3 µm?
>
> **SOLUTION:**
>
> a. From the "1,000 °C dry" curve in Fig. 3–4, it takes 2.5 h to grow 0.1 µm of oxide.
> b. In this part, use the "900 °C wet" curve only. First we determine that it would have taken 0.7 h to grow the 0.1 µm oxide at 900 °C in a wet ambient and 2.4 h to grow 0.3 µm of oxide from bare silicon. This means that it will take 2.4 − 0.7 h = 1.7 h in a wet 900 °C furnace to increase the oxide thickness from 0.1 to 0.3 µm. This is the correct answer regardless of how the first 0.1 µm oxide is produced (900 °C wet or 1,000 °C dry or any other condition). The answer is 1.7 h.

3.3 ● LITHOGRAPHY ●

How can we selectively remove oxide from those areas in which dopant atoms are to be introduced in Fig. 3–1b? Spatial selection is accomplished using a process called **photolithography** or **optical lithography**.

Major steps in the lithography process are illustrated in Fig. 3–5 using the patterning of an SiO_2 film as an example. The top surface of the wafer is first coated with an ultraviolet (UV) light sensitive material called **photoresist**. Liquid photoresist is placed on the wafer, and the wafer is spun at high speed to produce a thin, uniform coating. After spinning, a short bake at about 90 °C is performed to drive solvent out of the resist.

The next step is to expose the resist through a photomask and a high-precision reduction (for example 5 to 1 reduction) lens system using UV light as illustrated in Fig. 3–5b. The **photomask** is a quartz photoplate containing the patterns to be produced. Opaque regions on the mask block the UV light. Regions of the photoresist exposed to the light undergo a chemical reaction that varies with the type of resist being employed. In **negative resists**, the areas where the light strikes become polymerized and more difficult to dissolve in solvents. When placed in a developer (solvent), the polymerized regions remain, while the unexposed regions dissolve and wash away. The net result after development is pictured on the right-hand side of Fig. 3–5c. **Positive resists** contain a stabilizer that slows down the dissolution rate of the resist in a developer. This stabilizer breaks down when exposed to light, leading to the preferential removal of the exposed regions as shown on the left-hand side of Fig. 3–5c. Steps (a) through (c) make up the complete lithography process. To give a context for the purpose of lithography, we include step (d) for oxide removal. Buffered hydrofluoric acid (HF) may be used to dissolve unprotected regions of the oxide film. Lastly, the photoresist is removed in a step called **resist strip**. This is accomplished by using a chemical solution or by oxidizing or "burning" the resist in an oxygen plasma or a UV ozone system called an **asher**.

Optical diffraction limits the minimum feature size that can be resolved to k times the wavelength of the light used in the optical exposure system.

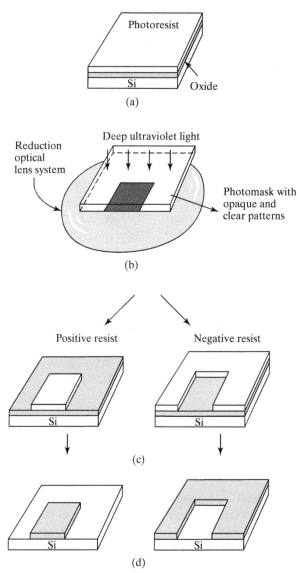

FIGURE 3–5 Major steps in the lithography process: (a) application of resist; (b) resist exposure through a mask and an optical reduction system; (c) after development of exposed photoresist; and (d) after oxide etching and resist removal. (After [2]. Reprinted by permission of Pearson Education, Inc., Upper Saddle River, NJ.)

$$\text{Lithography Resolution} = k\lambda \qquad (3.3.1)$$

A straightforward (but not easy) way to extend the resolution limit is to use UV light of shorter and shorter wavelengths that correspondingly reduce the resolution limit. Laser light sources of 248 and 193 nm (deep UV) are widely used. It is difficult to further reduce the wavelength (e.g., to 157 nm) owing to the lack of suitable transparent materials for lenses and mask plates at this wavelength. The factor k depends on the lens system and the photomask technology as described in the next paragraphs.

To obtain the best optical resolution, only a small area, about 10 cm², of the wafer is exposed in step (b). This area is called the **lithography field** and may contain a few to tens of IC chips. This exposure step is repeated for a neighboring area on the wafer and then another area by moving the wafer until the entire wafer has been exposed. For this reason, the lithography equipment is called a **stepper** for its **step-and-repeat** action.

Distortion of a pattern can result from the effect of the neighboring patterns surrounding it on the photomask. For example, a line may be successfully resolved but two lines close to each other may be bridged. This can be corrected by making the line slightly thinner on the photomask to begin with. This important technique is called **optical proximity correction** or **OPC**. Much computational resource is needed to perform OPC, i.e., to fine tune the photomask for a large IC pattern by pattern.

The k value in Eq. (3.3.1) can be reduced and the resolution limit can be pushed out with several other resolution enhancement techniques. For example, a **phase-shift photomask** might produce a 180° phase difference in the two clear regions on either side of a thin dark line by selective etching of the photomask substrate. Their diffractions into the dark region have electric fields of opposite signs (180° phase difference) and partially cancel each other out. As a result, thinner lines can be resolved. Some other examples of enhancement techniques are excluding certain ranges of the line-space pitch or allowing only certain ranges of it, shaped rather than uniform light source, and exposing only the vertical line patterns with one mask followed with exposing only the horizontal line patterns with another mask.

In addition to resolving small features, lithography technology also provides **alignment** between two lithography steps with an accuracy of about one-third the minimum feature size. *Lithography is the most difficult and expensive process among all the IC fabrication steps.* A typical IC fabrication flow applies the lithography technique over 20 times, each time using a different photomask.

3.3.1 Wet Lithography

Because of the difficulty of finding suitable materials for lenses and masks at wavelengths shorter than 193 nm, a clever technology has been developed to obtain better lithography resolution without requiring a shorter wavelength.

Figure 3–6a shows the objective lens of the optical lithography system and a wafer placed beneath it waiting to be exposed. The gap between the lens and the

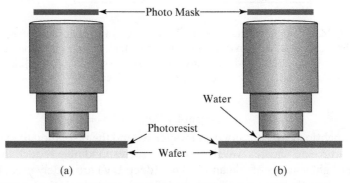

FIGURE 3–6 Schematics of (a) conventional dry lithography and (b) wet or immersion lithography. The wavelength of light source is 193 nm in both cases, but the effective wavelength in (b) is reduced by the refraction index of water, 1.43.

● **Extreme UV Lithography** ●

A bold extension of optical lithography, **extreme ultraviolet lithography** or **EUVL** technology, would use a 13-nm wavelength. This is a huge leap in the reduction of the light source wavelength and the theoretically achievable resolution. Because extreme ultraviolet light is strongly absorbed by all materials, an all-reflective optical system using mirrors instead of lenses is used as shown in Fig. 3–7. Even the photomask is based on reflection rather than transmission. The optical surfaces need to be flat and smooth to 0.25 nm (the size of an atom). The EUV light may be generated by zapping a stream of Xe gas with laser pulses.

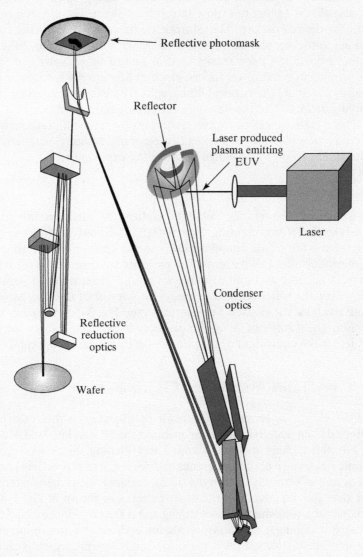

FIGURE 3–7 A schematic illustration of an extreme UV lithography system. (After Scott Hector, Motorola.)

wafer is a few millimeters. If this gap is filled with water as shown in Fig. 3–6b by immersing the system in water, we have the gist of **wet lithography** or **immersion lithography**. When light enters the water, its wavelength is reduced by the refraction index of water, 1.43, and therefore the lithography resolution is improved according to Eq. (3.3.1). Furthermore, the resolution can be improved even more by using a suitable liquid that has a larger index of refraction than water.

3.3.2 Electron Lithography

It is well known that electron microscopes have better resolution than optical microscopes. Electron lithography similarly is an alternative to optical lithography with resolution advantage. In **electron-beam lithography**, a focused stream of electrons delivers energy to expose the electron resist. The electron beam is scanned to write the desired pattern. The information necessary to guide the electron beam is stored in a computer and no mask is used. Electron-beam lithography has long been used to fabricate the photomasks used in optical lithography and for EUVL. For direct printing of patterns on wafers, electron lithography has slower exposure rates (in wafers per hour) than optical lithography. The exposure rate can be increased by employing multiple electron beams in each lithography machine.

There are schemes to expose a complex pattern simultaneously using a mask and a reduction electron-lens system (a carefully designed magnetic field), similar to optical lithography. This would improve the exposure rate.

3.3.3 Nanoimprint

High-resolution lithography, whether optical or electron lithography, is very expensive. Therefore, creating fine patterns without performing the expensive lithography is attractive. **Nanoimprint** is such a technique. Electron lithography is used to produce the fine patterns. The patterns are transferred (etched, see Section 3.4) into a suitable material to make a "stamp." This stamp is pressed into a soft coating over the wafer surface to create an *imprint* of the fine patterns. After the coating hardens, the desired fine patterns (see Fig. 3–5d) have been replicated on the wafer. The stamp can be used repeatedly to produce many wafers. In this sense, the stamp is the equivalent of the photomask in optical lithography.

3.4 ● PATTERN TRANSFER—ETCHING ●

After the pattern is formed in the resist by lithography, the resist pattern is often transferred to an underlying film, for example, the SiO_2 in Fig. 3–5d. If SiO_2 is removed with HF, this etching method is called **wet etching**. Since wet etching is usually **isotropic** (meaning without preference in direction, and proceeding laterally under the resist as well as vertically toward the silicon surface), the etched features are generally larger than the dimensions of the resist patterns as shown in Fig. 3–8a. **Dry etching** technique can overcome this shortcoming and is the dominant etching technology.

In dry etching, also known as **plasma etching** or **reactive-ion etching** or **RIE**, the wafer with patterned resist is exposed to a plasma, which is an almost neutral mixture of energetic molecules, ions, and electrons that is usually created by a radio frequency (RF) electric field as shown in Fig. 3–9a. The energetic species react

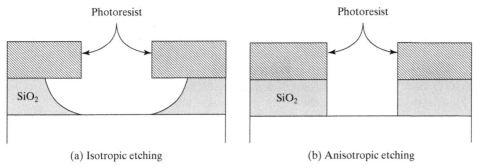

FIGURE 3–8 Comparison between (a) isotropic etching and (b) anisotropic etching.

chemically with the exposed regions of the material to be etched, while the ions in the plasma bombard the surface vertically and knock away films of the reaction products on the wafer surface. The latter action is directional so that the etching is preferentially vertical because the vertical surfaces can be covered with films of the reaction products. Hence the etch rate is **anisotropic**.

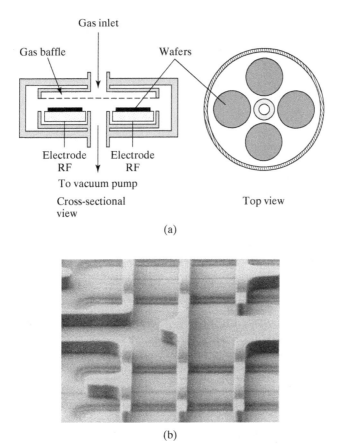

FIGURE 3–9 (a) A reactive-ion etching chamber and (b) scanning electron microscope view of a 0.16 μm pattern etched in polycrystalline silicon film. Excellent line width control is achieved even though the underlying surface is not flat [3].

By proper choices of the reactor design and etching chemistry, nearly vertical walls are produced in the etched material as shown in Fig. 3–9b. Low pressures and highly one-directional electric field tend to make etching anisotropic. Dry etching can also be designed to be isotropic or partially anisotropic if that is desired.

Suitable gas(es) is (are) introduced into the etch chamber based on the material to be etched. Silicon and its compounds can be etched by plasmas containing fluorine (F), whereas aluminum is etched with chlorine-containing plasmas.

The material **selectivity** of dry etching is usually not as high as that of wet etching. The material to be etched and the underlying material (e.g., SiO_2 and the underlying silicon) can both be significantly attacked during the etching process. Therefore, the dry etching process must be terminated as soon as the desired layer has been removed. This can be done with an **end-point detector**, which monitors the telltale light emission from the various etching products. There is often a trade-off between selectivity and anisotropy. For example, bromine (Br) provides better selectivity between Si and SiO_2 but poorer anisotropy than Cl.

Processing using plasma can potentially cause damage to the devices on the wafer. This is known as **plasma process-induced damage** or **wafer charging damage**. The main damage mechanism is the charging of conductors by the ions in the plasma, leading to an overly high voltage across a thin oxide and causing oxide breakdown. The worst condition is a small, thin oxide area connected to a large conductor, which collects a large amount of charge and current from the plasma and funnels them into the small-area oxide. The sensitivity of the damage to the size of the conductor is called the **antenna effect**.

Of course, pattern transfer is not limited to transferring a resist pattern onto another material. A pattern in an oxide may be transferred to Si, for example.

3.5 • DOPING •

The density profile of the dopant atoms in the silicon (**dopant profile**) is generally determined in two steps. First, the dopant atoms are placed on or near the surface of the wafer by **ion implantation**, **gas-source doping**, or **solid-source diffusion**. This step may be followed by an intentional or unintentional **drive-in diffusion** that transports the dopant atoms further into the silicon substrate.

3.5.1 Ion Implantation

Ion implantation is the most important doping method because of the precise control it provides. In ion implantation, an impurity is introduced into the semiconductor by creating ions of the impurity, accelerating the ions to high energies ranging from subkiloelectronvolt to megaelectronvolt, and then literally shooting the ions onto the semiconductor surface (Fig. 3–10). As one might suspect, the implanted ions displace semiconductor atoms along their paths into the crystal. Moreover, the ions themselves do not necessarily come to rest on lattice sites. A follow-up **anneal** (heating) of the wafer is therefore necessary for damage removal and for **dopant activation** (placing the dopant atoms on lattice sites as shown in Fig. 1–6) so that implanted impurities behave as donors and acceptors.

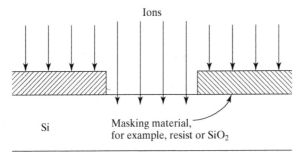

FIGURE 3–10 In ion implantation, a beam of high-energy ions penetrates into the unprotected regions of the semiconductor.

A schematic of an ion implantation system is presented in Fig. 3–11. Ions of the desired impurity are produced in the ion source shown at the extreme left. The ions are next accelerated into the mass analyzer where only the desired ions pass through a slit in the ion selection aperture. The resulting ion beam is then accelerated to the implantation energy, and finally the inch-size ion beam is scanned over the surface of the wafer, which is mounted on a massive metal plate. Scanning is accomplished by electrostatically scanning the ion beam, by mechanically moving the wafer, or by a combination of the two methods. An electrical contact to the wafer allows a flow of electrons to neutralize the implanted ions. A very precise determination of the total number of implanted ions per square centimeter (called the **implantation dose**, N_i) is

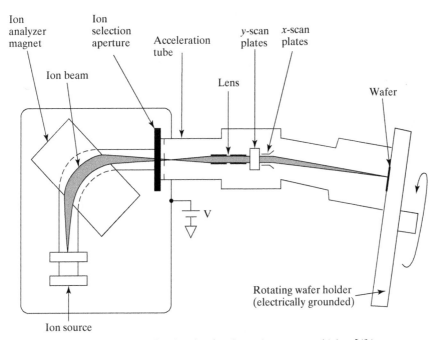

FIGURE 3–11 Simplified schematic of an ion implantation system. (After [4].)

obtained by simply integrating the beam current over the time of the implant. The concentration profile produced by ion implantation has the general form of a Gaussian function and is described by the peak location below the surface (R, called the **implantation range**), and the spread (ΔR, called **implantation straggle**).

$$N(x) = \frac{N_i}{\sqrt{2\pi}(\Delta R)} \cdot e^{-(x-R)^2/2\Delta R^2} \tag{3.5.1}$$

These parameters vary with the implant ion and substrate material and are roughly proportional to the ion energy as shown in Fig. 3–12. Computed distributions for phosphorus implanted into Si at various energies are shown in Fig. 3–13. Ion implantation processes can sometimes cause wafer charging damage. To alleviate this problem, electrons may be introduced near the wafer to neutralize the charge on the wafer.

3.5.2 Gas-Source Doping

In practice, gas-source doping is used to dope Si with phosphorus only. There are no convenient gas sources for As or B. It is carried out in a furnace similar to that used for oxidation (see Figs. 3–2 and 3–3). The N_2 carrier gas in Fig. 3–3 would pass through a bubbler containing phosphorus oxychloride ($POCl_3$, often pronounced "pockle") that is a liquid at room temperature. The N_2 carries the vapor of the source into the furnace tube. The reaction with Si or other gases liberates phosphorus atoms, which diffuse into the silicon.

3.5.3 Solid-Source Diffusion

In solid-source diffusion, the Si surface is first coated with a thin film (of a SiGe alloy, for example) containing dopants as deposited or due to subsequent implant of dopants into this film (and leave the damages in it). Dopants are diffused into Si. The SiGe film may be removed by wet etching.

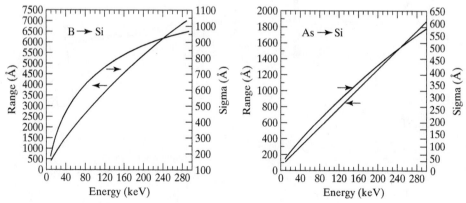

FIGURE 3–12 R and ΔR of implantation of (a) B and (b) As in silicon, versus energy [5].

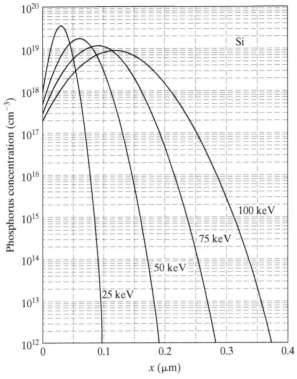

FIGURE 3–13 Computed implantation profiles of phosphorus assuming a constant dose of $10^{14}/cm^2$. ([6]. Reprinted by permission of Pearson Education, Inc., Upper Saddle River, NJ.)

3.6 • DOPANT DIFFUSION •

After dopant introduction by implantation or gaseous deposition, we may want to drive the dopant deeper into silicon. This is accomplished by **diffusion**. Unwanted diffusion also may occur during the post-implant anneal. The diffusion process is illustrated in Fig. 3–14. The dopant impurity diffuses with time at high temperature. If the diffusing dopant is of the opposite doping type to the substrate, as shown in Fig. 3–14, a line may be drawn to indicate the boundary where $N_a = N_d$. This structure is known as a **PN junction**, and the thickness of the diffusion layer is called the **junction depth**. For some applications, very deep junctions are desired. For other important applications, the shallowest possible junction is desired. Excessive diffusion is often the undesirable side effect of the necessary post-implantation anneal. In either case, it is important to control diffusion tightly.

Regardless of whether the shallow dopant addition is carried out by implantation or gaseous predeposition, the impurity concentration versus position inside the semiconductor after sufficient diffusion can be shown to be Gaussian [4].

$$N(x, t) = \frac{N_0}{\sqrt{\pi D t}} e^{-x^2/4Dt} \qquad (3.6.1)$$

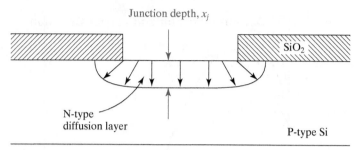

FIGURE 3–14 The basic diffusion process.

N_0 is the number of dopants per square centimeter and is determined by the dopant addition step, x is the distance into the semiconductor measured from the semiconductor surface, $N(x, t)$ is the impurity concentration at a depth x after a given time t, D is the **diffusivity** for the given impurity and furnace temperature, and t is the time for the diffusion step. Figure 3–15 shows the diffusivities of some common dopants in silicon. The diffusion rate increases with increasing temperature.

Diffusion is commonly performed in an open tube system similar in construction to that used for oxidation (Figs. 3–2 and 3–3). Diffusion temperatures range from roughly 900 °C to 1,200 °C. Sometimes the term **diffusion** refers to the combined process of gaseous dopant deposition and diffusion. The gaseous dopant deposition step is followed by a second step where the gaseous dopant source is

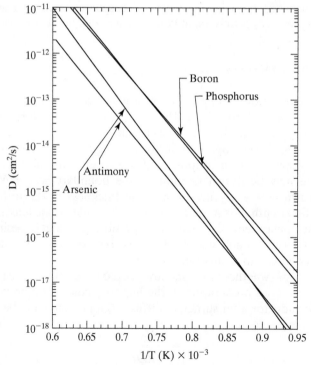

FIGURE 3–15 Diffusivity versus $1/T$ for Sb, As, B, and P in silicon. (From [5].)

● Dopant Diffusion and Carrier Diffusion ●

The dopant diffusivity has the same dimension as the electron or hole diffusion constant, square centimeter per second. Their values, however, differ by a huge factor. Even at a high temperature, dopants only diffuse a small distance in an hour. Fortunately, the dopant diffusivities are negligibly small at room temperature. Otherwise, the device structures would change with time after they have been fabricated!

shut off, and the impurities are driven deeper into the semiconductor. The portion of the process step with the source present is called the **predeposition**, and the latter portion with the source shut off is called the **drive-in**.

● Shallow Junctions and Rapid Thermal Annealing ●

High-performance devices often require that the junction depth (see Fig. 3–14) be kept shallow. This in turn requires that the Dt product in Eq. (3.6.1) be minimized. However, in order to activate the dopant and repair the crystal damage after ion implantation, thermal annealing is required. Unfortunately, **furnace annealing** may need 30 min in a furnace at 900 °C. This condition causes too much diffusion of the dopant, especially with B.

As it turns out, annealing can be completed at 1,050 °C in 20 s, which condition causes much less diffusion. In order to heat the silicon wafer up (and to cool it off) rapidly for short-duration annealing, a special heating technique is required. In **rapid thermal annealing (RTA)**, the silicon wafer is heated to high temperature in seconds by a bank of heat lamps. Cooling off is also fast because the thermal mass of the entire system is small. Similar systems can be used for **rapid thermal oxidation** and **rapid thermal chemical vapor deposition (CVD)** (see Section 3.7). Together, they are called **rapid thermal processing (RTP)**.

Pushing RTA further to 0.1 s annealing, one can obtain even shallower junctions. Such short annealing is called **flash annealing**. For even shorter durations (less than a microsecond) of heating, the silicon wafer can be heated with very short laser pulses. The process is called **laser annealing**, which may or may not involve melting a very thin layer of silicon.

As it turns out, crystal damage caused by ion implantation raises the dopant diffusivity at lower temperatures to values much larger than those shown in Fig. 3–15. This is called **transient enhanced diffusion** or **TED**. As a result, it is difficult to make ultra-shallow junctions using furnace annealing. The term *transient* denotes the fact that the enhancement of diffusion disappears after a short time during which the crystal damage is annealed out.

3.7 ● THIN-FILM DEPOSITION ●

Silicon nitride, silicon dioxide, Si, and many types of metal thin films are deposited during IC fabrication. Deposited films are usually not single crystalline.

● **Three Kinds of Solid** ●

A solid material may be **crystalline**, **polycrystalline**, or **amorphous**. They are illustrated in Fig. 3–16. A crystalline structure has nearly perfect periodic structure as described in Section 1.1. Silicon wafers and epitaxially deposited films (see Section 3.7.3) fall in this category as do high-quality gemstones such as ruby and sapphire (Al_2O_3 with impurities that produce the characteristic colors) as well as diamond.

Often, materials are **polycrystalline**, which means the material is made of densely packed crystallites or grains of single crystal. Each grain has a more or less random orientation. The interface between crystallites is called a **grain boundary**. Each grain may be 10–10,000 nm in size. Metal films and Si films deposited at higher temperatures fall in this category, as do all metal objects that we encounter in daily life. Because each grain contains a large number of atoms, *polycrystalline materials have basically the same properties as single crystalline materials. In particular, polycrystalline and crystalline silicon have qualitatively similar electronic properties.*

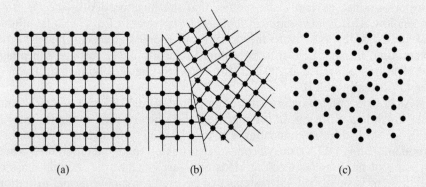

(a) (b) (c)

FIGURE 3–16 Crystalline material (a) has perfect ordering. Polycrystalline material (b) is made of tiny crystalline grains. (c) Amorphous material has no significant ordering.

An **amorphous** material has no atomic or molecular ordering to speak of. It may be thought of as a liquid with its molecules frozen in space. Thermally grown or deposited SiO_2, silicon nitride, and Si deposited at low temperature fall in this category. At high temperature, Si atoms have enough mobility to move and form crystallites on the substrate.

Carrier mobilities are lower in amorphous and polycrystalline Si than in single-crystalline Si. However, transistors of lower performance levels can be made of amorphous or polycrystalline Si, and are widely used in flat-panel computer monitors and other displays. They are called **thin-film transistors** or **TFTs**. The are also used in solar cells presented in Chapter 4.

3.7.1 Sputtering

Sputtering is performed in a vacuum chamber. The source material, called the **sputtering target**, and the substrate holding the Si wafer form opposing parallel plates connected to a high-voltage power supply. During deposition, the chamber is

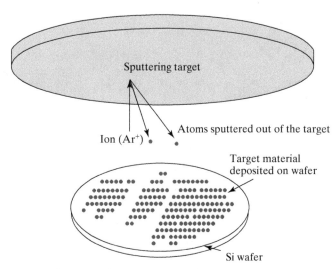

FIGURE 3–17 Schematic illustration of the sputtering process.

first evacuated of air and then a low-pressure amount of sputtering gas (typically Ar) is admitted into the chamber. Applying an interelectrode voltage ionizes the Ar gas and creates a plasma between the plates. The target is maintained at a negative potential relative to the substrate, and Ar ions are accelerated toward the sputtering target. The impacting Ar ions cause target atoms or molecules to be ejected from the target. The ejected atoms or molecules readily travel to the substrate, where they form the desired thin film. A simplified illustration of the sputtering process is shown in Fig. 3–17. A DC power supply can be used when depositing metals, but an RF supply is necessary when depositing insulating films. Sputtering may be combined with a chemical reaction in **reactive sputtering**. For example, when Ti is sputtered in a nitrogen-containing plasma, a TiN (titanium nitride) film is deposited on the Si wafer. Sputtering is the chief method of depositing Al and other metals. Sputtering is sometimes called a method of **physical vapor deposition (PVD)**.

3.7.2 Chemical Vapor Deposition (CVD)

While sputtering is a relatively simple and satisfactory way of depositing thin film over flat surfaces, it is directional and cannot deposit uniform films on the vertical walls of holes or steps in the surface topography. This is called a **step coverage** problem. CVD, on the other hand, deposits a much more *conformal* film, which covers the vertical and horizontal surfaces with basically no difference in the film thickness.

In CVD, the thin film is formed from gas-phase components. Either a compound decomposes to form the thin film or a reaction between gas components takes place to form it. A schematic of the CVD process is shown in Fig. 3–18. The CVD process is routinely used to deposit films of SiO_2, Si_3N_4 (a dielectric with excellent chemical and electrical stability), and polycrystalline silicon or poly-Si (see the sidebar "Three Kinds of Solid").

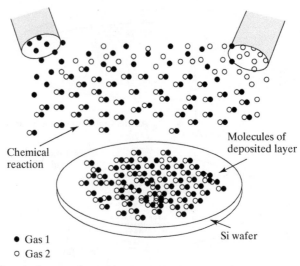

FIGURE 3–18 Chemical vapor deposition process.

These are some commonly used chemical reactors in the CVD deposition process:

Poly-Si: SiH_4 (silane) $\rightarrow Si + 2H_2$

Si_3N_4: $3SiH_2Cl_2$ (dichlorosilane) $+ 4NH_3 \rightarrow Si_3N_4 + 6HCl + 6H_2$

SiO_2: $SiH_4 + O_2 \rightarrow SiO_2 + 2H_2$

High-temperature SiO_2: $SiH_2Cl_2 + 2H_2O \rightarrow SiO_2 + 2HCl + 2H_2$

A **high-temperature oxide (HTO)** is particularly conformal because the high deposition temperature promotes particle movement on the surface so that even sidewall coverage is excellent. Commonly used CVD processes include **low-pressure chemical vapor deposition** or **LPCVD**, and **plasma-enhanced chemical vapor deposition** or **PECVD** processes. Low pressure offers better thickness uniformity and lower gas consumption. A simple LPCVD deposition system is illustrated in Fig. 3–19a. In PECVD, the electrons in the plasma impart energy to the reaction gases, thereby enhancing the reactions and permitting lower deposition temperatures. Figure 3–19b shows the schematic of a PECVD reactor.

Dopant species can be introduced during the CVD deposition of Si. This doping process is called *in situ* **doping** and is a method of heavily doping the Si film.

3.7.3 Epitaxy

Epitaxy is a very special type of thin-film deposition technology [7]. Whereas the deposition methods described in the preceding section yield either amorphous or polycrystalline films, **epitaxy** produces a crystalline layer over a crystalline substrate. The film is an extension of the underlying crystal. In a CVD reactor with special precautions to eliminate any trace of oxide at the substrate surface and at sufficiently high temperature, an arriving atom can move over the surface till it stops at a correct location to perfectly extend the lattice pattern of the substrate crystal. Figure 3–20a illustrates the epitaxy process. Selective epitaxy (Fig. 3–20b) is

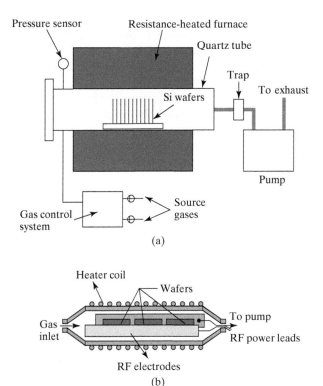

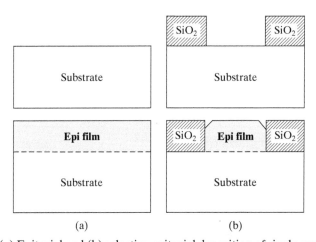

FIGURE 3–19 Schematic illustration of (a) an LPCVD system (after [1]) and (b) a PECVD reactor chamber with plasma generated radio-frequency power.

FIGURE 3–20 (a) Epitaxial and (b) selective epitaxial deposition of single crystalline film.

a variation of the basic epitaxy technology and has interesting device applications. In selective epitaxy deposition, an etching gas is introduced to simultaneously etch away the material. The net deposition rate is positive, i.e., atoms are deposited only over the single crystal substrate. There is no net deposition over the oxide mask because the deposition rate over the oxide is lower than the etching rate.

Epitaxy is useful when we want a lightly doped layer of crystal Si over a heavily doped substrate (see Fig. 8–22). Also, a different material may be epitaxially deposited over the substrate material as long as the film and the substrate have closely matched lattice constants (see Section 1.1). Epitaxially grown dissimilar materials are widely used in light-emitting diodes (see Fig. 4–30) and diode lasers (see Fig. 4–33). The interface between two different semiconductors is called a **heterojunction**. An application example of selective heterojunction epitaxial growth (of SiGe over Si) may be found in Fig. 7–1.

3.8 • INTERCONNECT—THE BACK-END PROCESS •

To build an IC, the individual devices must be interconnected by metal lines. This procedure is sometimes called **metallization**.

A basic **interconnect** is illustrated in Fig. 3–21a. First, the SiO_2 is removed from areas where a contact is to be made with the silicon. Then a layer of metal is deposited over the surface, typically by sputtering. The metal, perhaps aluminum, is then removed from areas where it is not desired (by lithography and dry etching). The metal interconnect in Fig. 3–21a performs the function of connecting the two diffusion regions.

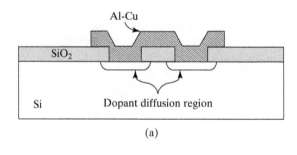

(a)

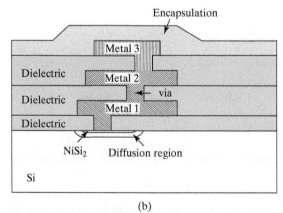

(b)

FIGURE 3–21 Schematic drawing of device interconnections: (a) a basic metallization example and (b) a multilevel metallization structure.

To build complex and dense circuits, the **multilevel metallization** structure shown in Fig. 3–21b is routinely employed. Up to about ten metal layers may be used. The metal thickness ranges from a small fraction of a micron to several microns. The thinner interconnects route signals while the thicker layers serve as power lines. The adjacent layers of metal are separated by **intermetal dielectric** layers. Electrical connection between the adjacent metal layers is made through a **via**. To reduce the contact resistance (see Section 4.21) between the metal and the N^+ or P^+ diffusion region, a silicide such as $NiSi_2$ is added. An interconnect structure with all the dielectric etched away is shown in Fig. 3–22.

From the first ICs, the interconnect metal has been aluminum, Al. Al interconnects suffer a potential reliability problem called **electromigration**. Electron flow in the metal line, over time, can cause the metal atoms to migrate along crystal grain boundaries or the metal/dielectric interfaces in a quasi-random manner. Voids may develop in the metal lines as a result and cause the line resistance to increase or even become open-circuited. Copper has replaced Al as the interconnect material in advanced ICs. Cu has excellent electromigration reliability and 40% lower resistance than Al. Copper may be deposited by plating or CVD. Because dry etching of Cu is difficult, copper patterns are commonly defined by a **damascene** process, which is illustrated in Fig. 3–23.

Because Cu diffuses rapidly in dielectrics, a barrier material such as TiN is deposited as a liner before Cu is deposited in Fig. 3–23c. Excess copper is removed by **chemical-mechanical polishing** or **CMP**. In CMP, a polishing pad and slurry are used to polish away material and leave a very flat surface.

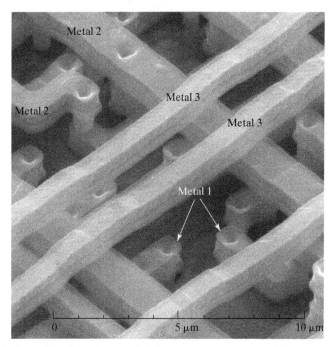

FIGURE 3–22 An example of a metal interconnect system. (Courtesy of Analytical Laboratory Services, Inc.)

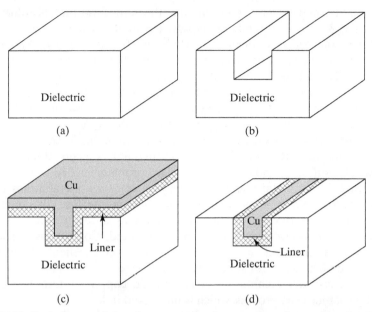

FIGURE 3–23 Basic steps of forming a copper interconnect line using the damascene process: (a) cover the wafer with a dielectric such as SiO_2; (b) etch a trench in the dielectric; (c) deposit a liner film and then deposit Cu; and (d) polish away the excess metal by CMP.

The dielectric material between the metal layers used to be SiO_2. It has been supplemented with **low-k dielectrics**, which often contain carbon or fluorine, and are designed to have much lower dielectric constants (k) than SiO_2. Lower k leads to lower capacitances between the interconnects. This is highly desirable because capacitance in a circuit slows down the circuit speed, raises power consumption (see Sections 6.7.2 and 6.7.3), and introduces **cross talk** between neighboring interconnect lines.

Since a large number of metal layers and process steps are involved, making the interconnects *consumes a large part of the IC fabrication budget*. This part of the fabrication process is called the **back-end process**. In contrast, the steps used to produce the transistors are called the **front-end process**.

● Planarization ●

A flat surface is highly desirable in IC processing because it greatly improves subsequent optical lithography (the whole surface is in focus) and etching. For this reason, CMP **planarization** may also be performed in the front-end process, for example, in the formation of the shallow trench isolation (see Fig. 6–1). Although there are several ways to perform planarization, CMP provides the best flatness.

3.9 ● TESTING, ASSEMBLY, AND QUALIFICATION ●

After the wafer fabrication process is completed, individual ICs are electrically probed on the wafer to determine which IC chips are functional. The rest are marked and will not be packaged.

After this preliminary functional testing, the wafer is diced into individual circuits or chips by sawing or laser cutting. Functional chips may be encased in plastic or ceramic packages or directly attached to circuit boards. Multiple chips may be put in one package to make **multi-chip modules**. The electrical connections between the chip and the package are made by automated wire bonding or through **solder bumps**. In the solder bump process, the metal pads on the IC chip are aligned with the matching pads on the ceramic package. All connections are simultaneously made by melting preformed solder bumps on the IC pads in what is called the **flip-chip** bonding process. Finally, the package is sealed with a ceramic or metal cover before it undergoes final at-speed testing. As the complexity of ICs increases, testing becomes more and more difficult and expensive. Ease of testing is an important consideration in circuit design.

The quality of manufacturing and the reliability of the technology are verified with a **qualification** routine performed on hundreds to thousands of product samples including an **operating life test** that lasts over one thousand hours. This process is long and onerous but the alternative, shipping unreliable parts, is unthinkable. To ensure a very high level of reliability, every chip may be subjected to **burn-in** at higher-than-normal voltage and temperature. The purpose is to accelerate failures in order to weed out the unreliable chips.

3.10 • CHAPTER SUMMARY—A DEVICE FABRICATION EXAMPLE •

Figure 3–24 illustrates how the individual fabrication steps are combined and sequenced to fabricate a simple PN diode. A typical IC fabrication process involves over one hundred steps.

The starting point is a flat, P-type single-crystal Si wafer. A preclean removes all particulates, organic film, and adsorbed metal from the semiconductor surface. Then a thermal oxide is grown. Step 2 is a lithography process performed to open a hole in the oxide that will eventually become the position of the PN junction.

The wafer is implanted with an appropriate dose of As at an appropriate energy (step 4). After annealing and diffusion, the junction is formed in step 5. Note that the junction edge is protected by the oxide. Some oxide may be formed in the diffusion process. This must be cleaned off before step 6, metallization.

Sputtering of Al deposits a thin metal film over the entire surface of the wafer as pictured in step 6. A lithography process (step 7) is then performed to pattern the metal. A low-temperature (≤ 450 °C) anneal is performed to produce a low-resistance contact between the metal and Si. In step 9, SiO_2 and Si_3N_4 films are deposited for encapsulation to protect the device from moisture and other contaminants. In step 10, an opening is made to access the Al for wire bonding. If electrical contact is to be made to the P-type substrate, the oxide grown on the back of the wafer in step 2 must be removed while the front of the wafer is protected with photoresist as shown in step 11. Gold (Au) is deposited at the back of the wafer for electrical contact in step 12. Finally, the wafer is diced into individual diode chips, and each chip is soldered to a package; a bond wire connects the Al to a second electrical lead. For a slide show of the device fabrication steps, see http://jas.eng.buffalo.edu/education/fab/pn/diodeframe.html.

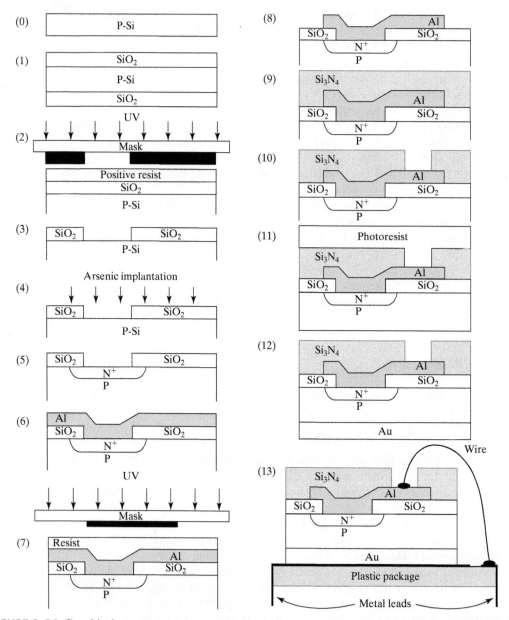

FIGURE 3–24 Graphical summary of the major processing steps in the formation of a PN junction diode. (0) Start; (1) oxidation; (2) lithography; (3) oxide etching; (4) As implantation; (5) annealing and diffusion; (6) sputtering Al; (7) lithography; (8) metal etching; (9) CVD nitride deposition; (10) lithography and bonding window etching; (11) removal of oxide from back side of wafer; (12) deposition of Au on back side; and (13) dicing and packaging. (After [6].)

PROBLEMS

Terminology and General Knowledge

3.1 Copy all the bold-faced terms in Chapter 3 Introduction and Sections 3.1–3.5. Give each of them a short definition or explanation (one word to two sentences), preferably in your own words.

3.2 Do Problem 3.1 for all the bold-faced terms in the remaining sections of Chapter 3.

3.3 Answer each of the following questions in one to three sentences.

(a) What is lithography field?

(b) What is misalignment in lithography?

(c) What is selectivity in an etching process?

(d) What is end-point detection in an etching process?

3.4 Answer the following questions.

(a) In an older MOSFET technology, the field oxide is a 1 μm thick thermal oxide. Would you grow it in a dry or wet ambient? Why?

(b) For etching a small feature with faithful replication of the resist pattern, is dry or wet etching technique preferred? Why?

(c) If the junction depth is to be kept as small as possible, which ion species would you use to make a P–N junction (for an ion implantation process on a P-type silicon substrate)? Give reasons to support your answer.

(d) If you want to deposit oxide at the lowest possible temperature, what processing technology would you use?

(e) What processing technology would you use to deposit aluminum? What is the processing technology you would use to etch a fine aluminum line? What chemicals are involved?

(f)

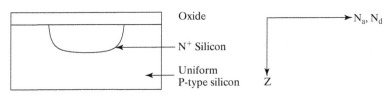

FIGURE 3–25

In the accompanying N_a N_d vs. Z coordinates (Fig. 3–25), quantitatively draw typical N_a and N_d profiles through the P–N junction and indicate the position of the junction. Assume the N⁺ dopant peak is at the Si–SiO₂ interface.

Oxidation

3.5 Why is wet oxidation faster than dry oxidation? Please speculate. One or two sentences will be sufficient.

3.6 Assume that the oxide thickness is T_{init} at time 0 and that the oxide thickness is given by $T_{ox}^2 + AT_{ox} = B(t + \tau)$, where $\tau \equiv \dfrac{T_{init}^2 + AT_{init}}{B}$. (For example, see 900°C wet oxidation curve in Fig. 3–4.)

(a) Calculate the final thickness of the silicon dioxide on a wafer that initially has 0.2 µm after an additional 3 h of 1,000°C dry oxidation ($A = 0.165$ µm and $B = 0.0117$ µm²/h at 1,000°C dry oxidation).

(b) There are two important limiting cases for this equation. For sufficiently thin oxides, the quadratic term is negligible. On the other hand, if the oxide is sufficiently thick, the linear term can be ignored. How much error is introduced if this question is answered with the linear approximation and the quadratic approximation?

● **Deposition** ●

3.7 Verify that chemical equations in Section 3.7.2 are balanced. If some are not balanced, correct them by providing the right coefficients.

● **Diffusion** ●

3.8 Assume $x_j = C\sqrt{Dt}$, where C may be assumed to be 1.
  (a) Show that additional diffusion with an increment $\Delta(Dt)$ would increase the junction depth by $\Delta(Dt)/2x_j$.
  (b) If a boron doped junction has a depth of $x_j = 0.1$ µm, by how much will x_j increase at 500 K in 10 years?

3.9 Assume $D = D_0 e^{-E_a/kT}$ is the diffusion coefficient of boron in silicon surface, where $D_0 = 10.5$ cm²/s and $E_a = 3.7$ eV. The substrate is N-type silicon doped to 10^{15} cm⁻³. $N_0 = 10^{15}$ cm⁻² of boron is introduced just below the silicon surface.
  (a) What is the junction depth after a 1-h drive-in at 1,100°C?
  (b) By how much will the junction depth change after 10^6 h (~100 years) of operation at 100°C?

● **Visualization** ●

3.10 For the following process steps, assume that you use a positive photoresist and that etch selectivity is infinite. A composite plot of four photomasks is given in Fig. 3–26. Assume that mask alignment is perfect. All contact sizes are 0.5 × 0.5 µm. The poly 1 and poly 2 areas are opaque, and the contact 1 and contact 2 areas are clear in the masks. Draw the cross section at the end of each process step along the cut line shown in the figure.

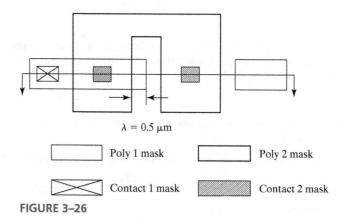

FIGURE 3–26

(a) Grow 1 μm thermal oxide on < 100 > bare Si wafer.

(b) Expose and develop photoresist with contact 1 mask. Assume that the resist thickness is 1 μm.

(c) Etch the 1 μm thermal oxide anisotropically. Assume the final oxide profile is perfectly vertical.

(d) Remove the photoresist with O_2 plasma.

(e) Implant phosphorus and anneal. Assume that the final junction depth is 0.3 μm.

(f) Deposit 1 μm *in situ* doped poly silicon by LPCVD. The thickness on the sidewalls is the same as that on the flat surface.

(g) Expose and develop the photoresist with poly 1 mask.

(h) Etch the 1 μm poly silicon anisotropically.

(i) Remove the photoresist with O_2 plasma.

(j) Deposit 1 μm oxide with PECVD. Again, the thickness on the sidewalls is the same as that on the flat surface.

(k) Expose and develop photoresist with contact 2 mask.

(l) Etch 0.2 μm of the PECVD oxide with HF. Assume the profile is cylindrical as shown in Fig. 3–8a.

(m) Etch the remaining 1.8 μm oxide anisotropically.

(n) Remove the photoresist with O_2 plasma.

(o) Implant phosphorus and anneal. Assume the junction depth is 0.3 μm and there is no additional dopant diffusion.

(p) Deposit 1 μm *in situ* doped poly silicon by LPCVD. The thickness on the sidewalls is the same as that on the flat surface.

(q) Expose and develop photoresist with poly 2 mask.

(r) Etch the 1.0 μm poly silicon anisotropically.

(s) Remove the photoresist with O_2 plasma.

(This is just an exercise. The structure does not have any known usefulness.)

3.11 Assume a negative resist is used instead of a positive resist in Problem 3.10 with the same contact 1 mask. Answer parts (a), (b), (c), and (d) of Problem 3.10. What changes does one have to make in order to obtain the same cross section as Problem 3.10 (d) with a negative resist?

● REFERENCES ●

1. Dance, B. "Europe Prepares Its Future Technology," *Semiconductor International* (1995), 125.

2. Jaeger, R. C. *Introduction to Microelectronic Fabrication*, Vol. 5, in *The Modular Series on Solid State Devices*, 2nd ed., G. W. Neudeck and R. F. Pierret. Reading, MA: Addison-Wesley, 2002, pp. 49, 24.

3. Warren, J. "Leaping into the Unknown with 0.18 mm," *Semiconductor International* (1998), 111.

4. Runyan, W. R., and K. E. Bean. *Semiconductor Integrated Circuit Processing Technology*, Reading, MA: Addison-Wesley, 1990.

5. Sze, S. M. *VLSI Technology*, 2nd ed. New York: McGraw-Hill Book Company, 1988.

6. Pierret, R. F. *Semiconductor Device Fundamentals.* Reading MA: Addison-Wesley Publishing Company Inc., 1996, 158 and 167.
7. Herman, M. A., W. Richter, and H. Sitter. *Epitaxy: Physical Foundation and Technical Implementation.* New York: Springer-Verlag, 2004.
8. Wolf, S., and R. N. Tauker. *Silicon Processing for the VLSI Era*, 2nd ed. Sunset Beach, CA: Lattice Press, 2000.
9. Burggraaf, P. "Stepping to Mix-and-Match I-line Lithography," *Semiconductor International* (1995), 47.

● GENERAL REFERENCES ●

1. Sze, S. M. *Semiconductor Devices: Physics and Technology*, 2nd ed. New York: John Wiley & Sons, 2002.
2. Wolf, S., and R. N. Tauker. *Silicon Processing for the VLSI Era*, 2nd ed. Sunset Beach, CA: Lattice Press, 2000.

4

PN and Metal–Semiconductor Junctions

CHAPTER OBJECTIVES

This chapter introduces several devices that are formed by joining two different materials together. PN junction and metal–semiconductor junction are analyzed in the forward-bias and reverse-bias conditions. Of particular importance are the concepts of the depletion region and minority carrier injection. Solar cells and light-emitting diode are presented in some detail because of their rising importance for renewable energy generation and for energy conservation through solid-state lighting, respectively. The metal–semiconductor junction can be a rectifying junction or an ohmic contact. The latter is of growing importance to the design of high-performance transistors.

PART I: PN JUNCTION

As illustrated in Fig. 4–1, a PN junction can be fabricated by implanting or diffusing (see Section 3.5) donors into a P-type substrate such that a layer of semiconductor is converted into N type. Converting a layer of an N-type semiconductor into P type with acceptors would also create a PN junction.

A PN junction has rectifying current–voltage (I–V or IV) characteristics as shown in Fig. 4–2. As a device, it is called a **rectifier** or a **diode**. The PN junction is the basic structure of solar cell, light-emitting diode, and diode laser, and is present in all types of transistors. *In addition, PN junction is a vehicle for studying the theory*

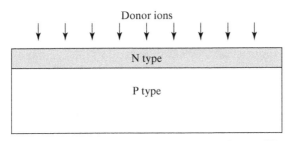

FIGURE 4–1 A PN junction can be fabricated by converting a layer of P-type semiconductor into N-type with donor implantation or diffusion.

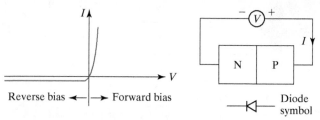

FIGURE 4–2 The rectifying *IV* characteristics of a PN junction.

of the depletion layer, the quasi-equilibrium boundary condition, the continuity equation, and other tools and concepts that are important to the understanding of transistors.

4.1 • BUILDING BLOCKS OF THE PN JUNCTION THEORY •

For simplicity, it is usually assumed that the P and N layers are uniformly doped at acceptor density N_a, and donor density N_d, respectively.[1] This idealized PN junction is known as a **step junction** or an **abrupt junction**.

4.1.1 Energy Band Diagram and Depletion Layer of a PN Junction

Let us construct a rough energy band diagram for a PN junction at equilibrium or zero bias voltage. We first draw a horizontal line for E_F in Fig. 4–3a because there is only one Fermi level at equilibrium (see Sec. 1.7.2). Figure 4–3b shows that far from the junction, we simply have an N-type semiconductor on one side (with E_c close to E_F), and a P-type semiconductor on the other side (with E_v close to E_F). Finally, in Fig. 4–3c we draw an arbitrary (for now) smooth curve to link the E_c from the N layer to the P layer. E_v of course follows E_c, being below E_c by a constant E_g.

QUESTION • *Can you tell which region (P or N) in Fig. 4–3 is more heavily doped? (If you need a review, see Section 1.8.2).*

Figure 4–3d shows that a PN junction can be divided into three layers: the neutral N layer, the neutral P layer, and a **depletion layer** in the middle. In the middle layer, E_F is close to neither E_v nor E_c. Therefore, both the electron and hole concentrations are quite small. For mathematical simplicity, it is assumed that

$$n \approx 0 \text{ and } p \approx 0 \quad \text{in the depletion layer} \quad (4.1.1)$$

The term *depletion layer* means that the layer is depleted of electrons and holes.

---

[1] N_d and N_a are usually understood to represent the *compensated* (see end of Section 1.9), or the net, dopant densities. For example, in the N-type layer, there may be significant donor *and* acceptor concentrations, and N_d is the former minus the latter.

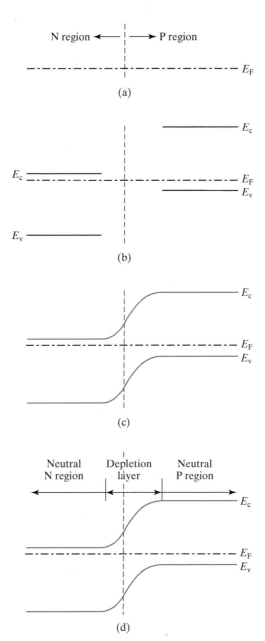

FIGURE 4–3 (a) and (b) Intermediate steps of constructing the energy band diagram of a PN junction. (c) and (d) The complete band diagram.

4.1.2 Built-In Potential

Let us examine the band diagram of a PN junction in Fig. 4–4 in greater detail. Figure 4–4b shows that E_c and E_v are not flat. This indicates the presence of a voltage differential. The voltage differential, ϕ_{bi}, is called the **built-in potential**. A built-in potential is present at the interface of any two dissimilar materials. We are usually

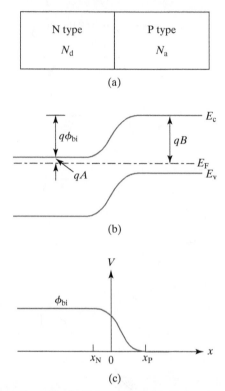

FIGURE 4–4 (a) A PN junction. The built-in potential in the energy band diagram (b) shows up as an upside down mirror image in the potential plot (c).

unaware of them because they are difficult to detect directly. For example, if one tries to measure the built-in potential, ϕ_{bi}, by connecting the PN junction to a voltmeter, no voltage will be registered because the net sum of the built-in potentials at the PN junction, the semiconductor–metal contacts, the metal to wire contacts, etc., in any closed loop is zero (see the sidebar, "Hot-Point Probe, Thermoelectric Generator and Cooler," in Sec. 2.1). However, the built-in voltage and field are as real as the voltage and field that one may apply by connecting a battery to a bar of semiconductor. For example, electrons and holes are accelerated by the built-in electric field exactly as was discussed in Chapter 2. Applying Eq. (1.8.5) to the N and P regions, one obtains

$$\text{N-region} \quad n = N_d = N_c e^{-qA/kT} \Rightarrow A = \frac{kT}{q} \ln \frac{N_c}{N_d}$$

$$\text{P-region} \quad n = \frac{n_i^2}{N_a} = N_c e^{-qB/kT} \Rightarrow B = \frac{kT}{q} \ln \frac{N_c N_a}{n_i^2}$$

$$\phi_{bi} = B - A = \frac{kT}{q} \left(\ln \frac{N_c N_a}{n_i^2} - \ln \frac{N_c}{N_d} \right)$$

$$\boxed{\phi_{bi} = \frac{kT}{q}\ln\frac{N_d N_a}{n_i^2}} \qquad (4.1.2)$$

The built-in potential is determined by N_a and N_d through Eq. (4.1.2). The larger the N_a or N_d is, the larger the ϕ_{bi} is. Typically, ϕ_{bi} is about 0.9 V for a silicon PN junction.

Since a lower E_c means a higher voltage (see Section 2.4), the N side is at a higher voltage or electrical potential than the P side. This is illustrated in Fig. 4–4c, which arbitrarily picks the neutral P region as the voltage reference. In the next section, we will derive $V(x)$ and $E_c(x)$.

4.1.3 Poisson's Equation

Poisson's equation is useful for finding the electric potential distribution when the charge density is known. In case you are not familiar with the equation, it will be derived from Gauss's Law here. Applying Gauss's Law to the volume shown in Fig. 4–5, we obtain

$$\varepsilon_s \mathscr{E}(x+\Delta x)A - \varepsilon_s \mathscr{E}(x)A = \rho \Delta x A \qquad (4.1.3)$$

where ε_s is the semiconductor permittivity and, for silicon, is equal to 12 times the permittivity of free space. ρ is the charge density (C/cm^3) and $\mathscr{E}$ is the electric field.

$$\frac{\mathscr{E}(x+\Delta x) - \mathscr{E}(x)}{\Delta x} = \frac{\rho}{\varepsilon_s} \qquad (4.1.4)$$

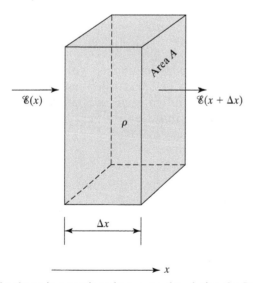

FIGURE 4–5 A small volume in a semiconductor, used to derive the Poisson's equation.

Taking the limit of $\Delta x \to 0$,

$$\frac{d\mathscr{E}}{dx} = \frac{\rho}{\varepsilon_s} \qquad (4.1.5)$$

$$\boxed{\frac{d^2V}{dx^2} = -\frac{d\mathscr{E}}{dx} = -\frac{\rho}{\varepsilon_s}} \qquad (4.1.6)$$

Equation (4.1.5) or its equivalent, Eq. (4.1.6), is *Poisson's equation*. It will be the starting point of the next section.

4.2 • DEPLETION-LAYER MODEL •

We will now solve Eq. (4.1.5) for the step junction shown in Fig. 4–6. Let's divide the PN junction into three regions—the neutral regions at $x > x_P$ and $x < -x_N$, and the **depletion layer** or **depletion region** in between, where $p = n = 0$ as shown in Fig. 4–6b. The charge density is zero everywhere except in the depletion layer where it takes the value of the dopant ion charge density as shown in Fig. 4–6c.

4.2.1 Field and Potential in the Depletion Layer

On the P side of the depletion layer ($0 \le x \le x_P$)

$$\rho = -qN_a \qquad (4.2.1)$$

Eq. (4.1.5) becomes

$$\frac{d\mathscr{E}}{dx} = -\frac{qN_a}{\varepsilon_s} \qquad (4.2.2)$$

Equation (4.2.2) may be integrated once to yield

$$\mathscr{E}(x) = -\frac{qN_a}{\varepsilon_s}x + C_1 = \frac{qN_a}{\varepsilon_s}(x_P - x) \qquad 0 \le x \le x_P \qquad (4.2.3)$$

C_1 is a constant of integration and is determined with the boundary condition $\mathscr{E} = 0$ at $x = x_P$. You may verify that Eq. (4.2.3) satisfies this boundary condition. The field increases linearly with x, having its maximum magnitude at $x = 0$ (see Fig. 4–6d).

On the N-side of the depletion layer, the field is similarly found to be

$$\mathscr{E}(x) = -\frac{qN_d}{\varepsilon_s}(x - x_N) \qquad x_N \le x \le 0 \qquad (4.2.4)$$

x_N is a negative number. The field must be continuous, and equating Eq. (4.2.3) and Eq. (4.2.4) at $x = 0$ yields

$$\boxed{N_a|x_P| = N_d|x_N|} \qquad (4.2.5)$$

$|x_N|$ and $|x_P|$ are the widths of the depletion layers on the two sides of the junction. They are inversely proportional to the dopant concentration; the more heavily doped side holds a smaller portion of the depletion layer. PN junctions are usually highly asymmetrical in doping concentration. A highly asymmetrical junction

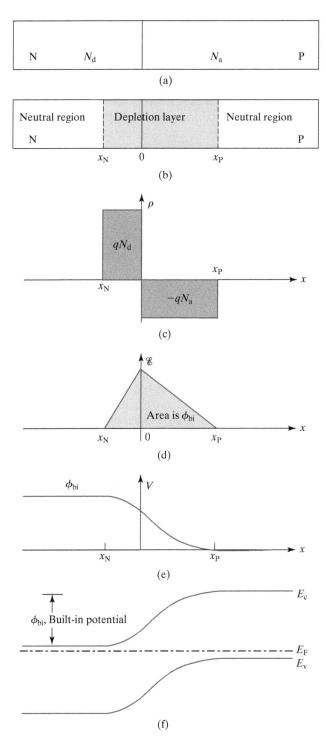

FIGURE 4–6 (a) Step PN junction; (b) depletion approximation; (c) space charge profile; (d) electric field from integration of ρ/ε_s (Poisson's equation); (e) electric potential from integrating $-\mathscr{E}$; and (f) energy band diagram.

is called a **one-sided junction,** either an **N⁺P junction** or a **P⁺N junction**, where N⁺ and P⁺ denote the heavily doped sides. *The depletion layer penetrates primarily into the lighter doping side, and the width of the depletion layer in the heavily doped material can often be neglected.* It may be helpful to think that a heavily doped semiconductor is similar to metal (and there is no depletion layer in metal).

Equation (4.2.5) tells us that the area density of the negative charge, $N_a|x_P|$ (C/cm²), and that of the positive charge, $N_d|x_N|$ (C/cm²), are equal (i.e., the net charge in the depletion layer is zero). In other words, the two rectangles in Fig. 4–6c are of equal size.

Using $\mathscr{E} = -dV/dx$, and integrating Eq. (4.2.3) yields

$$V(x) = \frac{qN_a}{2\varepsilon_s}(x_P - x)^2 \qquad 0 \leq x \leq x_P \qquad (4.2.6)$$

We arbitrarily choose the voltage at $x = x_P$ as the reference point for $V = 0$. Similarly, on the N-side, we integrate Eq. (4.2.4) once more to obtain

$$V(x) = D - \frac{qN_a}{2\varepsilon_s}(x - x_N)^2$$

$$= \phi_{bi} - \frac{qN_d}{2\varepsilon_s}(x - x_N)^2 \qquad x_N \leq x \leq 0 \qquad (4.2.7)$$

where D is determined by $V(x_N) = \phi_{bi}$ (see Fig. 4–6e and Eq. (4.1.2)). $V(x)$ is plotted in Fig. 4–6e. The curve consists of two parabolas (Eqs. (4.2.6) and (4.2.7)). Finally, we can quantitatively draw the energy band diagram, Fig. 4–6f. $E_c(x)$ and $E_v(x)$ are identical to $V(x)$, but inverted as explained in Section 2.4.

4.2.2 Depletion-Layer Width

Equating Eqs. (4.2.6) and (4.2.7) at $x = 0$ (because V is continuous at $x = 0$), and using Eq. (4.2.5), we obtain

$$x_P - x_N = W_{dep} = \sqrt{\frac{2\varepsilon_s \phi_{bi}}{q}\left(\frac{1}{N_a} + \frac{1}{N_d}\right)} \qquad (4.2.8)$$

$x_N + x_P$ is the total **depletion-layer width**, represented by W_{dep}.

If $N_a \gg N_d$, as in a P⁺N junction,

$$\boxed{W_{dep} \approx \sqrt{\frac{2\varepsilon_s \phi_{bi}}{qN_d}}} \approx |x_N| \qquad (4.2.9)$$

If $N_d \gg N_a$, as in an N⁺P junction,

$$\boxed{W_{dep} \approx \sqrt{\frac{2\varepsilon_s \phi_{bi}}{qN_a}}} \approx |x_P| \qquad (4.2.10)$$

$$|x_N| = |x_P|N_a/N_d \cong 0$$

EXAMPLE 4–1   A P^+N junction has $N_a = 10^{20} cm^3$ and $N_d = 10^{17} cm^{-3}$. What is (a) the built-in potential, (b) W_{dep}, (c) x_N, and (d) x_P?

SOLUTION:

a. Using Eq. (4.1.2),

$$\phi_{bi} = \frac{kT}{q} \ln \frac{N_d N_a}{n_i^2} \approx 0.026 \text{ V} \ln \frac{10^{20} \times 10^{17} cm^{-6}}{10^{20} cm^{-6}} \approx 1 \text{ V}$$

b. Using Eq. (4.2.9),

$$W_{dep} \approx \sqrt{\frac{2\varepsilon_s \phi_{bi}}{qN_d}} = \left(\frac{2 \times 12 \times 8.85 \times 10^{-14} \times 1}{1.6 \times 10^{-19} \times 10^{17}}\right)^{1/2}$$

$$= 1.2 \times 10^{-5} cm = 0.12 \text{ μm} = 120 \text{ nm} = 1200 \text{ Å}$$

c. In a P^+N junction, nearly the entire depletion layer exists on the N-side.

$$|x_N| \approx W_{dep} = 0.12 \text{ μm}$$

d. Using Eq. (4.2.5),

$$|x_P| = |x_N| N_d / N_a = 0.12 \text{ μm} \times 10^{17} cm^{-3} / 10^{20} cm^{-3} = 1.2 \times 10^{-4} \text{ μm}$$

$$= 1.2 \text{ Å} \approx 0$$

The point is that the heavily doped side is often hardly depleted at all. *It is useful to remember that $W_{dep} \approx 0.1$ μm for $N = 10^{17}$ cm^{-3}.* For more examples of the PN junction, see http://jas.eng.buffalo.edu/education/pn/pnformation2/pnformation2.html.

From Eqs. (4.2.9) and (4.2.10), we learn that *the depletion-layer width is determined by the lighter doping concentration.* Those two equations can be combined into

$$W_{dep} = \sqrt{2\varepsilon_s \phi_{bi}/qN} \qquad (4.2.11)$$

$$1/N = 1/N_d + 1/N_a \approx 1/\text{lighter dopant density} \qquad (4.2.12)$$

4.3 • REVERSE-BIASED PN JUNCTION •

When a positive voltage is applied to the N region relative to the P region, the PN junction is said to be **reverse-biased**. The zero-biased and reverse-biased PN junction energy diagrams are shown in Fig. 4–7. Under reverse bias, there is very little current since the bias polarity allows the flow of electrons from the P side to the N side and holes from the N side to the P side, but there are few electrons (minority carriers) on the P side and few holes on the N side. Therefore, the current is negligibly small. Since the current is small, the IR drop in the neutral regions is also negligible. All the reverse-bias voltage appears across the depletion layer. The potential barrier increases from $q\phi_{bi}$ in Fig. 4–7b to $q\phi_{bi} + qV_r$ in Fig. 4–7c.

98   Chapter 4 ● PN and Metal–Semiconductor Junctions

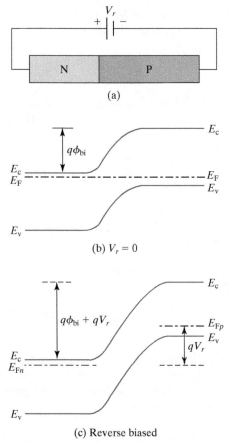

FIGURE 4–7 Reverse-biased PN junction (a) polarity of reverse bias; (b) energy band diagram without bias; and (c) energy band diagram under reverse bias.

The equations derived in the previous section for $V_r = 0$ are also valid under reverse bias if the ϕ_{bi} term is replaced with $\phi_{bi} + V_r$. The depletion layer width becomes

$$W_{dep} = \sqrt{\frac{2\varepsilon_s(\phi_{bi} + V_r)}{qN}} = \sqrt{\frac{2\varepsilon_s \times \text{potential barrier}}{qN}} \quad (4.3.1)$$

The depletion layer widens as the junction is more reverse biased. Under reverse bias, the depletion layer needs to widen in order to dissipate the larger voltage drop across it.

4.4 ● CAPACITANCE-VOLTAGE CHARACTERISTICS ●

The depletion layer and the neutral N and P regions in Fig. 4–8 may be viewed as an insulator and two conductors. Therefore, the PN junction may be modeled as a parallel-plate capacitor with capacitance

$$C_{dep} = A\frac{\varepsilon_s}{W_{dep}} \quad (4.4.1)$$

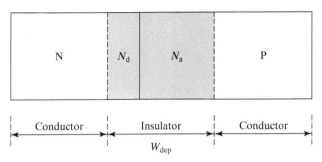

FIGURE 4–8 The PN junction as a parallel-plate capacitor.

where C_{dep} is the **depletion-layer capacitance** and A is the area. PN junction is prevalent in semiconductor devices and its capacitance is an unwelcome capacitive load to the devices and the circuits. C_{dep} can be lowered by reducing the junction area and increasing W_{dep} by reducing the doping concentration(s) and/or applying a reverse bias. *Numerically, $C \approx 1 fF/\mu m^2$ when $W_{dep} = 0.1 \mu m$.*

Using Eq. (4.4.1) together with Eq. (4.3.1), we obtain

$$\frac{1}{C_{dep}^2} = \frac{W_{dep}^2}{A^2 \varepsilon_s^2} = \frac{2(\phi_{bi} + V_r)}{qN\varepsilon_s A^2} \qquad (4.4.2)$$

Equation (4.4.2) suggests a linear relationship between $1/C_{dep}^2$ and V_r. Figure 4–9 illustrates the most common way of plotting the C–V data of a PN junction. From the slope of the line in this figure, one can determine N (or the lighter dopant concentration of a one-sided junction; see Eq. (4.2.12)). From the intercept with the horizontal axis, one can determine the built-in potential, ϕ_{bi}.

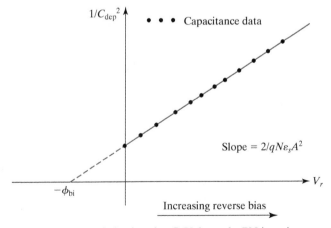

FIGURE 4–9 The common way of plotting the C–V data of a PN junction.

EXAMPLE 4-2 The slope of the line in Fig. 4–9 is 2×10^{23} F^{-2} V^{-1}, the intercept is 0.84 V, and the area of the PN junction is 1 μm². Find the lighter doping concentration, N_l, and the heavier doping concentration, N_h.

SOLUTION:

Using Eq. (4.4.2), the lighter doping concentration is

$$N_l = 2/(\text{slope} \times q\varepsilon_s A^2)$$
$$= 2/(2 \times 10^{23} \times 1.6 \times 10^{-19} \times 12 \times 8.85 \times 10^{-14} \times 10^{-8} \text{cm}^2)$$
$$= 6 \times 10^{15} \text{cm}^{-3}$$

(There is no way to determine whether the lightly doped side is the N side or the P side.) Now, using Eq. (4.1.2),

$$\phi_{bi} = \frac{kT}{q} \ln \frac{N_h N_l}{n_i^2}$$

$$N_h = \frac{n_i^2}{N_l} e^{q\phi_{bi}/kT} = \frac{10^{20} \text{cm}^{-6}}{6 \times 10^{15} \text{cm}^{-3}} e^{0.84/0.026} = 1.8 \times 10^{18} \text{cm}^{-3}$$

This example presents an accurate way to determine N_l, less so for N_h. If the intercept data has a small experimental error of 60 mV and the correct value is 0.78 V, Eq. (4.1.2) would have yielded $N_h = 1.8 \times 10^{17}$ cm^{-3}. *We should be aware of when a conclusion, though correct, may be sensitive to even small errors in the data.*

4.5 • JUNCTION BREAKDOWN[2] •

We have stated that a reverse-biased PN junction conducts negligibly small current. This is true until a critical reverse bias is reached and **junction breakdown** occurs as shown in Fig. 4–10a. There is nothing inherently destructive about junction breakdown. If the current is limited to a reasonable value by the external circuit so that heat dissipation in the PN junction is not excessive, the PN junction can be operated in reverse breakdown safely. A **Zener diode** is a PN junction diode designed to operate in the breakdown mode with a breakdown voltage that is tightly controlled by the manufacturer. A Zener protection circuit is shown in Fig. 4–10b. If the breakdown voltage of the Zener diode is 3.7 V, the maximum voltage that can appear across the integrated circuit (IC) leads C and D would be 3.7 V (even in the presence of a surge voltage of, say, 50 V, across the lines A and B). The resistance R is chosen to limit the current to a level safe for the Zener diode. Figure 4–10b can also represent a rudimentary voltage-reference circuit. In that case, a voltage supply (battery) with $V >$ 3.7 V is connected between A and B. The voltage that appears at C and D will be maintained at 3.7 V, within a tight range specified by the manufacturer, even if the battery voltage fluctuates with usage and temperature.

---

[2] This section may be omitted in an accelerated course.

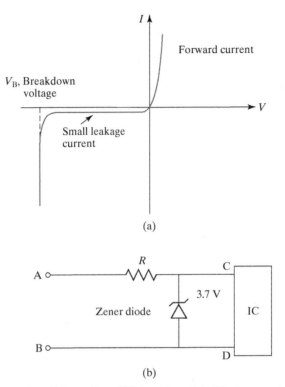

FIGURE 4–10 Reverse breakdown in a PN junction. (a) *IV* characteristics; (b) a Zener protection circuit or voltage-reference circuit.

4.5.1 Peak Electric Field

Junction breakdown occurs when the **peak electric field** in the PN junction reaches a critical value. Consider the N^+P junction shown in Fig. 4–11. Employing Eqs. (4.2.4) and (4.3.1) and evaluating the peak electric field at $x = 0$, we obtain

$$\mathscr{E}_p = \mathscr{E}(0) = \left[\frac{2qN}{\varepsilon_s}(\phi_{bi} + V_r)\right]^{1/2} \tag{4.5.1}$$

As the reverse bias voltage V increases in Fig. 4–11, $\mathscr{E}_p$ increases with it. When $\mathscr{E}_p$ reaches some critical value, $\mathscr{E}_{crit}$, breakdown occurs. Equating $\mathscr{E}_p$ in Eq. (4.5.1) with $\mathscr{E}_{crit}$ allows us to express the breakdown voltage in terms of the doping concentration. Remember, N is the lighter dopant density in a one-sided junction. In general, N is an average of N_a and N_d [see Eq. (4.2.12)].

$$V_B = \frac{\varepsilon_s \mathscr{E}_{crit}^2}{2qN} - \phi_{bi} \tag{4.5.2}$$

4.5.2 Tunneling Breakdown

When a heavily doped junction is reverse biased, as shown in Fig. 4–12, only a small distance separates the large number of electrons in the P-side valence band and the empty states in the N-side conduction band. Therefore, tunneling of electrons can

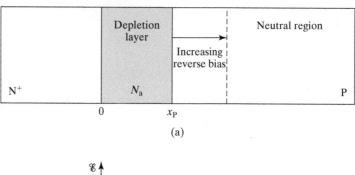

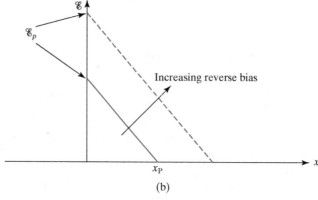

FIGURE 4–11 The field distribution in a one-side PN junction. (a) $N^+ P$ junction with $x_N \approx 0$ and (b) electric field profile.

occur (see Fig. 4–12b). The tunneling current density has an exponential dependence on $1/\mathscr{E}$ [1]:

$$J = Ge^{-H/\mathscr{E}_p} \quad (4.5.3)$$

where G and H are constants for a given semiconductor. The *IV* characteristics are shown in Fig. 4–12c. This is known as **tunneling breakdown.** The critical electric field for tunneling breakdown is proportional to H, which is proportion to the 3/2 power of E_g and 1/2 power of the effective mass of the tunneling carrier. The critical field is about 10^6 V/cm for Si. V_B is given in Eq. (4.5.2). Tunneling is the dominant breakdown mechanism when N is very high and V_B is quite low (below a few volts). Avalanche breakdown, presented in the next section, is the mechanism of diode breakdown at higher V_B.

4.5.3 Avalanche Breakdown

With increasing electric field, electrons traversing the depletion layer gain higher and higher kinetic energy. Some of them will have enough energy to raise an electron from the valence band into the conduction band, thereby creating an electron–hole pair as shown in Fig. 4–13. This phenomenon is called **impact ionization**. The electrons and holes created by impact ionization are themselves also accelerated by the electric field. Consequently, they and the original carrier can create even more carriers by impact ionization. The result is similar to a snow

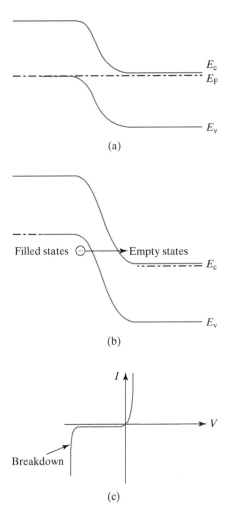

FIGURE 4–12 Tunneling breakdown. (a) Heavily doped junction at zero bias; (b) reverse bias with electron tunneling from valence band to conduction band; and (c) *IV* characteristics.

avalanche on a mountainside. (Furthermore, holes accelerate to the left and generate electrons *upstream*, thus providing positive feedback.) When $\mathscr{E}_p$ reaches $\mathscr{E}_{crit}$, the carrier creation rate and the reverse current rise abruptly. This is called **avalanche breakdown.** If the ϕ_{bi} term in Eq. (4.5.2) is ignored,

$$V_B = \frac{\varepsilon_s \mathscr{E}_{crit}^2}{2qN} \quad (4.5.4)$$

$\mathscr{E}_{crit}$ is about 5×10^5 V/cm at $N = 10^{17}$ cm^{-3}, and is approximately proportional to $N^{0.2}$ [2]. Therefore Eq. (4.5.2) may be reduced to

$$V_B(V) \approx 15 \times \left(\frac{10^{17}}{N}\right)^{0.6} \propto 1/N^{0.6} \quad (4.5.5)$$

V_B is about 15 V at $N = 10^{17}$ cm^{-3}.

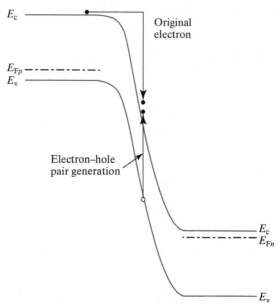

FIGURE 4–13 Electron–hole pair generation by impact ionization. The incoming electron gives up its kinetic energy to generate an electron–hole pair.

In general, it is necessary and effective to reduce the junction doping concentration(s) if a larger breakdown voltage is desired. The energy that a carrier must possess in order to initiate impact ionization increases (and therefore, $\mathscr{E}_{crit}$ also increases) with increasing band gap, E_g. Thus, the breakdown voltage, for given N_a and N_d, progressively increases from Ge to Si to GaAs diodes, due to increasing E_g.

• Applications of High-Voltage Devices •

While IC devices typically have junction breakdown voltages under 20 V, silicon **power devices** can operate at 100–1,000 V because their junction breakdown voltages can be that high. They are used to control gasoline-electric hybrid cars, diesel-electric trains, urban subway trains, and industrial processes. They are even used in the **HVDC** or **high-voltage DC** utility power transmission systems where giga-watt power may be converted from AC to DC at the generator end (e.g., in Northwestern U.S.), transmitted as HVDC power, and converted back to AC for voltage down-transformation and distribution at the user-market end (e.g., Southern California). Compared to AC transmission, HVDC improves the power grid stability and reduces power transmission loss.

QUESTION: Estimate the order of magnitude of the doping density, N, required to achieve a 1,000 V junction breakdown voltage.

SOLUTION: Using Eq. (4.5.5), we estimate that N should be 10^{14} cm^{-3} or smaller.

4.6 • CARRIER INJECTION UNDER FORWARD BIAS— QUASI-EQUILIBRIUM BOUNDARY CONDITION

Let us now examine the PN junction under forward bias. As shown in Fig. 4–14, a forward bias of V reduces the barrier height from ϕ_{bi} to $\phi_{bi} - V$. This reduces the drift field and upsets the balance between diffusion and drift that exists at zero bias. Electrons can now diffuse from the N side into the P side. This is called **minority-carrier injection**. Similarly, holes are *injected from the P side into the N side*. Figure 4–15 presents a way of visualizing carrier injection. As the barrier is reduced, a larger portion of the "Boltzmann tail" of the electrons on the N side can move into the P side (see Fig. 1–20). More electrons are now present at x_P and more holes appear at $-x_N$ than when the barrier is higher.

On the N side, $E_c - E_{Fn}$ is of course determined by N_d [Eq. (2.8.1)]. Let us assume that E_{Fn} remains constant through x_P because the depletion layer is narrow.[3] Therefore, at the edge of the neutral P region,

$$n(x_P) = N_c e^{-(E_c - E_{Fn})/kT} = N_c e^{-(E_c - E_{Fp})/kT} e^{(E_{Fn} - E_{Fp})/kT}$$

$$= n_{P0} e^{(E_{Fn} - E_{Fp})/kT} = n_{P0} e^{qV/kT} \quad (4.6.1)$$

n_{P0} is the equilibrium (denoted by subscript 0) electron concentration of the P region (denoted by the subscript P), simply n_i^2/N_a. The minority carrier density has been raised by $e^{qV/kT}$.

A similar equation may be derived for $p(x_N)$.

$$\boxed{\begin{aligned} n(x_P) &= n_{P0} e^{qV/kT} = \frac{n_i^2}{N_a} e^{qV/kT} \\ p(x_N) &= p_{N0} e^{qV/kT} = \frac{n_i^2}{N_d} e^{qV/kT} \end{aligned}} \quad (4.6.2)$$

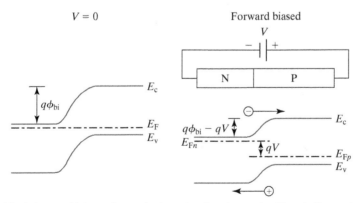

FIGURE 4–14 A forward bias reduces the junction barrier to $\phi_{bi} - V$ and allows electrons and holes to be injected over the reduced barrier.

---

[3] It can also be shown that dE_{Fn}/dx is very small in the depletion layer [1].

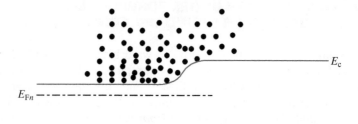

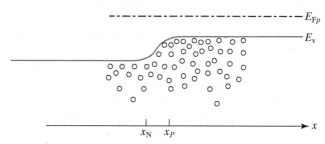

FIGURE 4–15 n at x_P (electron density at the edge of the neutral P region) is determined by $E_c - E_{Fn}$. Similarly, p at x_N is determined by $E_v - E_{Fp}$.

Equation (4.6.2) is called the **quasi-equilibrium boundary condition** or the **Shockley boundary condition**. It states that *a forward bias, V, raises the minority carrier densities at the edges of the depletion layer by the factor* $e^{qV/kT}$. This factor is 10^{10} for a moderate forward bias of 0.6 V. The derivation of Eq. (4.6.2) is also valid for a reverse bias, i.e., V can be either positive (forward bias) or negative (reverse bias). When V is a large negative number, $n(x_P)$ and $p(x_N)$ become essentially zero. This situation is sometimes called **minority carrier extraction** as opposed to carrier injection.

Another version of the quasi-equilibrium boundary condition expresses the **excess minority carrier** concentrations.

$$\boxed{\begin{aligned} n'(x_P) \equiv n(x_P) - n_{P0} &= n_{P0}(e^{qV/kT} - 1) \\ p'(x_N) \equiv p(x_N) - p_{N0} &= p_{N0}(e^{qV/kT} - 1) \end{aligned}} \qquad (4.6.3)$$

Commit Eq. (4.6.3) to memory. The next two sections will analyze what happens in the neutral regions.

EXAMPLE 4–3 Carrier Injection

A PN junction has $N_a = 10^{19}$ cm^{-3} and $N_d = 10^{16}$ cm^{-3}. (a) With V = 0, what are the minority carrier densities at the depletion region edges? Assume V = 0.6 V for (b)–(d). (b) What are the minority carrier densities at the depletion region edges? (c) What are the excess minority carrier densities? (d) What are the majority carrier densities? (e) Under the reverse bias of 1.8 V, what are the minority carrier concentrations at the depletion region edges?

a. On the P-side $\quad n_{P0} = n_i^2/N_a = 10^{20}/10^{19} = 10$ cm^{-3}

On the N-side $\quad p_{N0} = 10^{20}/10^{16} = 10^4$ cm^{-3}

b. $n(x_P) = n_{P0}e^{qV/kT} = 10 \times e^{0.6/0.026} = 10^{11} \text{cm}^{-3}$

$p(x_N) = p_{N0}e^{qV/kT} = 10^4 \times e^{0.6/0.026} = 10^{14} \text{cm}^{-3}$

We see that a moderate forward bias can increase the minority carrier densities dramatically.

c. $n'(x_P) = n(x_P) - n_{P0} = 10^{11} - 10 = 10^{11} \text{cm}^{-3}$

$p'(x_N) = p(x_N) - p_{N0} = 10^{14} - 10^4 = 10^{14} \text{cm}^{-3}$

Carrier injection into the heavily doped side is negligible when compared with injection into the lightly doped side.

d. $n' = p'$ due to charge neutrality [Eq. (2.6.2)]

On the P-side: $p(x_P) = N_a + p' = N_a + n'(x_P) = 10^{19} + 10^{11} = 10^{19} \text{cm}^{-3}$

On the N-side: $n(x_N) = N_d + n' = N_d + p'(x_N) = 10^{16} + 10^{14}$

$= 1.01 \times 10^{16} \text{cm}^{-3}$

e. $n(x_P) = n_{P0}e^{qV/kT} = 10 \times e^{-1.8/0.026} = 10^{-29} \text{cm}^{-3}$

$p(x_N) = p_{N0}e^{qV/kT} = 10^4 \times e^{-1.8/0.026} = 10^{-26} \text{cm}^{-3}$

These are meaninglessly small concentrations. We conclude that $n = p = 0$ at the junction edge under reverse bias.

4.7 • CURRENT CONTINUITY EQUATION •

In the interiors of the neutral N and P regions the minority carrier densities will be determined by the equation developed in this section. Consider the box shown in Fig. 4–16. $A \cdot \Delta x \cdot p$ is the number of holes in the box and $A \cdot J_p/q$ is the number of holes

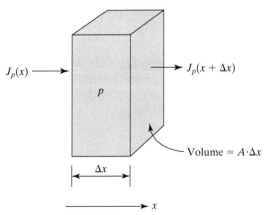

FIGURE 4–16 In steady state, the number of holes flowing into the box per second is equal to the number of holes flowing out per second plus the number of holes lost to recombination in the box per second.

flowing into the box per second. In steady state, the number of holes flowing into the box per second = number of holes flowing out of the box per second + number of holes recombining in the box per second.

$$A \cdot \frac{J_p(x)}{q} = A \cdot \frac{J_p(x + \Delta x)}{q} + A \cdot \Delta x \cdot \frac{p'}{\tau} \tag{4.7.1}$$

$$-\frac{J_p(x + \Delta x) - J_p(x)}{\Delta x} = q\frac{p'}{\tau} \tag{4.7.2}$$

Taking the limit of $\Delta x \to 0$,

$$-\frac{dJ_p}{dx} = q\frac{p'}{\tau} \tag{4.7.3}$$

τ is the recombination lifetime of the carriers. Equation (4.7.3) says that dJ_p/dx is zero only if there is no recombination. $J_p(x)$ must be larger than $J_p(x + \Delta x)$ in order to supply the holes lost to recombination in the box.

Equation (4.7.3) is valid for both the majority and minority carriers. However, it is particularly easy to apply it to the minority carriers. The minority carrier current (but not the majority carrier current) is dominated by diffusion and the drift component can be ignored. Appendix III at the end of this book provides a self-consistency proof. Let us apply Eq. (4.7.3) to the N side of the PN junction by substituting J_p, the minority carrier current, with the diffusion current [Eq. (2.3.3)].

$$qD_p\frac{d^2p}{dx^2} = q\frac{p'}{\tau_p} \tag{4.7.4}$$

A subscript p is given to τ to indicate the recombination lifetime in the N type semiconductor, in which the minority carriers are the holes.

$$\boxed{\frac{d^2p'}{dx^2} = \frac{p'}{D_p\tau_p} = \frac{p'}{L_p^2}} \tag{4.7.5}$$

$$\boxed{L_p \equiv \sqrt{D_p\tau_p}} \tag{4.7.6}$$

In Eq. (4.7.5), we replaced d^2p/dx^2 with d^2p'/dx. This assumes that the equilibrium hole concentration, p_0, is not a function of x. In other words, N_a is assumed to be uniform. (If N_a is not uniform, the mathematics becomes complicated but the result is qualitatively the same as those presented here.) Similarly, for electrons,

$$\boxed{\frac{d^2n'}{dx^2} = \frac{n'}{L_n^2}} \tag{4.7.7}$$

$$\boxed{L_n \equiv \sqrt{D_n\tau_n}} \tag{4.7.8}$$

L_n and L_p have the dimension of length. They are called the hole and electron **diffusion lengths**. They vary from a few μm to hundreds of μm depending on τ. Equations (4.7.6) and (4.7.8) are only valid for the minority carriers. Fortunately, we will not have to solve the continuity equations for the majority carriers.

4.8 • EXCESS CARRIERS IN FORWARD-BIASED PN JUNCTION •

In Fig. 4–17, we put the P side (positively biased side) on the left so that the current will flow in the positive x direction. The drawing shows x_P and x_N close to $x = 0$ because the length scale involved in this section is usually two orders of magnitude larger than W_{dep}. On the N side of the PN junction, we analyze the movement of the minority carriers (holes) by solving

$$\frac{d^2 p'}{dx^2} = \frac{p'}{L_p^2} \tag{4.7.5}$$

for the boundary conditions

$$p'(\infty) = 0$$
$$p'(x_N) = p_{N0}(e^{qV/kT} - 1) \tag{4.6.3}$$

The general solution of Eq. (4.7.5) is

$$p'(x) = Ae^{x/L_p} + Be^{-x/L_p} \tag{4.8.1}$$

The first boundary condition demands $A = 0$. The second [Eq. (4.6.3)] determines B and leads to

$$p'(x) = p_{N0}(e^{qV/kT} - 1)e^{-(x-x_N)/L_p}, \quad x > x_N \tag{4.8.2}$$

Similarly, on the P side,

$$n'(x) = n_{P0}(e^{qV/kT} - 1)e^{(x-x_P)/L_n}, \quad x < x_P \tag{4.8.3}$$

Figure 4–18 presents sample minority carrier profiles. They are simple exponential functions of x. The characteristic lengths are the **diffusion lengths,** L_n and L_p. The carrier concentrations at the depletion-layer edges are determined by the quasi-equilibrium boundary conditions (i.e., by N_a or N_d and V). Carriers are

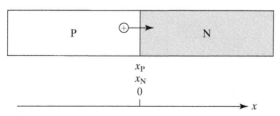

FIGURE 4–17 PN diode structure for analyzing the motion of holes after they are injected into the N side.

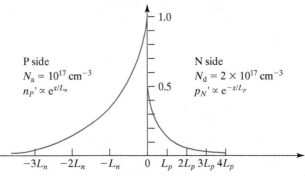

FIGURE 4–18 Normalized n' and p'. $n'(0) = 2p'(0)$ because $N_d = 2N_a$. $L_n = 2L_p$ is assumed.

mostly injected into the lighter doping side. From the depletion-layer edges, the injected minority carriers move outward by diffusion. As they diffuse, their densities are reduced due to recombination, thus creating the exponentially decaying density profiles. Beyond a few diffusion lengths from the junction, n' and p' have decayed to negligible values.

EXAMPLE 4–4 Minority and Majority Carrier Distribution and Quasi-Fermi Level

N-type	P-type
$N_d = 5 \times 10^{17}$ cm^{-3}	$N_a = 10^{17}$ cm^{-3}
$D_p = 12$ cm^2/s	$D_n = 36.4$ cm^2/s
$\tau_p = 1$ µs	$\tau_n = 2$ µs

A 0.6 V forward bias is applied to the diode. (a) What are the diffusion lengths on the N side and the P side? (b) What are the injected excess minority carrier concentrations at the junction edge? (c) What is the majority carrier profile on the N side? (d) Sketch the excess carrier densities, $p'(x)$ and $n'(x)$. (e) Sketch the energy diagram including $E_{Fp}(x)$ and $E_{Fn}(x)$, the quasi-Fermi levels. Note that $W_{dep} < 0.1$ µm may be assumed to be zero in the diagram.

SOLUTION:

a. On the N side,

$$L_p \equiv \sqrt{D_p \tau_p} = \sqrt{12 \text{cm}^2/\text{s} \times 10^{-6}\text{s}} = \sqrt{12 \times 10^{-6} \text{cm}^2} = 3.5 \times 10^{-3} \text{cm} = 35 \, \mu\text{m}$$

On the P side,

$$L_n \equiv \sqrt{D_n \tau_n} = \sqrt{36 \times 2 \times 10^{-6}} = 8.5 \times 10^{-3} \text{cm} = 85 \, \mu\text{m}$$

Please note that these diffusion lengths are much larger than the dimensions of IC devices, which are less than 1 µm.

b. Using Eqs. (4.8.2) and (4.8.3),

Excess hole density on the N-side, $p'(x_N) = p_{N0}(e^{qV/kT} - 1)$

$$= \frac{n_i^2}{N_d}(e^{qV/kT} - 1) = \frac{10^{20}}{5 \times 10^{17}} e^{0.6/0.026} = 2 \times 10^{12} \text{cm}^{-3}$$

Excess electron density on the P-side, $n'(x_P) = n_{P0}(e^{qV/kT} - 1)$

$$= \frac{n_i^2}{N_a}(e^{qV/kT} - 1) = \frac{10^{20}}{10^{17}} e^{0.6/0.026} = 10^{13} \text{cm}^{-3}$$

c. Charge neutrality requires $p'(x) = n'(x)$. If charge neutrality is not maintained, the net charge would create an electric field (see Eq. 4.1.5) that would drive the majority carriers to redistribute themselves (by drift) until charge neutrality is achieved. On the N side, electrons are the majority carriers,

$$n_N(x) = n_{N0} + n'(x) = N_d + p'(x) = 5 \times 10^{17} + p'(x_N)e^{-x/L_p}$$
$$= 5 \times 10^{17} \text{cm}^{-3} + 2 \times 10^{12} e^{-x/35} \text{ μm}$$

Here $x_N \approx 0$ is assumed.

The excess carrier density is often much smaller than the doping density as is the case here. This is called **low-level injection**.

The reader is requested to write down the answer for $p_P(x)$.

d.

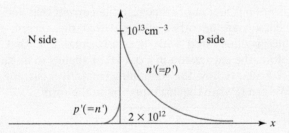

FIGURE 4–19 (a) Clearly, more carriers are injected into the lower-doping side than the higher-doping side. The ratio $p'(x_N)/n'(x_P)$ is equal to N_a/N_d. Majority carriers adjust themselves such that $n' \approx p'$ to maintain charge neutrality.

e. On the N side,

$$N_v e^{-(E_{Fp} - E_v)/kT} = p(x) = p_{N0} + p'(x)$$

$$E_{Fp}(x) = E_v + kT \ln \frac{N_v}{200 + 2 \times 10^{12} \exp(x/35 \text{ μm})}$$

Similarly, on the P side,

$$E_{Fn}(x) = E_c - kT \ln \frac{N_c}{10^3 + 10^{13} \exp(-x/85 \text{ μm})}$$

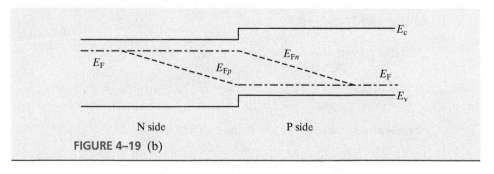

FIGURE 4–19 (b)

4.9 • PN DIODE IV CHARACTERISTICS •

Using Eqs. (4.8.2) and (4.8.3), J_p on the N side and J_n on the P side are

$$J_{pN} = -qD_p \frac{dp'(x)}{dx} = q\frac{D_p}{L_p} p_{N0}(e^{qV/kT} - 1)e^{-(x-x_N)/L_p} \quad (4.9.1)$$

$$J_{nP} = qD_n \frac{dn'(x)}{dx} = q\frac{D_n}{L_n} n_{P0}(e^{qV/kT} - 1)e^{(x-x_P)/L_n} \quad (4.9.2)$$

The two current components are shown in Fig. 4–20 a. Since both J_n and J_p are known at $x \approx 0$, the total current, J, can be determined at that location.

$$\text{Total current} = J_{pN}(x_N) + J_{nP}(x_P) = \left(q\frac{D_p}{L_p}p_{N0} + q\frac{D_n}{L_n}n_{P0}\right)(e^{qV/kT} - 1) \quad (4.9.3)$$

= Total current at all x

We know that J is not a function of x because the current that goes into one end of the diode must come out the other end and that the current is continuous in between. This fact is illustrated with the horizontal line that represents J in Fig. 4–20a. Therefore, the expression in Eq. (4.9.3) applies to all x. Once the total J is known, Fig. 4–20b shows how the remaining (majority) current components can be determined. We can rewrite the last equation in the form

$$I = I_0(e^{qV/kT} - 1) \quad (4.9.4)$$

$$I_0 = Aqn_i^2\left(\frac{D_p}{L_pN_d} + \frac{D_n}{L_nN_a}\right) \quad (4.9.5)$$

A is the diode area and I_0 is known as the **reverse saturation current** because the diode current saturates at $-I_0$ at large reverse bias (negative V). The concepts of carrier injection, carrier recombination, and reverse leakage current are illustrated with interactive animation at http://jas.eng.buffalo.edu/education/pn/biasedPN/index.html.

Several diode *IV* curves are plotted in Fig. 4–21. Note that the diodes have relatively sharp turn-on characteristics. It is often said that *Si PN diodes have a turn-on voltage of about 0.6 V at room temperature*. The turn-on voltage is lower at higher temperature.

The diode *IV* model and data can be examined more quantitatively and over a wide range of current when plotted in a semi-log plot as shown in Fig. 4–22. The diode

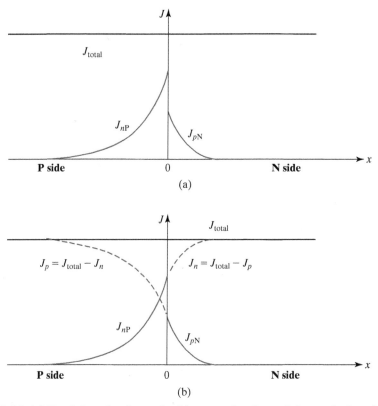

FIGURE 4–20 (a) Total J can be determined by summing J_{nP} and J_{pN} at the junction where both are known; (b) the other majority current components can now be determined.

model provides excellent fit to the measured diode current over a large current range under **forward bias**. The model predicts a straight line with a slope that is $(\ln 10)kT/q$ or *60 mV per decade* at room temperature. Equation (4.9.4) is used in circuit simulators to represent a diode. The I_0 for this purpose, however, is usually determined by fitting the forward IV data rather than calculated using Eq. (4.9.5) because N_a, N_d, τ_n, and τ_p are generally not accurately known and can vary with x.

● **The PN Junction as a Temperature Sensor** ●

Equation (4.9.4) is plotted in Fig. 4–21 for a Si diode at several temperatures. The figure shows that at a higher temperature it takes a smaller V for the diode to conduct a given I, due to a larger n_i in Eq. (4.9.5). This effect can be used to make a simple IC-based thermometer. A problem at the end of this chapter asks you to analyze dV/dT.

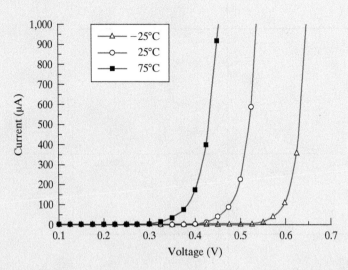

FIGURE 4–21 The *IV* curves of the silicon PN diode shift to lower voltages with increasing temperature.

4.9.1 Contributions from the Depletion Region

Equation (4.9.3) assumes that J_p and J_n do not change between x_N and x_P. As a result, Eq. (4.9.5) assumes that electrons and holes neither recombine, nor get generated, in the depletion region. In reality, there is net carrier recombination or generation in the depletion region, which contributes to the forward current and the reverse current. This contribution is called the **SCR current** for **space-charge region current**. **Space-charge region** is just another name for the depletion region. The rate of recombination (generation) in the SCR may be understood this way. Inside the depletion region, multiplying Eqs. (2.8.1) and (2.8.2) leads to

$$pn = N_c N_v e^{-(E_c - E_v)/kT} e^{(E_{Fn} - E_{Fp})/kT} = n_i^2 e^{\frac{qV}{kT}} \quad (4.9.6)$$

In the last step, Eq. (1.8.10) and $E_{Fn} - E_{Fp} = qV$ (see Fig. 4–14) were used. Equation (4.6.2) is a special case of Eq. (4.9.6) when p is known (N_a) or n is known (N_d). In the SCR, neither is known. In fact, n and p vary through the depletion-layer width. However, recombination requires the presence of both electrons and holes as shown in Fig. 2–12. It stands to reason that the recombination rate is the largest where $n \approx p$. This reasoning together with Eq. (4.9.6) suggests that the recombination rate is the largest where

$$n \approx p \approx n_i e^{\frac{qV}{2kT}} \quad (4.9.7)$$

$$\text{Net recombination (generation) rate per unit volume} = \frac{n_i}{\tau_{dep}}\left(e^{\frac{qV}{2kT}} - 1\right) \quad (4.9.8)$$

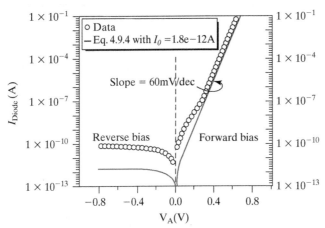

FIGURE 4–22 A semi-log plot of measured diode *IV* curve normalized to 1 cm². Eq. (4.9.4) represented by the color curves is accurate for the forward current except for very low current region. Reverse current is raised by thermal generation in the depletion layer.

τ_{dep} is the generation/recombination lifetime in the depletion layer. The –1 term ensures the recombination/generation rate is zero at $V = 0$ (equilibrium). When the rate is negative, there is net generation. The carriers so generated are swept by the field into the N and P regions as an additional current component to Eq. (4.9.4).

$$I = I_0\left(e^{\frac{qV}{kT}} - 1\right) + A\frac{qn_iW_{dep}}{\tau_{dep}}\left(e^{\frac{qV}{2kT}} - 1\right) \quad (4.9.9)$$

The second term is the SCR current. Under forward bias, it is an extra current with a slope corresponding to 120 mV/decade as shown in Fig. 4–22 below 10^{-7} A. This **nonideal current** is responsible for the low gain of bipolar transistor at low current (see Section 8.4). The reverse leakage current, i.e., Eq. (4.9.9) under reverse bias, is

$$I_{leakage} = I_0 + A\frac{qn_iW_{dep}}{\tau_{dep}} \quad (4.9.10)$$

Junction leakage current is a very important issue in DRAM (dynamic random access memory) technology (see Section 6.15.2) and generates noise in imager devices (see Section 5.10). Manufacturing these devices requires special care to make the generation/recombination lifetime long with super-clean and nearly crystal-defects free processing to minimize the density of recombination traps (see Fig. 2–12).

4.10 • CHARGE STORAGE[4]

Figure 4–19 shows that excess electrons and holes are present in a PN diode when it is forward biased. This phenomenon is called **charge storage.** Clearly the **stored charge** is proportional to $p_N'(0)$ and $n_P'(0)$ (i.e., to $e^{qV/kT} - 1$). Therefore, the

---

[4] This section may be omitted in an accelerated course. The charge storage concept is presented more thoroughly in Section 8.7.

stored charge, Q (Coulombs), is proportional to I, which is also proportional to $e^{qV/kT} - 1$.

$$Q \propto I \quad (4.10.1)$$

There is a simple explanation to this proportionality. I is the rate of minority charge injection into the diode. In steady state, this rate must be equal to the rate of charge recombination, which is Q/τ_s (see Section 2.6).

$$\boxed{\begin{aligned} I &= Q/\tau_s \\ Q &= I\tau_s \end{aligned}} \quad (4.10.2)$$

τ_s is called the **charge-storage time**. In a one-sided junction, τ_s is the recombination lifetime on the lighter-doping side, where charge injection and recombination take place. In general, τ_s is an average of the recombination lifetimes on the N side and the P side. *In any event, I and Q are simply linked through a charge-storage time.*

4.11 ● SMALL-SIGNAL MODEL OF THE DIODE[5] ●

In a class of circuits called **analog circuits**, the diode is biased at a DC current, I_{DC}, and the circuit behavior is determined by how the diode reacts to small changes in the diode voltage or current.

The diode appears to the circuit as a parallel RC circuit as shown in Fig. 4–23. The conductance is, using Eq. (4.9.4) and assuming $qV/kT \gg 1$,

$$G \equiv \frac{1}{R} = \frac{dI}{dV} = \frac{d}{dV}I_0(e^{qV/kT} - 1) \approx \frac{d}{dV}I_0 e^{qV/kT}$$

$$= \frac{q}{kT}I_0 e^{qV/kT} = I_{DC}/\frac{kT}{q} \quad (4.11.1)$$

At room temperature, $G = I_{DC}/26$ mV. The small-signal conductance and resistance can be altered by adjusting the bias current, I_{DC}. The small-signal capacitance is, using Eq. (4.10.2),

$$C = \frac{dQ}{dV} = \tau_s \frac{dI}{dV} = \tau_s G = \tau_s I_{DC}/\frac{kT}{q} \quad (4.11.2)$$

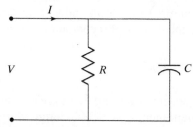

FIGURE 4–23 The small-signal equivalent circuit of a PN diode.

---

[5] This section may be omitted in an accelerated course.

The diode RC delay is therefore just the charge storage time, τ_s. Measuring the diode capacitance provides a convenient way to determine τ_s. This capacitance is often called the **diffusion capacitance** or the **charge-storage capacitance** because it is related to charge storage, which is related to the diffusion process. To be more accurate, one can add to Eq. (4.11.2) a term $A\varepsilon_s/W_{dep}$, representing the depletion-layer capacitance, C_{dep} [see Eq. (4.4.1)]. Under strong forward bias, the diffusion capacitance usually overwhelms C_{dep}. Both diffusion capacitance and depletion capacitance are undesirable because they slow down the devices and the circuits.

PART II: APPLICATION TO OPTOELECTRONIC DEVICES

4.12 • SOLAR CELLS •

A PN rectifier is useful, for example, to convert AC utility power to the DC power for powering the electronic equipment. Several other useful and even more interesting devices are also based on the PN junction. They are all **optoelectronic devices**. The first is the solar cell.

4.12.1 Solar Cell Basics

Commonly made of silicon, **solar cells**, also known as **photovoltaic cells**, can covert sunlight to electricity with 15 to 30% energy efficiency. Panels of solar cells are often installed on rooftops or open fields to generate electricity. A solar cell's structure is identical to a PN junction diode but with finger-shaped or transparent electrodes so that light can strike the semiconductor.

• **Earth's Energy Reserves** •

"Semiconductor Devices Save the Earth." This unusually brazen title of a scientific paper [3] called attention to the potential of solar cells as one solution to the global energy and climate warming problems. Suppose the rate of world consumption of energy does not grow with time but stays at its present level. The confirmed global oil reserve will last about 35 years. It is 60 years for natural gas, 170 years for coal, and 60 years for uranium. If our consumption rate increases by only 3% a year, the projected energy consumption (Fig. 4–24) will deplete the fossil fuel reserves even sooner. Although new reserves may be discovered from time to time, there is the additional problem of global warming (greenhouse effect) caused by the emission of carbon dioxide from burning fossil fuels. It would be foolhardy not to aggressively conserve energy and develop **renewable energy sources** such as solar cells.

Solar energy can be converted into electricity through many means besides photovoltaics. For example, it is converted into heat that drives a thermal engine that drives an electric generator in a solar thermal-electric system. Wind electricity generation harnesses the energy of the wind, which is created by solar heating of the earth. Growing plants and then burning them to generate electricity is another way. They all generate electricity without net emission of carbon dioxide.

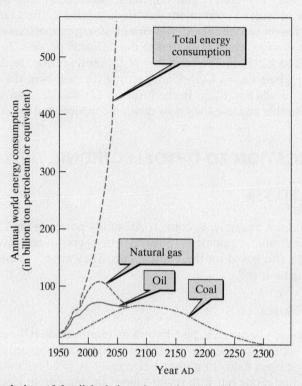

FIGURE 4–24 Depletion of fossil-fuel deposits and recent history and projection of world energy consumption assuming 3% annual growth. (From [3]. © 1992 IEEE.)

We have already seen that a voltage, without light, can produce a diode current

$$I = I_0(e^{qV/kT} - 1) \tag{4.9.4}$$

Light is also a driving force that can produce a diode current without voltage, i.e., with the two diode terminals short-circuited so that $V = 0$. Figure 4–25 shows that when light shines on the PN junction, the minority carriers that are generated by light within a diffusion length (more or less) from the junction can diffuse to the junction, be swept across the junction by the built-in field, and cause a current to flow out of the P terminal through the external short circuit and back into the N-terminal. This current is called the **short-circuit current**, I_{sc}. I_{sc} is proportional to the light intensity and to the cell area, of course.

The total diode (solar cell) current is the sum of the current generated by the voltage and that generated by light.

$$I = I_0(e^{qV/kT} - 1) - I_{sc} \tag{4.12.1}$$

The negative sign indicates that the direction of I_{sc} (see Fig. 4–25a) is opposite to that of the voltage-generated current (see Fig. 4–2). The solar cell *IV* curve is shown in Fig. 4–25b. Solar cell operates in the fourth quadrant of the *IV* plot. Since *I* and *V* have opposite signs, solar cell *generates* power. Each silicon solar cell produces

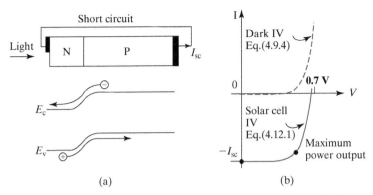

FIGURE 4–25 (a) Light can produce a current in PN junction at $V = 0$. (b) Solar cell IV product is negative, indicating power generation. (After [4].)

about 0.6 V. Many cells are connected in series to obtain the desired voltage. Many such series strings are connected in parallel into a solar cell panel. There is a particular operating point on the IV curve (see Fig. 4–25b) that maximizes the output power, $|I \times V|$. A load-matching circuit ensures that the cell operates at that point.

Unfortunately, solar energy is diffuse. Each square meter of solar cells can produce about 25 W of continuous power when averaged over day, night, and sunny/cloudy days. The average electricity consumption of the world is over 10^{12} W. If all the electricity is to be provided by solar cells, the cells need to cover a huge 200 km by 200 km area. Therefore, low cost and high **energy conversion efficiency**, defined as the ratio of the electric energy output to the solar energy input, are important. Solar cells can be made of amorphous or polycrystalline (see the sidebar in Section 3.7) as well as single-crystalline semiconductors. The first two types are less expensive to manufacture but also less efficient in electricity generation.

4.12.2 Light Penetration Depth—Direct-Gap and Indirect-Gap Semiconductors

The sunlight spans a range of wavelength or photon energy mostly from infrared to violet. The photon energy and the wavelength, λ, are related by

$$\text{Photon energy(eV)} = \frac{hc}{\lambda} = \frac{1.24}{\lambda}(\mu m) \quad (4.12.2)$$

Photons with energy less than E_g are not absorbed by the semiconductor as shown in Fig. 4–26. Photons with energy larger than E_g are absorbed but some photons may travel a considerable distance in the semiconductor before being absorbed. The light intensity decreases exponentially with the distance of travel x

$$\text{Light intensity}(x) \propto e^{-\alpha x} \quad (4.12.3)$$

α is called the **absorption coefficient**. $1\backslash\alpha$ may be called the light **penetration depth**. A solar cell must have a thickness significantly larger than the light penetration depth in order to capture nearly all the photons. Figure 4–26 shows that

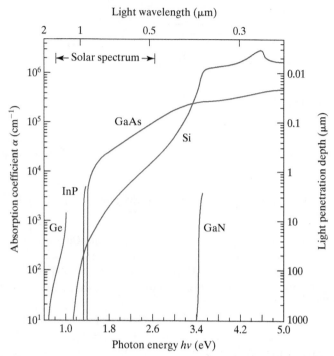

FIGURE 4–26 Light absorption coefficient as a function of photon energy. Si and Ge are indirect-gap semiconductors. InAs, GaAs, and GaN are direct-gap semiconductors, which exhibit steeply rising absorption coefficients.

a Si solar cell should be no thinner than 50 µm in order to absorb most of the photons with energy above E_g. On the other hand, a GaAs solar cell needs to be only 1 µm thick because its absorption coefficient rises steeply when the photon energy exceeds E_g. Si and Ge have lower absorption coefficients because of a characteristic of their energy band structures.

Electrons have both particle and wave properties. An electron has energy E and wave vector k. k represents the direction and the wavelength of the electron wave ($k = 2\pi$/electron wavelength). The upper branch of Fig. 4–27a shows the schematic energy versus k plot of the electrons near the bottom of the GaAs conduction band. The lower branch shows the E versus k plot of the electrons near the top of its valence band. This E–k relationship is the solution of the Schrödinger equation [see Equation (1.5.3)] of quantum mechanics for the GaAs crystal. Because the bottom of the conduction band and the top of the valence band occur at the same k, a characteristic shared also by InP and GaN, they are called **direct-gap semiconductors**. "Gap" refers to the energy band gap. A photon can move an electron from the valence band to the conduction band efficiently because k conservation (the equivalent of momentum conservation) is satisfied. Hence, the absorption coefficient is large.

Figure 4–27b shows a schematic E–k plot of Si and Ge. They are called **indirect-gap semiconductors.** Light absorption is inefficient because assistance by phonons is required to satisfy k conservation.

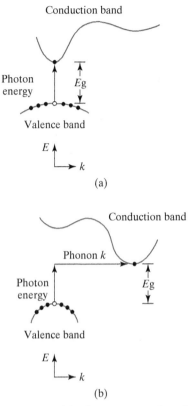

FIGURE 4–27 The *E–k* diagrams of (a) direct-gap semiconductor and (b) indirect-gap semiconductor.

Everything else being equal, direct-gap semiconductors are preferred for solar cell applications because only a very thin film and therefore a small amount of semiconductor material is needed. This has positive cost implications. In reality, silicon is the most prevalent solar cell material by far because silicon is inexpensive. Nonetheless, intense research is going on to search for inexpensive inorganic or **organic semiconductor** materials to drive thin-film solar cell system cost, including packaging and installation, lower than the silicon solar cell system cost.

A related property of the direct-gap semiconductors makes them the only suitable materials for LEDs and diode lasers (see Sections 4.13 and 4.14).

4.12.3 Short-Circuit Current and Open-Circuit Voltage

If light shines on the semiconductor in Fig. 4–16 and generates holes (and electrons) at the rate of G s^{-1}cm^{-3}, an additional term $G \cdot A \cdot \Delta x$ should be added to the left-hand side of Eq. (4.7.1). As a result, Eq. (4.7.5) is modified to

$$\frac{d^2 p'}{dx^2} = \frac{p'}{L_p^2} - \frac{G}{D_p} \qquad (4.12.4)$$

Assume that the P$^+$N solar cell in Fig. 4–28a has a very thin P$^+$ layer for simplicity and that all the electron–hole pairs are generated in the N region at a uniform rate

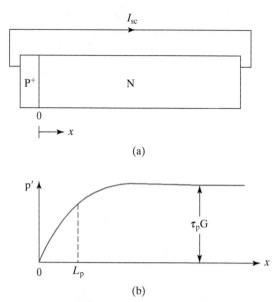

FIGURE 4–28 (a) A P$^+$N solar cell under the short-circuit condition and (b) the excess carrier concentration profile. Effectively only the carriers generated within $x < L_p$ can diffuse to the junction and contribute to the short-circuit current.

of G pairs s^{-1}cm^{-3}. Under the short-circuit condition ($V = 0$), the boundary condition at $x = 0$ is (see Eq. 4.6.3)

$$p'(0) = 0 \qquad (4.12.5)$$

At $x = \infty$, p' should reach a constant value and therefore the left-hand side of Eq. (4.12.4) should be zero.

$$p'(\infty) = L_p^2 \frac{G}{D_p} = \tau_p G \qquad (4.12.6)$$

The solution of Eq. (4.12.4) is

$$p'(x) = \tau_p G(1 - e^{-x/L_p}) \qquad (4.12.7)$$

One can easily verify that Eq. (4.12.7) satisfies Eqs. (4.12.4), (4.12.5), and (4.12.6) by substitution. $p'(x)$ is plotted in Fig. 4–28b.

$$J_p = -qD_p \frac{dp'(x)}{dx} = q\frac{D_p}{L_p} \tau_p G e^{-x/L_p} \qquad (4.12.8)$$

$$I_{sc} = A J_p(0) = A q L_p G \qquad (4.12.9)$$

There is an insightful interpretation of Eq. (4.12.9). Only the holes generated within a distance L_p from the junction, i.e., in the volume $A \cdot L_p$, are collected by the PN junction and contribute to the short-circuit current. The carriers generated farther from the junction are lost to recombination. The conclusion is that a large minority carrier diffusion length is good for the solar cell current. This is always true, although the relationship between I_{sc} and the diffusion length is more complex in a realistic solar cell structure than the simple proportionality in Eq. (4.12.9) derived under the assumption of uniform light absorption.

The solar cell material should have a large carrier diffusion length, i.e., a long carrier recombination time. In other words, the material should be quite free of defects and impurities that serve as recombination centers (see Fig. 2–12). This is particularly important for indirect-gap materials. For direct-gap semiconductors, light does not penetrate deep. In that case, all carriers are generated in a narrow region and the carrier diffusion length does not have to be large if the junction is properly positioned in this narrow region to collect the carriers.

Substituting Eqs. (4.12.9) and Eq. (4.9.5) (assuming a P$^+$N solar cell) into Eq. (4.12.1) leads to

$$I = Aq\frac{n_i^2 D_p}{N_d L_p}(e^{qV/kT} - 1) - AqL_p G \tag{4.12.10}$$

By setting $I = 0$, we can solve for the open-circuit voltage V_{oc} (assuming $e^{qV_{oc}/kT} \gg 1$ for simplicity).

$$0 = \frac{n_i^2 D_p}{N_d L_p}e^{qV_{oc}/kT} - L_p G \tag{4.12.11}$$

$$V_{oc} = \frac{KT}{q}\ln(\tau_p G N_d/n_i^2) \tag{4.12.12}$$

4.12.4 Output Power

Equation (4.12.10) is sketched in Fig. 4–25b. There is a particular operating point on the solar cell IV curve that maximizes the output power, $|I \times V|$. A load-matching circuit is usually employed to ensure that the cell operates at that point.

$$\text{Output Power} = I_{sc} \times V_{oc} \times FF \tag{4.12.13}$$

where FF (called the **fill factor**) is simply the ratio of the maximum $|I \times V|$ to $I_{sc} \times V_{oc}$. FF is typically around 0.75. The short-circuit current, I_{sc}, is proportional to the light intensity as shown in Eq. (4.12.9).

Increasing N_d can raise V_{oc} according to Equation (4.12.12). Solar cells should therefore be doped fairly heavily. Large carrier generation rate, G, is good for V_{oc}. Using optical concentrators to focus sunlight on a solar cell can raise G and improve V_{oc}. Besides reducing the solar cell area and cell cost, light concentration can thus increase the cell efficiency provided that the cell can be effectively cooled. If the cell becomes hot, n_i^2 increases and V_{oc} drops. A larger band-gap energy, E_g, reduces n_i^2 exponentially (see Eq. 1.8.12). V_{oc} therefore increases linearly with E_g. On the other hand, if E_g is too large, the material would not absorb the photons in a large long-wavelength (red and infrared) portion of the solar spectrum (see Fig. 4–26) and I_{sc} drops. The best solar cell efficiency (~24%) is obtained with E_g values between 1.2 and 1.9 eV. Commercial rooftop silicon solar-cell panels have conversion efficiencies between 15 and 20%. **Tandem solar cells** can achieve very high (over

30%) energy conversion efficiency by using two or more semiconductor materials. One material with a larger E_g absorbs and converts the short-wavelength portion of the solar radiation to electricity and another smaller E_g material, positioned behind the first, does the same to the solar radiation that is not absorbed by the first material.

4.13 • LIGHT-EMITTING DIODES AND SOLID-STATE LIGHTING •

LEDs or **light-emitting diodes** of various colors are used for such applications as traffic lights, indicator lights, and video billboards. LEDs can also provide space lighting at much higher energy efficiency than the incandescent lamps. The electrons and holes recombine by emitting photons (light) with $h\nu \approx E_g$. By adjusting the composition of the semiconductor, E_g can be altered to make blue, green, yellow, red, infrared, and UV LEDs possible. They are made of compound semiconductors involving In, Ga, Al, As, P, and N.

Figure 4–29 shows a basic LED. A PN junction made of an appropriate semiconductor is forward biased to inject minority carriers. When the injected minority carriers recombine with the majority carriers, photons are emitted. The light is emitted to all directions. To reduce the reflection of light at the semiconductor and air interface (back into the semiconductor) and therefore project more light into the forward direction, the semiconductor surface may be textured or a dome shaped lens may be provided.

4.13.1 LED Materials and Structures

Direct-gap semiconductors such as GaN (gallium nitride) are much better for LED applications than indirect-gap semiconductors such as Si. The electrons and holes in direct-gap materials have matching wave vectors (see Section 4.12.2) and can recombine easily. Figure 4–27 illustrates how a photon generates an electron–hole pair in the two types of semiconductors. If the arrows were reversed, this figure also explains how a photon is generated when an electron

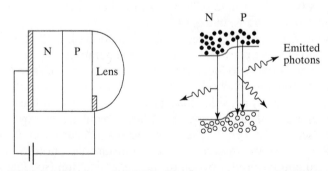

FIGURE 4–29 Schematic drawing of an LED. Photons are generated when the electrons and holes injected by the PN junction recombine.

and a hole recombine. The photon generation process, called radiative recombination, is straightforward and fast in direct-gap semiconductors with nanosecond lifetime. Therefore, the radiative recombination process is the dominant recombination process, i.e., a high percentage of the injected carriers generate photons. This percentage is known as the **quantum efficiency**. The quantum efficiency of photon generation is much lower in indirect-gap semiconductors because the radiative recombination is slow with millisecond lifetime. As a result, the recombination-through-traps process (see Fig. 2–13), which generates phonons rather than photons, is the faster and dominant process of recombination.

The next consideration is the band-gap energy. From Eq. (4.12.2)

$$\text{LED wavelength } (\mu m) = \frac{1.24}{\text{photon energy}} \approx \frac{1.24}{E_g(eV)} \quad (4.13.1)$$

Table 4–1 shows the semiconductors commonly used in LEDs. Materials with band gaps in the infrared range such as InP and GaAs are popular for LEDs used in optical communication applications (See Section 4.14). There are a few suitable semiconductors for visible-light applications. For example, GaP has a band-gap energy corresponding to yellow light. Mixing GaP and GaAs at varying ratios, i.e., using $GaAs_{1-x}P_x$, we can make LEDs that emit yellow ($x \approx 1$), orange, red, and infrared ($x \approx 0$) light.

GaAs and GaP are called **binary semiconductors**, which are made of two chemical elements. $GaAs_{1-x}P_x$, containing three elements, is a **ternary semiconductor**. They are all known as **compound semiconductors**.

There is a third consideration—the substrate material. High-efficiency LEDs are made of thin semiconductor films grown epitaxially (see Section 3.7.3) over a substrate wafer having a crystal lattice constant closely matched to that of the films. The lattice constants are given in the last column of Table 4–1. High-quality wafers of InP, GaAs, GaP, and Al_2O_3 are available at reasonable costs.

TABLE 4–1 Optoelectronic-device materials.

	$E_g(eV)$	Wavelength (μm)	Color	Lattice constant (Å)
InAs	0.36	3.44		6.05
InN	0.65	1.91	infrared	3.45
InP	1.36	0.92		5.87
GaAs	1.42	0.87		5.66
GaP	2.26	0.55	red yellow	5.46
AlP	3.39	0.51	blue violet	5.45
GaN	2.45	0.37		3.19
AlN	6.20	0.20	UV	3.11

TABLE 4–2 Some common LEDs.

Spectral Range	Material System	Substrate	Example Applications
Infrared	InGaAsP	InP	Optical communication
Infrared-Red	GaAsP	GaAs	Indicator lamps. Remote control
Red-Yellow	AlInGaP	GaP	Optical communication. High-brightness traffic signal lights
Green-Blue	InGaN	Sapphire	High-brightness signal lights. Video billboards
Blue-UV	AlInGaN	GaN or sapphire	Solid-state lighting
Red-Blue	Organic semiconductors	Glass	Displays

For example, GaP film on GaAs substrate is used to produce yellow LEDs. Unfortunately, GaP does not have the same lattice constant as the GaAs substrate. The mismatch in lattice constants creates crystal defects in the expitaxial GaP film (see Section 3.7.3) and degrades LED efficiency and reliability. GaP can be mixed with InP to obtain a lattice constant matching that of the GaAs substrate. However, that mixture has an E_g that is too low for the desired visible-light emission. Fortunately, both E_g and the lattice constant can be independently tuned if one mixes GaP, InP, and AlP in varying proportions. In other words, **quaternary semiconductors**, which contain four chemical elements such as AlInGaP, can provide a range of E_g (wavelengths) while meeting the requirement of matching the lattice constant of the substrate material.

Table 4–2 summarizes some important LED material systems. GaP substrate, replacing GaAs, improves the efficiency of red and yellow LEDs because GaP is transparent while GaAs absorb light at these wavelengths. Sapphire (Al_2O_3) with an epitaxial GaN overlayer is an excellent substrate for blue and UV LEDs because it is transparent to UV light.

With well-chosen materials, LED can have internal quantum efficiency close to unity, i.e., most of the injected carriers generate photons. Care must be taken to extract the light. Figure 4–30a shows a red LED with its substrate shaped into a truncated pyramid to collected light with reflector on the back and total internal reflection from the sides. The GaP substrate is transparent to red light.

Figure 4–30b illustrates the concept of energy well or **quantum well**. Because the AlInGaP film has a smaller band gap than GaP, it forms a well, called a quantum well, between the GaP on both sides. The concentrations of both electrons and holes are high in the well, a condition favorable for recombination and light emission. Often **multiple quantum wells** are used with several repeated alternating layers of different semiconductors.

Another class of LEDs is the **organic light-emitting diodes (OLEDs)**. Certain organic compounds have semiconductor properties and can be made into PN junction diodes, LEDs, and transistors [5]. Different colors can be obtained by modifying the molecules or adding phosphorescent materials that emit longer wavelength light when excited with a shorter wavelength light. OLED single and multicolor displays on glass substrates compete with LCD displays in some applications such as car dashboard displays.

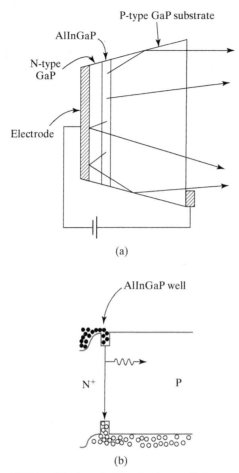

FIGURE 4–30 (a) A red LED with sloped sides for better light extraction; and (b) energy band diagram showing the quantum well.

4.13.2 Solid-State Lighting

Solid-state lighting refers to space lighting using LEDs in lieu of traditional light sources such as incandescent light bulbs and fluorescent lamps, which involve gas or vacuum. About 25% of the world electricity usage is attributable to lighting (more if the air-conditioning load due to heat generated by lamps is included). Improving the energy efficiency of light can significantly reduce energy consumption and greenhouse gas emission.

Lumen (lm) is a measure of the visible light energy (normalized to the sensitivity of the human eye at different wavelengths). A 100 W incandescent light bulb produces about 1,700 lm, therefore the **luminous efficacy** of the light bulb is 17 lm/W. Table 4–3 compares the efficacy of white LED with other light sources and the theoretical limit. The efficacy of LED continues to improve through material and device innovations.

TABLE 4–3 Luminous efficacy of lamps in lumen/watt.

Incandescent lamp	Compact fluorescent lamp	Tube fluorescent lamp	White LED	Theoretical limit at peak of eye sensitivity (λ = 555 nm)	Theoretical limit (white light)
17	60	50-100	90-?	683	~340

White light may be obtained by packaging red, green, and blue LEDs in one lamp housing. An economically attractive alternative is to use a UV LED and phosphors that convert the UV light into red, green, and blue, i.e., white light. Efficient UV LEDs can be fabricated with AlInGaN/InGaN/AlInGaN quantum wells grown on sapphire substrate (see Table 4–2). Light is extracted from the UV-transparent sapphire substrate.

White LED efficacy up to ten times that of the incandescent lamps are achievable. The technical challenge is to reduce the cost including the cost of the substrate, epitaxial film growth, and packaging. One approach of cost reduction is to use OLED. Organic semiconductors are polymers and can be printed on large sheets of glass or even flexible substrates at low cost. Different polymers can generate light of different colors. Using several layers of different materials or adding fluorescent dopants can produce white light. OLEDs, however, have lower efficacy than nitride and aluminide based LEDs. For solid-state lighting, both high efficacy and low cost are required. White LEDs are also used as the back lighting source for LCD displays.

4.14 • DIODE LASERS •

Lasers can be made with many materials from gas to glass, and energy may be supplied to the lasers by many means such as electric discharge in gas or intense light from flash lamps. **Diode lasers, powered by the diode currents,** are by far the most compact and lowest-cost lasers. Their basic structure has a PN junction under forward-bias. Diode lasers shine in many applications from fiber-optic communications to DVD and CD-ROM readers. The applications make use of one or both of these characteristics of the laser light: single frequency and ability to be focused to a small spot size. Diode lasers, as LEDs, employ direct-gap semiconductors for efficient light emission.

4.14.1 Light Amplification

Laser operation requires light amplification. Figure 4–31 illustrates how to achieve light amplification. Part (a) shows a photon generating an electron–hole pair as in solar-cell operation. Part (b) shows that when an electron falls from the conduction band to the valence band, a photon is emitted in a random direction as in the operation of an LED. This is called **spontaneous emission**. If a photon of a suitable energy comes along as shown in Fig. 4–31c, the incident photon can stimulate the electron, causing it to fall and emit a second photon. This is called **stimulated emission**. In light-wave terms, the amplitude of the incident light wave is amplified

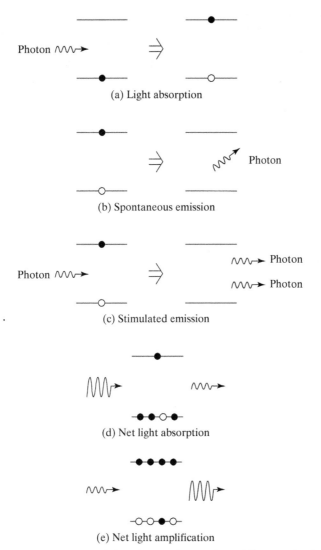

FIGURE 4–31 (a–c) Three types of light–electron interactions; (d) normally, light is absorbed in the semiconductor; and (e) under population inversion, light (wave amplitude) is amplified in the semiconductor.

just as an electrical signal waveform can be amplified. The amplified light wavefrom is identical to the incident light in frequency and direction of travel but has larger amplitude. This is the reason for the wavelength purity and directional tightness of the laser light. Stimulated emission is the foundation of laser operation.

In general, there are electrons in both the upper and the lower energy levels. If there is a higher probability of electron occupation in the lower states as shown in Fig. 4–31d (the normal case according to the Fermi function of Section 1.7), there is a higher rate of absorption than stimulated emission and the light is absorbed by the semiconductor. If there is a higher probability of electron occupation in the upper states as shown in Fig. 4–31e, a condition called **population inversion**, the light is amplified. Population inversion is a necessary condition for laser operation.

Consider a light beam traveling along the P$^+$N$^+$ junction as shown in Fig. 4–32a. At zero bias voltage, shown in Fig. 4–32b, there is a higher probability of electron occupation near E_v than E_c and the light beam is attenuated by absorption. Population inversion can be achieved by applying a large forward bias voltage to the P$^+$N$^+$ junction as shown in Fig. 4–32c. There is now a higher probability of electron occupation of states near E_c than E_v and the light beam is amplified. Population inversion is achieved when

$$E_{Fn} - E_{Fp} > E_g \qquad (4.14.1)$$

Equation (4.14.1) can be satisfied, i.e., population inversion achieved, more easily when a quantum well is inserted in or near the PN junction as shown in Fig. 4–32d and e. A quantum well or energy well is created when a thin layer of a narrower-gap semiconductor is sandwiched between two wider-gap semiconductors. The quantum well in Fig. 4–32e confines population inversion to a narrow region.

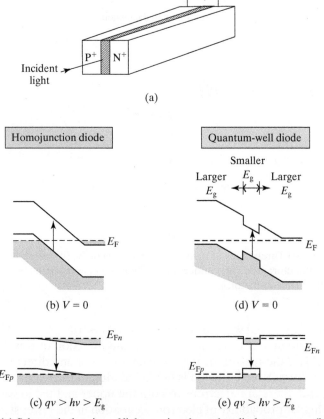

FIGURE 4–32 (a) Schematic drawing of light passing through a diode structure; (b) and (d) light is absorbed if the diode is at equilibrium. Energy states below E_F are basically filled with electrons, and those above E_F are empty. (c) and (e) under population inversion, light is amplified by stimulated emission. The arrows indicate the electron transitions caused by the light.

Fewer excess carriers are needed to achieve population inversion in the smaller volume of a narrow quantum well and the external circuitry does not need to inject carriers at a high rate, i.e., the "**threshold current**" of lasing is low.

Light amplification by stimulated emission is performed many times in a long-distance optical fiber communication line to compensate for light absorption in the fiber. To make a laser, light amplification must be paired with optical feedback as shown in the next section.

4.14.2 Optical Feedback

The operation of an electronic oscillator circuit depends on signal amplification (gain) and feedback. A laser is an optical oscillator. Besides optical amplification, it needs optical feedback. A simple way to provide optical feedback is to cleave or polish the end faces of the laser diode (see Fig. 4–33a) such that they are

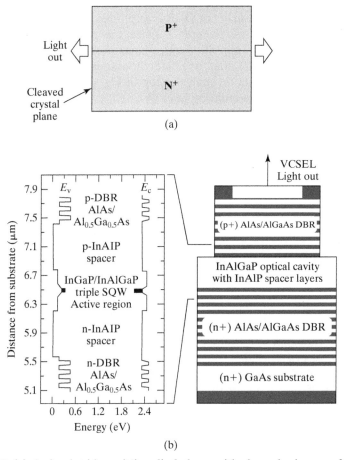

FIGURE 4–33 (a) A simple side-emitting diode laser with cleaved mirror surface. (b) The complex structure of a red-light VCSEL. Left half shows the energy band diagram of a few of the many layers of semiconductors. The energy axis is the x axis, not the usual y axis. (From [6].)

perpendicular to the PN junction. A part of the light incident on the semiconductor–air interface is reflected and amplified while traveling back through the laser diode before arriving at the other face of the diode, where a part is again reflected. The condition for oscillation is that the net round-trip gain be equal to or greater than unity.

$$R_1 \times R_2 \times G \geq 1 \qquad (4.14.2)$$

Where R_1 and R_2 are the reflectivities of the two ends and G is the light amplification factor (gain) for a round-trip travel of the light from one end to the other and back. When Eq. (4.14.2) is satisfied, the internal light intensity grows until it can stimulate emissions at a sufficient rate to consume the carriers injected by the external circuitry. That determines the steady-state internal laser light intensity, a fraction of which is emitted through the end reflectors.

Often a diode laser uses a series of alternating layers of two different semiconductors of the proper thicknesses that create constructive interference and function as a reflector for a particular wavelength. The series of layers form a **distributed Bragg reflector (DBR)** and provide **distributed feedback**. An example is shown in Fig. 4–33b. The advantage of DBR is improved wavelength purity because it only reflects light whose wavelength is twice the period of the layer series. Another advantage is its compatibility with planar processing techniques (mirror polish and cleaving are not) that produce thousands of lasers on one wafer substrate. The laser light exits through the top in Fig. 4–33b and this structure is called a **vertical-cavity surface-emitting laser (VCSEL)**.

The left-hand side of Fig. 4–33b is the energy-band diagram of the core region of the laser. The optical gain is provided only by the thin quantum wells in the middle of the sketch.

4.14.3 Diode Laser Applications

Red diode lasers are used in CD and DVD readers. The laser beam is focused to a tiny spot to read the indentations embossed into the plastic disks. Blue diode laser beams can be focused into even smaller spots because of the shorter wavelength. They are used in high-density or **Blu-ray** DVD readers. For **writable optical storage**, the focused laser melts a thin film and thus changes its optical reflectance for later reading.

Diode lasers are used in the highly important **fiber-optic communication** systems. The **optical fiber** is a thin (~20 μm) flexible fused quartz fiber that is extraordinarily transparent to light. Its transparency is the greatest at the 1.55 μm wavelength. Pulses of laser light of that wavelength are generated by modulating the laser diode current, the light pulse can travel more than 10 km in the fiber before losing half the intensity through absorption and scattering. With optical amplification every certain distance, optical fibers carry data between cities and nations. There are several hundred thousand kilometers of fiber-optic cables that crisscross the ocean floor to link the continents.

1.55 μm wavelength infrared diode lasers are constructed with InGaAsP materials on InP substrate. For short distance links, InGaAsP LEDs may be used (see Table 4–2). Light of different wavelengths travels at different speeds in the glass fiber. LEDs emit light of many different wavelengths, which arrives at the destination after different times of travel. Consequently, a short LED pulse at the originating point would arrive at the destination as a longer broadened pulse. For this reason, lasers, with their extraordinary purity of wavelength, are the light source of choice for long-distance high data rate links.

4.15 ● PHOTODIODES ●

Figure 4–25 shows that a reverse current flows through a diode when illuminated with light and the current is proportional to the light intensity. A reverse-biased PN diode can thus be used to detect light, and the device is called a **photodiode**. If the photodiode is biased near the avalanche breakdown voltage, photo-generated carriers are multiplied by impact ionization as they travel through the depletion layer (see Fig. 4–13) and thereby the sensitivity of the detector is increased. This device is called an **avalanche photodiode**. Photodiodes are used for optical communication, DVD reader, and other light-sensing applications.

PART III: METAL–SEMICONDUCTOR JUNCTION

There are two kinds of **metal–semiconductor junction**. The junctions between metal and lightly doped semiconductors exhibit rectifying *IV* characteristics similar to those of PN junctions. They are called **Schottky diodes** and have some interesting applications. The junction between metal and heavily doped semiconductors behaves as low-resistance **ohmic contacts** (basically electrical shorts). Ohmic contacts are an important part of semiconductor devices and have a significant influence on the performance of high-speed transistors.

4.16 ● SCHOTTKY BARRIERS ●

The energy diagram of a metal–semiconductor junction is shown in Fig. 4–34. The Fermi level, E_F, is flat because no voltage is applied across the junction. Far to the right of the junction, the energy band diagram is simply that of an N-type silicon sample. To the left of the junction is the energy band diagram of a metal—with the energy states below E_F almost totally filled and the states above E_F almost empty. The most striking and important feature of this energy diagram is the energy barrier at the metal–semiconductor interface. It is characterized by the **Schottky barrier height**, ϕ_B. ϕ_B is a function of the metal and the semiconductor. Actually, there are two energy barriers. In Fig. 4–34a, $q\phi_{Bn}$ is the barrier against electron flow between the metal and the N-type semiconductor.[6] In Fig. 4–34b, $q\phi_{Bp}$ is the barrier against hole flow between the metal and the P-type semiconductor. In both figures, there is clearly a depletion layer adjacent to the semiconductor–metal interface, where E_F is close to neither E_c nor E_v (such that $n \approx 0$ and $p \approx 0$).

---

[6] The hole flow in Fig. 4–34a is usually insignificant because there are few holes in the N-type semiconductor.

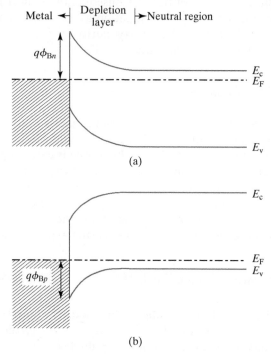

FIGURE 4–34 Energy band diagram of a metal–semiconductor contact. The Schottky barrier heights depend on the metal and semiconductor materials. (a) ϕ_{Bn} is the barrier against electron flow between the metal and the N-type semiconductor; (b) ϕ_{Bp} is the barrier against hole flow between the metal and the P-type semiconductor.

TABLE 4–4 Measured Schottky barrier heights for electrons on N-type silicon (ϕ_{Bn}) and for holes on P-type silicon (ϕ_{Bp}). (From [7].)

Metal	Mg	Ti	Cr	W	Mo	Pd	Au	Pt
ϕ_{Bn} (V)	0.4	0.5	0.61	0.67	0.68	0.77	0.8	0.9
ϕ_{Bp} (V)		0.61	0.50		0.42		0.3	
Work Function ψ_M (V)	3.7	4.3	4.5	4.6	4.6	5.1	5.1	5.7

It will become clear later that ϕ_B is the single most important parameter of a metal–semiconductor contact. Table 4–4 presents the approximate ϕ_{Bn} and ϕ_{Bp} for several metal–silicon contacts. Please note that the sum of $q\phi_{Bn}$ and $q\phi_{Bp}$ is approximately equal to E_g (1.12 eV), as suggested by Fig. 4–35.

$$\phi_{Bn} + \phi_{Bp} \approx E_g \qquad (4.16.1)$$

Why does ϕ_{Bn} (and ϕ_{Bp}) vary with the choice of the metal? Notice that Table 4–4 is arranged in ascending order of ϕ_{Bn}. There is a clear trend that ϕ_{Bn} increases with increasing metal work function (last row in Table 4–4). This trend may be partially explained with Fig. 4–2a.

$$\phi_{Bn} = \psi_M - \chi_{Si} \qquad (4.16.2)$$

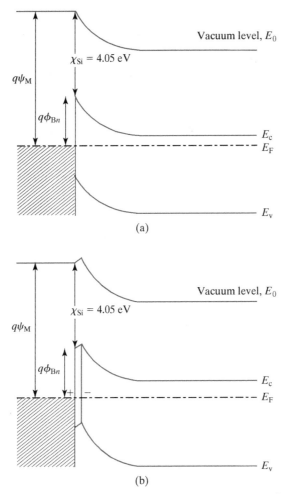

FIGURE 4–35 (a) An "ideal" metal–semiconductor contact and (b) in a real metal–semiconductor contact, there is a dipole at the interface.

ψ_M is the metal work function and χ_{Si} is the silicon electron affinity. See Sec. 5.1 for more discussion of these two material parameters. Equation (4.16.2) suggests that ϕ_{Bn} should increase with increasing ψ_M (in *qualitative* agreement with Table 4–4) by 1 eV for each 1 eV change in ψ_M (not in *quantitative* agreement with Table 4–4). The explanation for the quantitative discrepancy is that there are high densities of energy states in the band gap at the metal–semiconductor interface.[7] Some of these energy states are acceptor like and may be neutral or negative. Other energy states are donor like and may be neutral or positive. The net charge is zero when the Fermi level at the interface is around the middle of the silicon band gap. In other words, Eq. (4.16.2) is only correct for ψ_M around 4.6V, under which condition there is little interface charge. At any other ψ_M, there is a dipole at the interface as shown in

---

[7] In a three-dimensional crystal, there are no energy states in the band gap. Not so at the metal–semiconductor interface.

TABLE 4–5 Measured Schottky barrier heights of metal silicide on Si.

Silicide	ErSi$_{1.7}$	HfSi	MoSi$_2$	ZrSi$_2$	TiSi$_2$	CoSi$_2$	WSi$_2$	NiSi$_2$	Pd$_2$Si	PtSi
ϕ_{Bn} (V)	0.28	0.45	0.55	0.55	0.61	0.65	0.67	0.67	0.75	0.87
ϕ_{Bp} (V)			0.55	0.55	0.49	0.45	0.43	0.43	0.35	0.23

Fig. 4–35b and it prevents ϕ_{Bn} from moving very far from around 0.7 V. This phenomenon is known as **Fermi-level pinning**. Table 4–4 can be approximated with

$$\phi_{Bn} = 0.7 \text{ V} + 0.2(\psi_M - 4.75) \qquad (4.16.3)$$

The factor of 0.2 in Eq. (4.16.3) is determined by the polarizability of Si and the energy state density at the metal–silicon interface [8].

• Using C–V Data to Determine ϕ_B •

In Fig. 4–36a, ϕ_{bi} is the built-in potential across the depletion layer.

$$q\phi_{bi} = q\phi_{Bn} - (E_c - E_F) = q\phi_{Bn} - kT\ln\frac{N_c}{N_d} \qquad (4.16.4)$$

The depletion-layer thickness is [see Eq. (4.3.1)]

$$W_{dep} = \sqrt{\frac{2\varepsilon_s(\phi_{bi} + V)}{qN_d}} \qquad (4.16.5)$$

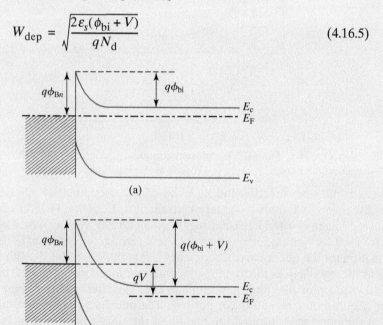

FIGURE 4–36 The potential across the depletion layer at the Schottky junction. (a) No voltage applied; (b) a negative voltage (reverse bias) is applied to the metal.

$$C = A\frac{\varepsilon_s}{W_{dep}} \qquad (4.16.6)$$

$$\frac{1}{C^2} = \frac{2(\phi_{bi} + V)}{qN_d\varepsilon_s A^2} \qquad (4.16.7)$$

Figure 4–37 shows how Eq. (4.16.7) allows us to determine ϕ_{bi} using measured C–V data. Once ϕ_{bi} is known, ϕ_{Bn} can be determined using Eq. (4.16.4).

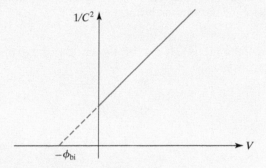

FIGURE 4–37 ϕ_{bi} (and hence ϕ_B) can be extracted from the C–V data as shown.

Much more prevalent in IC technology than metal–Si contacts are the silicide–Si contacts. Metals react with silicon to form metal like silicides at a moderate temperature. Silicide–Si interfaces are more stable than the metal–Si interfaces and free of native silicon dioxide. After the metal is deposited on Si by sputtering or CVD (Chemical Vapor Deposition) (see Chapter 3), an annealing step is applied to form a silicide–Si contact. *The term metal–silicon contact is understood to include silicide–silicon contacts.* Table 4–5 shows some available data of ϕ_{Bn} and ϕ_{Bp} of silicide–silicon contacts.

4.17 • THERMIONIC EMISSION THEORY •

Figure 4–38 presents the energy band diagram of a Schottky contact with a bias V applied to the metal. Let us analyze the current carried by the electrons flowing from Si over the energy barrier into metal, $J_{S \to M}$. This current can be predicted quite accurately by the thermionic emission theory.

In the **thermionic emission** theory, we assume that E_{Fn} is flat all the way to the peak of the barrier, the electron concentration at the interface (using Eqs. (1.8.5) and (1.8.6)) is

$$n = N_c e^{-q(\phi_B - V)/kT} = 2\left[\frac{2\pi m_n kT}{h^2}\right]^{3/2} e^{-q(\phi_B - V)/kT} \qquad (4.17.1)$$

The x-component of the average electron velocity is of course smaller than the *total* thermal velocity, $\sqrt{3kT/m_n}$ [Eq. (2.1.3)], and only half of the electrons travel

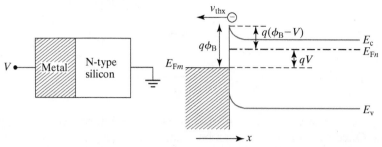

FIGURE 4–38 Energy band diagram of a Schottky contact with a forward bias V applied between the metal and the semiconductor.

toward the left (the metal). It can be shown that the average velocity of the left-traveling electrons is

$$v_{thx} = -\sqrt{2kT/\pi m_n} \qquad (4.17.2)$$

Therefore,

$$J_{S \to M} = -\frac{1}{2}qnv_{thx} = \frac{4\pi q m_n k^2}{h^3} T^2 e^{-q\phi_B/kT} e^{qV/kT} \qquad (4.17.3)$$

$$\equiv J_0 e^{qV/kT} \qquad (4.17.4)$$

Equation (4.17.4) carries two notable messages. First $J_0 \approx 100 e^{-q\phi_B/kT}$ (A/cm^2) is larger if ϕ_B is smaller. Second, $J_{S \to M}$ is only a function of $\phi_B - V$ (see Fig. 4–38). The shape of the barrier is immaterial as long as it is narrow compared to the carrier mean free path. $\phi_B - V$ determines how many electrons possess sufficient energy to surpass the peak of the energy barrier and enter the metal.

4.18 • SCHOTTKY DIODES •

At zero bias (Fig. 4–39a), the net current is zero because equal (and small) numbers of electrons on the metal side and on the semiconductor side have sufficient energy to cross the energy barrier and move to the other side. The probability of finding an electron at these high-energy states is $e^{-(E-E_c)/kT} = e^{-q\phi_B/kT}$ on both sides of the junction, as shown in Fig. 4–39a. Therefore, the net current is zero.[8] In other words, $I_{S \to M} = I_0$ and $I_{M \to S} = -I_0$, where $I_{S \to M}$ and $I_{M \to S}$ (see Fig. 4–39a) represent

---

[8] What if the densities of states are different on the two sides of the junction? Assume that the density of states at E on the metal side is twice that on the silicon side. There would be twice as many electrons on the metal side attempting to cross the barrier as on the Si side. On the other hand, there would be twice as many empty states on the metal side to *receive* the electrons coming from the Si side. Therefore, in a more detailed analysis, the net current is still zero.

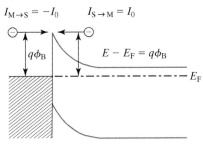
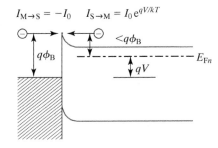
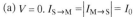

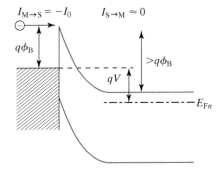
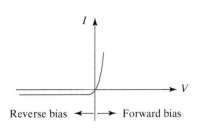
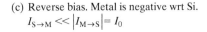

FIGURE 4–39 Explanation of the rectifying *IV* characteristics of Schottky diodes. The arrows in the subscripts indicate the direction of electron flows.

the electron current flowing from Si to metal and from metal to Si, respectively. According to the thermionic emission theory,

$$I_0 = AKT^2 e^{-q\phi_B/kT} \tag{4.18.1}$$

A is the diode area and

$$K = \frac{4\pi q m_n k^2}{h^3} \tag{4.18.2}$$

$K \approx 100 \text{ A}/(\text{cm}^2/\text{K}^2)$ is known as the **Richardson constant**. In Fig. 4–39b, a positive bias is applied to the metal. $I_{M \rightarrow S}$ remains unchanged at $-I_0$ because the barrier against $I_{M \rightarrow S}$ remains unchanged at ϕ_B. $I_{S \rightarrow M}$, on the other hand, is enhanced by $e^{qV/kT}$ because the barrier is now smaller by qV. Therefore,

$$I_{S \rightarrow M} = AKT^2 e^{-(q\phi_B - qV)/kT} = AKT^2 e^{-q\phi_B/kT} e^{qV/kT} = I_0 e^{qV/kT} \tag{4.18.3}$$

$$I = I_{S \rightarrow M} + I_{M \rightarrow S} = I_0 e^{qV/kT} - I_0 = I_0(e^{qV/kT} - 1) \tag{4.18.4}$$

In summary,

$$I = I_0(e^{qV/kT} - 1) \quad (4.18.5)$$

$$I_0 = AKT^2 e^{-q\phi_B/kT} \quad (4.18.6)$$

Equation (4.18.5) is applicable to the $V < 0$ case (reverse bias, Fig. 4–39c) as well. For a large negative V, Eq. (4.18.5) predicts $I = -I_0$. Figure 4–39c explains why: $I_{S \to M}$ is suppressed by a large barrier, while $I_{M \to S}$ remains unchanged at $-I_0$. Equation (4.18.5) is qualitatively sketched in Fig. 4–39d. I_0 may be extracted using Eq. (4.18.5) and the IV data. From I_0, ϕ_B can be determined using Eq. (4.18.6).

The similarity between the Schottky diode IV and the PN junction diode IV is obvious. The difference will be discussed in Section 4.19.

4.19 • APPLICATIONS OF SCHOTTKY DIODES •

Although Schottky and PN diodes follow the same IV expression

$$I = I_0(e^{qV/kT} - 1), \quad (4.19.1)$$

I_0 of a silicon Schottky diode can be 10^3–10^8 times larger than a typical PN junction diode, depending on ϕ_B (i.e., the metal employed). A smaller ϕ_B leads to a larger I_0. A larger I_0 means that a smaller forward bias, V, is required to produce a given diode current as shown in Fig. 4–40.

This property makes the Schottky diode the preferred rectifier in low-voltage and high-current applications where even a ~0.8 V forward-voltage drop across a PN junction diode would produce an undesirably large power loss. Figure 4–41 illustrates the switching power supply as an example. After the utility power is rectified, a 100 kHz pulse-width modulated (square-wave) AC waveform is produced so that a small (lightweight and cheap) high-frequency transformer can down-transform the voltage. This low-voltage AC power is rectified with Schottky diode (~0.3 V forward voltage drop) and filtered to produce the 50 A, 1 V, 50 W DC

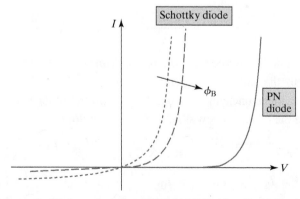

FIGURE 4–40 Schematic IV characteristics of PN and Schottky diodes having the same area.

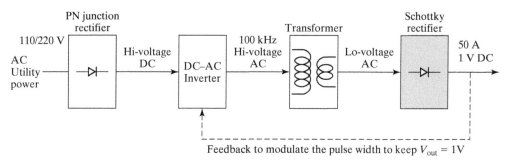

FIGURE 4–41 Block diagram of a switching power supply for electronic equipment such as PCs.

output. If a PN diode with 0.8 V forward voltage drop is used, it would consume 40 W (50 A × 0.8 V) of power and require a larger fan to cool the equipment.

For this application, a Schottky contact with a relatively small ϕ_B would be used to obtain a large I_0 and a small forward voltage drop. However, ϕ_B cannot be too small, or else the large I_0 will increase the power loss when the diode is reverse biased and can cause excessive heat generation. The resultant rise in temperature will further raise I_0 [Eq. (4.18.1)] and can lead to **thermal runaway**.

● **The Transistor as a Low Voltage-Drop Rectifier** ●

Even a Schottky diode's forward voltage may be too large when the power-supply output voltage is, say, 1V. One solution is to replace the diode with a MOSFET transistor [9]. A MOSFET is essentially an on–off switch as shown in Fig. 6–2. A low-power circuit monitors the voltage polarity across the transistor and generates a signal to turn the switch (transistor) on or off. In this way, the transistor, with the control circuit, functions as a rectifier and is called a **synchronous rectifier**. The MOSFET in this application would have a very large channel width in order to conduct large currents. The important point to note is that a MOSFET is not subjected to the same trade-off between the reverse leakage current and forward voltage drop as a diode [Eq. (4.18.5)].

The second difference between a Schottky diode and a PN junction diode is that the basic Schottky diode operation involves only the majority carriers (only electrons in Fig. 4–39, for example). There can be negligible minority carrier injection at the Schottky junction (depending on the barrier height). Negligible injection of minority carriers also means negligible storage of excess minority carriers (see Section 4.10). Therefore, Schottky diodes can operate at higher frequencies than PN junctiondiodes.

Schottky junction is also used as a part of a type of GaAs transistor as described in Section 6.3.2.

4.20 ● QUANTUM MECHANICAL TUNNELING ●

Figure 4–42 illustrates the phenomenon of **quantum mechanical tunneling**. Electrons, in quantum mechanics, are represented by traveling waves. When the electrons arrive at a potential barrier with potential energy (V_H) that is higher than the electron

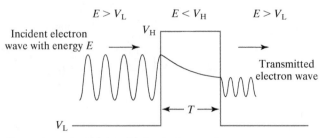

FIGURE 4–42 Illustration of quantum mechanical tunneling.

energy (E), the electron wave becomes a decaying function. Electron waves will emerge from the barrier as a traveling wave again but with reduced amplitude. In other words, there is a finite probability for electrons to tunnel through a potential barrier. The **tunneling probability** increases exponentially with decreasing barrier thickness [10] as

$$P \approx \exp\left(-2T\sqrt{\frac{8\pi^2 m}{h^2}(V_H - E)}\right) \qquad (4.20.1)$$

where m is the effective mass and h is the Planck's constant. This theory of tunneling will be used to explain the ohmic contact in the next section.

4.21 • OHMIC CONTACTS •

Semiconductor devices are connected to each other in an integrated circuit through metal. The semiconductor to metal contacts should have sufficiently low resistance so that they do not overly degrade the device performance. Careful engineering is required to reach that goal. These low-resistance contacts are called **ohmic contacts**. Figure 4–43 shows the cross-section of an ohmic contact. A surface layer of a heavily doped semiconductor diffusion region is converted into a silicide such as $TiSi_2$ or $NiSi_2$ and a dielectric (usually SiO_2) film is deposited.

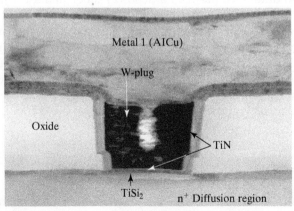

FIGURE 4–43 A contact structure. A film of metal silicide is formed before the dielectric-layer deposition and contact-hole etching. (From [11]. © 1999 IEEE.)

Lithography and plasma etching are employed to produce a contact hole through the dielectric reaching the silicide. A thin conducting layer of titanium nitride (TiN) is deposited to prevent reaction and interdiffusion between the silicide and tungsten. Tungsten is deposited by CVD to fill the contact hole. Figure 4–43 also shows what goes on top of the W plug: another layer of TiN and a layer of AlCu as the interconnect metal material.

An important feature of all good ohmic contacts is that the semiconductor is very heavily doped. The depletion layer of the heavily doped Si is only tens of Å thin because of the high dopant concentration.

When the potential barrier is very thin, the electrons can pass through the barrier by tunneling with a larger tunneling probability as shown in Fig. 4–44. The tunneling barrier height, $V_H - E$ in Eq. (4.20.1) is simply ϕ_{Bn}. The barrier thickness T may be taken as

$$T \approx W_{dep}/2 = \sqrt{\varepsilon_s \phi_{Bn}/(2qN_d)} \qquad (4.21.1)$$

$$P \approx e^{-H\phi_{Bn}/\sqrt{N_d}} \qquad (4.21.2)$$

$$H \equiv \frac{4\pi}{h}\sqrt{(\varepsilon_s m_n)/q} \qquad (4.21.3)$$

At $V = 0$, $J_{S \rightarrow M}$ and $J_{M \rightarrow S}$ in Fig. 4–44a are equal but of opposite signs so that the net current is zero.

$$J_{S \rightarrow M}(=-J_{M \rightarrow S}) \approx \frac{1}{2}qN_d v_{thx} P \qquad (4.21.4)$$

Only half of the electrons in the semiconductor, with density $N_d/2$, are in thermal motion toward the junction. The other half are moving away from the junction. v_{thx} may be found in Eq. (4.17.2). Assuming that $N_d = 10^{20}$ cm^3, P would be about 0.1 and $J_{S \rightarrow M} \approx 10^8$ A/cm^2. (This is a very large current density.) If a small voltage is applied across the contact as shown in Fig. 4–44b, the balance between $J_{S \rightarrow M}$ and $J_{M \rightarrow S}$ is broken. The barrier for $J_{M \rightarrow S}$ is reduced from ϕ_{Bn} to $(\phi_{Bn} - V)$.

$$J_{S \rightarrow M} = \frac{1}{2}qN_d v_{thx}\, e^{-H(\phi_{Bn} - V)/\sqrt{N_d}} \qquad (4.21.5)$$

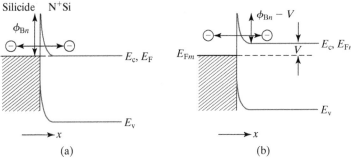

FIGURE 4–44 (a) Energy band diagram of metal–N$^+$ Si contact with no voltage applied and (b) the same contact with a voltage, V, applied to the contact.

At small V, the net current density is

$$J \approx \left.\frac{dJ_{S \to M}}{dV}\right|_{V=0} \cdot V = V \cdot \frac{1}{2} q v_{thx} H \sqrt{N_d} e^{-H\phi_{Bn}/\sqrt{N_d}} \quad (4.21.6)$$

$$R_c \equiv \frac{V}{J} = \frac{2 \cdot e^{H\phi_{Bn}/\sqrt{N_d}}}{q v_{thx} H \sqrt{N_d}} \quad (4.21.7)$$

$$\propto e^{H\phi_{Bn}/\sqrt{N_d}} \quad (4.21.8)$$

R_c is the **specific contact resistance** (Ω cm^2), the resistance of a 1 cm^2 contact. Of course, Eq. (4.21.8) is applicable to P$^+$ semiconductor contacts if ϕ_{Bn}, m_n, and N_d are replaced by ϕ_{Bp}, m_p, and N_a. Figure 4–45 shows the IV characteristics of a silicide–Si contact. The IV relationship is approximately linear, or ohmic in agreement with Eq. (4.21.6). The resistance decreases with increasing temperature in qualitative agreement with Eq. (4.21.7), due to increasing thermal velocity, v_{thx}. The contact resistance is 140 Ω and $R_c \approx 10^7$ Ω cm^2. The R_c model embodied in Eq. (4.2.7) is qualitatively accurate, but B and H are usually determined experimentally[9] [11]. R_c calculated from a more complex model is plotted in Fig. 4–46. If we want to keep the resistance of a 30 nm diameter contact below 1 kΩ, R_c should be less than 7×10^{-10} Ω cm^2. This will require a very high doping concentration and a low ϕ_B. Perhaps two different silicides will be used for N$^+$ and P$^+$ contacts, since a single metal cannot provide a low ϕ_{Bn} *and* a low ϕ_{Bp}.

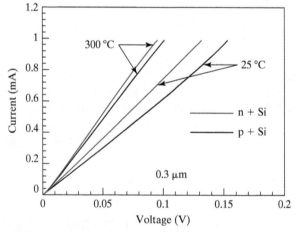

FIGURE 4–45 The IV characteristics of a 0.3 μm (diameter) TiSi$_2$ contact on N$^+$-Si and P$^+$-Si. (From [11]. ©1999 IEEE.)

● **Boundary Condition at an Ohmic Contact** ●

The voltage across an ideal ohmic contact is zero. This means that the Fermi level cannot deviate from its equilibrium position, and therefore $n' = p' = 0$ at an ideal ohmic contact.

---

[9] The electron effective mass in Eq. (4.21.2) is not equal to m_n (effective mass of electron in the conduction band) while it is tunneling under the barrier (in the band gap). Also Eq. (4.17.2) overestimates v_{thx} for a heavily doped semiconductor, for which the Boltzmann approximation is not valid.

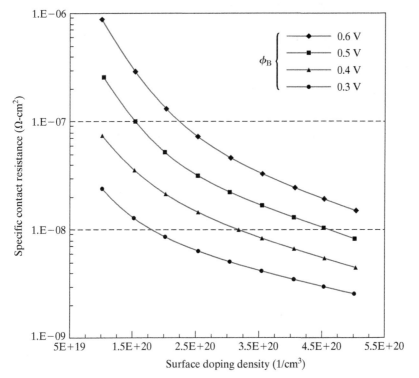

FIGURE 4–46 Theoretical specific contact resistance. (After [12].)

4.22 • CHAPTER SUMMARY

PART I: PN JUNCTION

It is important to know how to draw the energy band diagram of a PN junction. At zero bias, the potential barrier at the junction is the *built-in potential*,

$$\phi_{bi} = \frac{kT}{q} \ln \frac{N_d N_a}{n_i^2} \quad (4.1.2)$$

The potential barrier increases beyond ϕ_{bi} by 1V if a 1V reverse bias is applied and decreases by 0.1V if a 0.1V forward bias is applied.

The width of the depletion layer is

$$W_{dep} = \sqrt{\frac{2\varepsilon_s \times \text{potential barrier}}{qN}} \quad (4.3.1)$$

N is basically the smaller of the two doping concentrations. The main significance of W_{dep} is that it determines the *junction capacitance*.

$$C_{dep} = A \frac{\varepsilon_s}{W_{dep}} \quad (4.4.1)$$

In general, C_{dep} should be minimized because it contributes to the capacitive loading that slows down the circuit. The way to reduce C_{dep} is to reduce the capacitor area or doping concentrations. Applying a reverse bias will also reduce C_{dep} because W_{dep} increases.

Under forward bias, electrons are injected from the N side to the P side and holes are injected from the P side to the N side. This is called *minority carrier injection*. E_{Fn} is flat from the N region through the depletion layer up to the beginning of the neutral P region. This and similar consideration for E_{Fp} lead to the quasi-equilibrium boundary condition of minority carrier densities:

$$n(x_P) = n_{P0}e^{qV/kT}$$
$$p(x_N) = p_{N0}e^{qV/kT} \qquad (4.6.2)$$

"Quasi-equilibrium" refers to the fact that E_{Fn} and E_{Fp} are flat across the depletion layer so that the electrons and the holes are separately at equilibrium within each species. Equation (4.6.2) states that *more minority carriers are injected into the lighter-doping side.*

The steady-state *continuity equations* for minority carriers are

$$\frac{d^2p'}{dx^2} = \frac{p'}{L_p^2}, \qquad L_p \equiv \sqrt{D_p\tau_p} \qquad (4.7.5)$$

$$\frac{d^2n'}{dx^2} = \frac{n'}{L_n^2}, \qquad L_n \equiv \sqrt{D_n\tau_n} \qquad (4.7.7)$$

The injected minority carriers diffuse outward from the edges of the depletion layer and decay exponentially with distance due to recombination in the manner of $e^{-|x|/L_p}$ and $e^{-|x|/L_n}$. L_p and L_n are the diffusion lengths.

$$I = I_0(e^{qV/kT} - 1) \qquad (4.9.4)$$

$$I_0 = Aqn_i^2\left(\frac{D_p}{L_pN_d} + \frac{D_n}{L_nN_a}\right) \qquad (4.9.5)$$

The charge storage concept can be expressed as

$$Q = I\tau_s \qquad (4.10.2)$$

The storage charge gives rise to a diffusion capacitance under a forward current, I_{DC},

$$C = \tau_s G \qquad (4.11.2)$$

where the small-signal conductance, G, is

$$G = I_{DC}/\frac{kT}{q} \qquad (4.11.1)$$

PART II: APPLICATION TO OPTOELECTRONIC DEVICES

Solar cells convert light into electricity through a simple PN junction. To make the photovoltaic technology more competitive against the fossil-fuel based and other renewable energy technologies, its energy conversion efficiency and cost should be improved.

Low E_g semiconductors can collect larger portions of the solar spectrum and produce larger currents, while large E_g semiconductors can produce larger voltages. The highest theoretical energy conversion efficiency of around 24% is obtained with E_g in the range of 1.2 eV to 1.9 eV. Tandem solar cells stack multiple cells made of different E_gs can achieve even higher efficiency.

The low cost of silicon makes it a favorite solar cell material. Silicon is an indirect-gap semiconductor. Direct-gap semiconductors can collect light in a thin layer of materials and offer two potential cost advantages. First, a smaller quantity (thinner layer) of the semiconductor is needed. Second, the material purity requirement may be lower since a long diffusion length is not needed to collect the carriers generated by light at distances far from the PN junction. Low-cost organic or inorganic solar cells with high conversion efficiency and low installation cost would be an ideal renewable and carbon-emission-free electricity source.

LED generates light with photon energies about equal to the band gap energy when the injected carriers recombine in a forward biased PN junction diode. LEDs are used in signal lights, optical data links, and back lighting for LCD displays. Their potentially most important application may be space lighting replacing the incandescent lamps that are up to 10 times less efficient and fluorescent lamps that contain mercury.

In the most advanced LED, nearly every electron–hole pair recombination produces a photon, i.e., the internal quantum efficiency is 100%. This is achieved with the use of direct-gap semiconductors in which the radiative recombination lifetime is much shorter than the nonradiative recombination lifetime. The external quantum efficiency is raised by employing transparent substrates and reflectors in the back and sides of LED. The PN junction is produced in a thin film of a semiconductor having the desired band gap, which determines the emission wavelength or color. The thin film is epitaxially grown over a low cost and preferably transparent substrate. The suitable substrate materials are few. Low defect epitaxial growth requires the matching of crystal lattice constants of the substrate and the thin film. The thin film is often a quaternary compound semiconductor. Varying the composition of the compound can achieve the goals of tuning its band gap and tuning its lattice constant.

Lasers are optical oscillators. They are based on optical amplification and optical feedback. Both optical amplification and feedback can be achieved in a compact PN diode structure. A large forward bias voltage that exceeds E_g/q produces *population inversion* in a PN junction. A light wave passing through the diode under population inversion is amplified through *stimulated emission*. The amplified light retains the exact wavelength and direction of the original light wave. Population inversion can be achieved with a small forward current using the *quantum well* structure with a lower band gap semiconductor sandwiched between two wider band gap materials. The optical feedback can be provided with multi-layer *Bragg reflectors*. Diode lasers are widely used in CD and DVD readers and writers and fiber-optic communication systems.

PART III: METAL–SEMICONDUCTOR JUNCTION

The Fermi level at the metal–silicon interface is located at 0.3–0.9 eV below E_c, depending on the metal material. A low work function metal or silicide provides a low Schottky barrier height for electrons, ϕ_{Bn}, and a large barrier height for holes, ϕ_{Bp}.

A junction between metal and lightly doped silicon usually has rectifying *IV* characteristics, and is called a **Schottky diode**. The sense of the forward/reverse bias of a metal/N-type Si diode is the same as that of a P^+N junction diode. A metal on a P-silicon contact has the same forward/reverse sense as an N^+P junction diode. Compared with PN junction diodes, typical Schottky diodes have much larger reverse saturation currents, which are determined by the Schottky barrier height.

$$I_0 = AKT^2 e^{-q\phi_B/kT} \qquad (4.18.6)$$

where $K = 100$ A/cm^2/K^2. Due to the larger I_0, Schottky diodes require lower forward voltages to conduct a given current than PN junction diodes. They are often used as rectifiers in low-voltage power supplies.

Low-resistance ohmic contacts are critical to the performance of high-current devices. The key ingredient of an ohmic contact is heavy doping of the semiconductor. A low Schottky barrier height is also desirable. The carrier transport mechanism is quantum-mechanical tunneling. The contact resistance of a metal/N-Si ohmic contact is

$$R_c \propto e^{-\left(\frac{4\pi}{h}\phi_B \sqrt{(\varepsilon_s m_n)/(qN_d)}\right)} \qquad (4.21.8)$$

● **PROBLEMS** ●

Part I: PN Junction

● Electrostatics of PN Junctions ●

4.1 Applying the depletion approximation to a linearly graded junction with $N_d - N_a = ax$, derive expressions for
   (a) the electric field distribution,
   (b) the potential distribution,
   (c) the built-in potential, and
   (d) the depletion-layer width.

4.2 Consider a silicon PN step junction diode with $N_d = 10^{16} \text{cm}^{-3}$ and $N_a = 5 \times 10^{15} \text{cm}^{-3}$. Assume $T = 300$ K.

(a) Calculate the built-in potential ϕ_{bi}.

(b) Calculate the depletion-layer width (W_{dep}) and its length on the N side (x_n) and P side (x_p).

(c) Calculate the maximum electric field.

(d) Sketch the energy band diagram, electric potential, electric field distribution, and the space-charge profile.

(e) Now let $N_a = 10^{18} \text{cm}^{-3}$. Repeat (a), (b), and (c). Compare these to the previous results. How have the depletion widths changed?

4.3 Consider the silicon PN junction in Fig. 4–47.

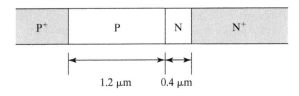

FIGURE 4–47

(a) If $N_a = 5 \times 10^{16} \text{cm}^{-3}$ in the P region and $N_d = 1 \times 10^{17} \text{cm}^{-3}$ in the N region, under increasing reverse bias, which region (N or P) will become completely depleted first? What is the reverse bias at this condition? (Hint: use $N_a x_p = N_d x_n$. The doping densities of P^+ and N^+ are immaterial).

(b) Repeat part (a) with $N_a = 1 \times 10^{16} \text{cm}^{-3}$ and $N_d = 1 \times 10^{17} \text{cm}^{-3}$.

(c) What are the small-signal capacitances (F/cm²) at the bias conditions in (a) and (b)?

4.4 A silicon sample maintained at 300 K is characterized by the energy band diagram in Fig. 4–48:

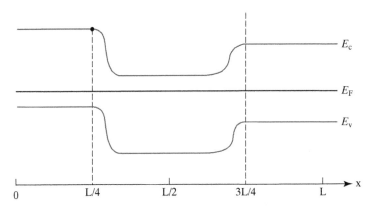

FIGURE 4–48

(a) Does the equilibrium condition prevail? How do you know?

(b) Roughly sketch n and p versus x.

(c) Sketch the electrostatic potential (Φ) as a function of x.

(d) Assume that the carrier pictured on Fig. 4–48 by the dot may move without changing its total energy. Sketch the kinetic and potential energies of the carrier as a function of its position x.

4.5

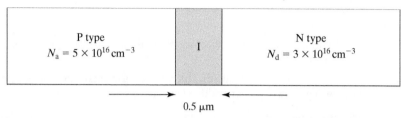

FIGURE 4–49

Consider the P-I-N structure shown in Fig. 4–49. The I region is intrinsic. Determine the quantities in (a) and (c). Assume that no bias is applied. (Hint: It may be helpful to think of the I region as a P or N and then let the doping concentration approach zero. That is, $N_d \cong N_a \cong 0$.)

(a) Find the depletion-layer width (W_{dep}) and its widths on the N side (x_n) and the P side.
(b) Calculate the maximum electric field.
(c) Find the built-in potential.
(d) Now assume that a reverse bias is applied. If the critical field for breakdown in silicon is 2×10^5 V/cm, compare the breakdown voltages between the P-I-N structure and a P-N structure (without the I region) with the doping levels shown above.

If interested, you can find more P-I-N diode examples at http://jas.eng.buffalo.edu/education/pin/pin2/index.html.

● Diffusion Equation ●

4.6 Consider a piece of infinitely long semiconductor sample shown in Fig. 4–50.

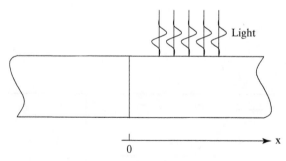

FIGURE 4–50

The $x > 0$ portion is illuminated with light. The light generates $G_L = 10^{15}$ electron–hole pairs per cm^2 per s uniformly throughout the bar in the region $x > 0$. G_L is 0 for $x < 0$. Assume that the steady-state conditions prevail, the semiconductor is made of silicon, $N_d = 10^{18}$ cm^{-3}, $\tau = 10^{-6}$ s, and $T = 300$ K.

(a) What is the hole concentration at $x = \infty$? Explain your answer.
(b) What is the hole concentration at $x = +\infty$? Explain your answer.

(c) Do low-level injection conditions prevail? Explain your answer.

(d) Determine $p'(x)$ for all x, where $p'(x)$ is the excess minority carrier concentration. (Hint: Solve the continuity equation.)

4.7

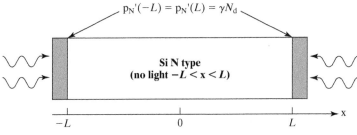

FIGURE 4–51

The two ends of a uniformly doped N-type silicon bar of length 2L are simultaneously illuminated (Fig. 4–51) so as to maintain $p' = \gamma N_d$ excess hole concentration at both $x = -L$ and $x = L$. L is the hole diffusion length. The wavelength and intensity of the illumination are such that no light penetrates into the interior ($-L < x < L$) of the bar and $\gamma = 10^{-3}$. Assume the steady-state conditions, $T = 300$ K, $N_d >> n_i$, and minority carrier lifetime of τ.

(a) Is the silicon bar at thermal equilibrium near $x = 0$? Why or why not?

(b) What are the excess concentrations of holes (p_N') and electron (n_N') at $x = -L$? What are the total electron and hole concentrations at $x = -L$?

(c) Do low-level injection conditions prevail inside the bar? Explain your answer.

(d) Write down the differential equation that you need to solve to determine $p_N'(x)$ inside the bar.

(e) Write down the general form of the $p_N'(x)$ solution and the boundary condition(s) appropriate for this particular problem.

4.8 Consider a P$^+$N junction diode with $N_d = 10^{16}$ cm^{-3} in the N region.

(a) Determine the diffusion length L on the N-type side.

(b) What are the excess hole density and excess electron density at the depletion-layer edge on the N-type side under (a) equilibrium and (b) forward bias $V = 0.4$ V?

4.9 Consider an ideal, silicon PN junction diode with uniform cross section and constant doping on both sides of the junction. The diode is made from 1 Ωcm P-type and 0.2 Ωcm N-type materials in which the recombination lifetimes are $\tau_n = 10^{-6}$ s and $\tau_p = 10^{-8}$ s, respectively.

(a) What is the value of the built-in voltage?

(b) Calculate the density of the minority carriers at the edge of the depletion layer when the applied voltage is 0.589 V (which is $23 \times kT/q$).

(c) Sketch the majority and minority carrier current as functions of distance from the junction on both sides of the junction, under the bias voltage of part (b).

(d) Calculate the location(s) of the plane (or planes) at which the minority carrier and majority carrier currents are equal in magnitude.

4.10

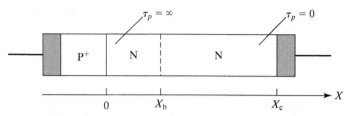

FIGURE 4–52

Consider the silicon P$^+$N junction diode pictured in Fig. 4–52. $\tau_p = \infty$ for $0 \leq x \leq x_b$ and $\tau_p = 0$ for $x_b \leq x \leq x_c$. Excluding biases that would cause high-level injection or breakdown, develop an expression for the IV characteristic of the diode. Assume the depletion-layer width (W_{dep}) never exceeds x_b for all biases of interest. The tinted regions are simply the metal contacts.

● **Proof of Minority Drift Current Being Negligible** ●

4.11 Consider an ideal, long-base, silicon abrupt P$^+$N junction diode with uniform cross section and constant doping on either side of the junction. The diode is made from a heavily doped P-type material and 0.5Ωcm N-type materials in which the minority carrier lifetime is $\tau_p = 10^{-8}$ s.

Answer the following questions on the n side of the junction only.

(a) Calculate the density of the minority carriers as a function of x (distance from the junction) when the applied voltage is 0.589 V (which is $23 \times kT/q$).

(b) Find and sketch the majority and minority carrier currents as functions of x (distance from the junction), under the applied bias voltage of part (a).

(c) What is the majority carrier diffusion current as a function of x?

The purpose of the following questions is to show that the minority drift current is negligible.

(d) Use the results of parts (b) and (c) to find the majority carrier drift current, J_{ndrift}. Then find electric field $\mathscr{E}(x)$, and finally the minority drift current J_{pdrift}. Is $J_{pdrift} \ll J_{pdiff}$? Sketch J_{pdrift} and J_{pdiff} in the same graph.

(e) Justify the assumption of $n' = p'$.

● **Temperature Effect on IV** ●

4.12 The forward-bias voltage (V) required to maintain a PN diode current (I) is a function of the temperature (T).

(a) Derive an expression for $\delta V/\delta T$.

(b) What is a typical value for a silicon diode?

(c) Compare the result of (b) with a numerical value extracted from Fig. 4–21.

4.13 Equation (4.9.5) can be interpreted this way: The minority carriers that are thermally generated within the diffusion length from the reverse-biased junction are collected by the junction and are responsible for the reverse-leakage current. Rewrite Eq. (4.9.5) in a way that justifies this interpretation.

4.14 Assume that the neutral regions of a PN diode present a series resistance R such that the voltage across the PN junction is not V but $V-RI$.

(a) How should Eq. (4.9.4) be modified?

(b) Find an expression of V as a function of I.

(c) Sketch a typical I–V curve without R for I from 0 to 100 mA. Sketch a second I–V curve in this figure for R = 200 Ω without using a calculator.

● **Charge Storage** ●

4.15 A PN diode with lengths much larger than the carrier diffusion length such as shown in Fig. 4–18 is called a long-base diode. A short-base diode has lengths much shorter than the diffusion lengths, and its excess carrier concentration is similar to that shown in Fig. 8–6. A uniformly doped short-base Si diode has $N_d = 10^{17}$ cm^{-3} and $N_a = 10^{16}$ cm^{-3}, $\tau_p = \tau_n = 1$ μs, $D_p = 10$ cm^2/s, $D_n = 30$ cm^2/s, and cross-sectional area = 10^{-5} cm^{-2}. The length of the quasi-neutral N-type and P-type regions $W_E' = W_B' = 1$ μm. The diode is at room temperature under applied forward bias of 0.5 V. Answer the following questions:

(a) Show that the total current and the sum of the charge stored on both N and P sides of the junction are proportional to each other: $Q_t = I_t \tau_s$.

Find the expression for τ_s. Use the short-base approximation, i.e., assume that the excess minority carrier concentration decreases linearly from its maximum value at the edge of the depletion region to zero at the ohmic contacts at either end of the diode.

(b) τ_s is called the charge-storage time. Show that it is significantly smaller than τ_p and τ_n.

(c) Which diode can operate at a higher frequency, short-based or long-based?

Part II: Application to Optoelectronic Devices

● **IV of Photodiode/Solar Cell** ●

4.16 Photodiodes and solar cells are both specially designed PN junction diodes packaged to permit light to reach the vicinity of the junction. Consider a P$^+$N step junction diode where incident light is uniformly absorbed throughout the N region of the device producing photogeneration rate of G_L electron–hole pairs/cm^{3}s. Assume that low-level injection prevails so that the minority drift current is negligible.

(a) What is the excess minority carrier concentration on the N side at a large distance ($x \to \infty$) from the junction? [Note: $p'(x \to \infty) \neq 0$. Far away from the junction, the recombination rate is equal to the photocarrier generation rate].

(b) The usual quasi-equilibrium boundary conditions still hold at the edges of the depletion layer. Using those and the boundary condition established in part (a), device an expression for the IV characteristic of the P$^+$N diode under the stated conditions of illumination. Ignore all recombination/generation, including photogeneration, occurring in the depletion layer.

(c) Sketch the general form of the IV characteristics for $G_L = 0$ and $G_L = G_{L0}$. Indicate the voltage developed across the diode when the diode is an open circuit, i.e., $I = 0$. What is the current that will flow when the diode is short circuited, i.e., $V = 0$?

Part III: Metal–Semiconductor Junction

● **Ohmic Contacts and Schottky Diodes** ●

4.17 Sketch the energy band diagram and comment on whether a very heavy doping is important, unimportant, or unacceptable for

(a) an ohmic contact between P$^+$-type silicon and TiSi$_2$ at equilibrium,

(b) an ohmic contact between N$^+$-type silicon and TiSi$_2$, and

(c) a rectifying contact between P-type silicon and TiSi$_2$ under 2 V reverse bias.

4.18 **(a)** Draw the energy band diagram for a metal–semiconductor contact (including the vacuum level) under 0.4 V applied forward bias. The metal has a work function φ_M of 4.8 eV, and the semiconductor is N-type Si with uniform doping concentration of 10^{16} cm^{-3}. Label clearly $q\varphi_M$, $q\phi_{Bn}$, $q(\phi_{bi} + V)$, and χ_{Si} on your sketch. Assume no surface states are present. Find the numerical values for $q\varphi_M$, $q\phi_{Bn}$, $q(\phi_{bi} + V)$.

(b) Sketch the charge density ρ, electric field ε, and potential ϕ, for the device in (a). For each diagram, draw two curves: one for equilibrium case and one for $V = 0.4$ V. No numbers or calculations are required.

4.19 Consider a Schottky diode with the doping profile shown in Fig. 4–53. Assume that the built-in potential ϕ_{bi} is 0.8 V.

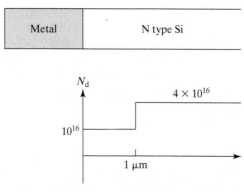

FIGURE 4–53

(a) Sketch $1/C^2$ vs. V (the reverse bias voltage) qualitatively. Do not find numerical values for C.

(b) Sketch the electric field profile for the bias condition when $W_{dep} = 2$ μm. Again, do not find numerical values for the electric field.

(c) What is the potential drop across the junction in part (b)?

(d) Derive an expression of C as a function of V for $W_{dep} > 1$ μm.

● **Depletion-Layer Analysis for Schottky Diodes** ●

4.20 **(a)** Calculate the small signal capacitance at zero bias and 300 K for an ideal Schottky barrier [see Eq. (4.16.2)] between platinum (work function 5.3 eV) and silicon doped with $N_d = 10^{16}$ cm^{-3}. The area of the Schottky diode is 10^{-5} cm^2.

(b) Calculate the reverse bias at which the capacitance is reduced by 25% from its zero-bias value.

4.21 The doping profile inside the semiconductor of a Schottky diode is linearly graded, i.e., $N_d(x) = ax$.

Derive expressions for ρ, ε, V, and W_{dep} inside the semiconductor.
Indicate how ϕ_{bi} is to be determined and computed.
Establish an expression for the junction (depletion layer) capacitance.

4.22 A metal/N-type semiconductor Schottky diode has the CV characteristic given in Fig. 4–54.

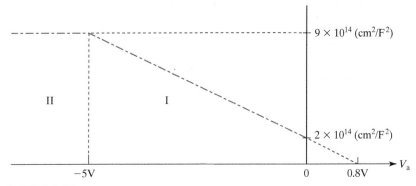

FIGURE 4–54

(a) What is the built-in voltage of the diode (from Region I data)?

(b) Find the doping profile of the N-type semiconductor.

● **Comparison Between Schottky Diodes and PN Junction Diodes** ●

4.23 (a) Qualitatively hand-sketch $\log(I)$ vs. V for a Schottky diode and a PN diode in the same figure. Comment on the similarity and difference.

(b) Calculate the I_0 of a 1 mm² $MoSi_2$ on N-type Si Schottky diode. Compare it with the I_0 of a 1 mm² P^+N diode with $N_d = 10^{18}$ cm^{-3} and $\tau_p=1$ μs.

(c) Compare the forward voltage of the two diodes in (b) at a forward current of 50 A.

(d) Besides increasing the diode area (cost), what can one do to reduce the forward voltage drop of the Schottky diode?

(e) What prevents one from using a Schottky diode having a much smaller ϕ_{Bn}?

● **Ohmic Contacts** ●

4.24 Consider an aluminum Schottky barrier on silicon having a constant donor density N_d. The barrier height $q\phi_B$ is 0.65 eV. The junction will be a low-resistance ohmic contact and can pass high currents by tunneling if the barrier presented to the electrons is thin enough. We assume that the onset of efficient tunneling occurs when the Fermi level extrapolated from the metal meets the edge of the conduction band (E_c) at a distance no larger than 10 nm from the interface.

(a) What is the minimum N_d such that this condition would be met at equilibrium?

(b) Draw a sketch of the energy band diagram under the condition of (a).

(c) Assume that N_d is increased four times from (a). By what factor is the tunneling distance (W_{dep}) reduced? And by what factor is R_c reduced?

4.25 A PN diode conducting 1 mA of current has an ohmic contact of area 0.08 μm² and surface density of 1×10^{20} cm^{-3}.

(a) What specific contact resistance can be allowed if the voltage drop at the ohmic contact is to be limited to 50 mV?

(b) Using Eq. (4.21.7), estimate the Φ_{Bn} that is allowed. Is that the maximum or minimum allowable Φ_{Bn}?

(c) Repeat (b), but this time use Fig. 4–46 to estimate Φ_{Bn}. (Fig. 4–46 is based on a more detailed model than Eq. 4.21.7)

4.26 Use Fig. 4-46, which is applicable to ohmic contacts to both N^+ and P^+ silicon, for this problem. Assume the doping concentration is 1.5×10^{20} cm^{-3}.

(a) Estimate the R_c of NiSi contact on N^+ and P^+ silicon.

(b) Estimate the R_c of PtSi contact on P^+ silicon.

(c) Estimate the R_c of ErSi$_{1.7}$ contact on N^+ silicon.

(d) If a contact resistance of $R_c \leq 4 \times 10^{-9} \Omega$ cm^2 is required for contact on both N^+ and P^+ silicon, what silicide(s) and doping concentration(s) would you have to use?

● REFERENCES ●

1. Sze, S. M. *Physics of Semiconductor Devices*, 2nd ed. New York: John Wiley & Sons, 1981, Ch. 2.
2. Muller, R. S., and T. I. Kamins. *Device Electronics for Integrated Circuits*, 2nd ed. New York: John Wiley & Sons, 1986, 194.
3. Kuwano, Y., S. Okamoto, and S. Tsuda. "Semiconductor Devices Save the Earth," *Technical Digest of International Electron Devices Meeting*, (1992), 3–10.
4. Hu, C., and R. M. White. *Solar Cells*. New York: McGraw Hill, 1983.
5. Kalinowski, J. *Organic Light-Emitting Diodes: Principles, Characteristics & Processes.* New York: Marcel Dekker, 2005.
6. Schneider, R. P., and J. A. Lott. "Cavity Design for Improved Electrical Injection in AlGaInP/AlGaAs Visible (639–661 nm) VCSEL Diodes," *Applied Physics Letter* 63 (1993), 917–919.
7. Beadle, W. E., J. C. Tsai, and R. D. Plummer. *Quick Reference Manual for Silicon Integrated Circuit Technology*. New York: Wiley-Interscience, 1985.
8. Monch, W. "Role of Virtual Gap States and Defects in Metal–Semiconductor Contacts," *Physics Review Letter*, 58 (12), (1987), 1260.
9. Kagen, R., M. Chi, and C. Hu. Improving Switching Power Supply Efficiency by Using MOSFET Synchronous Rectifiers. *Proceedings of Powercon*, 9, (July 1982), 5.
10. Choi, Y.-K., et al. "Ultrathin-body SOI MOSFET for deep-sub-tenth micron era," *IEEE Electron Device Letters*, 21(5), (2000), 254–255.
11. Banerjee, K., A. Amerasekera, G. Dixit, and C. Hu. "Temperature and Current Effect on Small-Geometry-Contact Resistance," *Technical Digest of International Electron Devices Meeting*, (1999), 115.
12. Ozturk, M. C. "Advanced Contact Formation," *Review of SRC Center for Front End Processes* 1999.

● GENERAL REFERENCES ●

1. Muller, R. S., T. I. Kamins, and M. Chen. *Device Electronics for Integrated Circuits*, 3rd ed. New York: John Wiley & Sons, 2003.
2. Streetman, B., and S. K. Banerjee. *Solid State Electronic Devices*, 6th ed. Upper Saddle River, NJ: Prentice Hall, 2006.
3. Sze, S. M. *Semiconductor Devices: Physics and Technology*, 2nd ed. New York: John Wiley & Sons, 2002.

5

MOS Capacitor

CHAPTER OBJECTIVES

This chapter builds a deep understanding of the modern MOS (metal–oxide–semiconductor) structures. The key topics are the concepts of surface depletion, threshold, and inversion; MOS capacitor C–V; gate depletion; inversion-layer thickness; and two imaging devices—charge-coupled device and CMOS (complementary MOS) imager. This chapter builds the foundation for understanding the MOSFETs (MOS Field-Effect Transistors).

The acronym **MOS** stands for **metal–oxide–semiconductor**. An MOS capacitor (Fig. 5–1) is made of a semiconductor body or substrate, an insulator film, such as SiO_2, and a metal electrode called a **gate**. The oxide film can be as thin as 1.5 nm. One nanometer is equal to 10 Å, or the size of a few oxide molecules.

Before 1970, the gate was typically made of metals such as Al (hence the M in MOS). After 1970, heavily doped polycrystalline silicon (see the sidebar, Three Kinds of Solid, in Section 3.7) has been the standard gate material because of its ability to

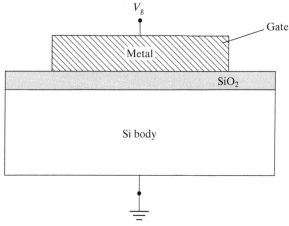

FIGURE 5–1 The MOS capacitor.

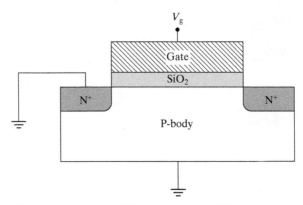

FIGURE 5–2 An MOS transistor is an MOS capacitor with PN junctions at two ends.

withstand high temperature without reacting with SiO_2. But the MOS name stuck. Unless specified otherwise, you may assume that the gate is made of heavily doped, highly conductive, polycrystalline silicon, or poly-Si for short. After 2008, the trend is to reintroduce metal gate and replace SiO_2 with more advanced dielectrics for the most advanced transistors (see Section 7.4).

The MOS capacitor is not a widely used device *in itself*. However, it is part of the MOS transistor—the topic of the next two chapters. The MOS transistor is by far the most widely used semiconductor device. An MOS transistor (Fig. 5–2) is an MOS capacitor with two PN junctions flanking the capacitor. This transistor structure is often a better structure for studying the MOS *capacitor* properties than the MOS capacitor itself as explained in Section 5.5.

5.1 • FLAT-BAND CONDITION AND FLAT-BAND VOLTAGE •

It is common to draw the energy band diagram with the oxide in the middle and the gate and the body on the left- and right-hand sides as shown in Fig. 5–3. The band diagram for $V_g = 0$ (Fig. 5–3b) is quite complex.

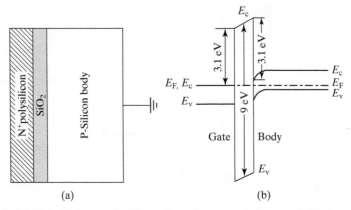

FIGURE 5–3 (a) Polysilicon-gate/oxide/semiconductor capacitor and (b) its energy band diagram with no applied voltage.

It is a good strategy to first study the energy band diagram for a special bias condition called the **flat-band condition**. Flat band is the condition where the energy band (E_c and E_v) of the substrate is flat at the Si–SiO$_2$ interface as shown in Fig. 5–4. This condition is achieved by applying a negative voltage to the gate in Fig. 5–3b, thus raising the band diagram on the left-hand side. (See Section 2.4 for the relation between voltage and the band diagram.) When the band is flat in the body as in Fig. 5–4, the surface electric field in the substrate is zero. Therefore the electric field in the oxide is also zero[1], i.e., E_c and E_v of SiO$_2$ are flat, too. E_c and E_v of SiO$_2$ are separated by 9 eV, the E_g of SiO$_2$. E_0, the **vacuum level**, is the energy state of electrons outside the material. E_0 of SiO$_2$ is above E_c by 0.95 eV. The difference between E_0 and E_c is called the **electron affinity**, another material parameter just as E_g is a material parameter. Si has an electron affinity equal to 4.05 eV. E_0 must be continuous at the Si–SiO$_2$ interface as shown in Fig. 5–4 (otherwise the electric field would be infinite). Therefore, E_c of SiO$_2$ is 3.1 eV higher than E_c of Si. This 3.1 eV is the Si–SiO$_2$ **electron energy barrier**. The **hole energy barrier** is 4.8 eV in Fig. 5–4. Because of these large energy barriers, electrons and holes normally cannot pass through the SiO$_2$ gate dielectric. E_c in the poly-silicon gate is also lower than the E_c of SiO$_2$ by 3.1 eV (the Si–SiO$_2$ energy barrier). Finally, E_F of the N$^+$poly-Si may be assumed to coincide with E_c for simplicity. In SiO$_2$, the exact position of E_F has no significance. If we place E_F anywhere around the middle of the SiO$_2$ band gap,

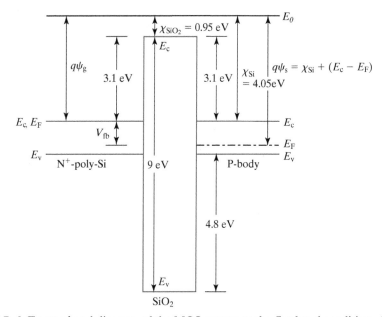

FIGURE 5–4 Energy band diagram of the MOS system at the flat-band condition. A voltage equal to V_{fb} is applied between the N$^+$-poly-Si gate and the P-silicon body to achieve this condition. ψ_g is the gate-material work function, and ψ_s is the semiconductor work function. E_0 is the vacuum level.

---

[1] According to Gauss's Law, with no interface charge, $\varepsilon_s \mathscr{E}_s = \varepsilon_{ox} \mathscr{E}_{ox}$ where $\mathscr{E}_s$ and $\mathscr{E}_{ox}$ are the body surface field and the oxide field.

$n = N_c \exp[(E_c - E_F)/kT]$ would be a meaninglessly small number such as 10^{-60} cm^{-3}. Therefore, the position of E_F in SiO$_2$ is immaterial.

The applied voltage at the flat-band condition, called V_{fb}, the **flat-band voltage**, is the difference between the Fermi levels at the two terminals.

$$V_{fb} = \psi_g - \psi_s \qquad (5.1.1)$$

ψ_g and ψ_s are the gate work function and the semiconductor work function, respectively, in volts. The work function is the difference between E_0 and E_F. For an N$^+$-poly-Si gate, $\psi_g = 4.05$ V.[2] For the P-Si body, $\psi_s = 4.05$ V $+ (E_c - E_F)/q$. For the example at hand, Eq. (5.1.1) and Fig. 5–4 indicate a negative V_{fb}, about –0.7 V.

5.2 • SURFACE ACCUMULATION •

How would Fig. 5–4 change if a more negative V_g than V_{fb} is applied? The band diagram on the gate side would be pushed upward (see Section 2.4). The result is shown in Fig. 5–5. Note that Fig. 5–5 is not drawn to scale (e.g., 3.1 eV is not about three times the silicon band gap) for the economy of page space. Such not-to-scale drawings are the norm. When $V_g \neq V_{fb}$, ϕ_s (surface voltage) and V_{ox} (oxide voltage) will be non-zero in general. $q\phi_s$ is the band bending in the substrate. Because the substrate is the voltage reference, ϕ_s is negative if E_c bends upward toward the surface as shown in Fig. 5–5 and positive if E_c bends downward. If this discussion of the sign of ϕ_s sounds strange, please review Sec. 2.4. V_{ox} is the voltage across the oxide. Again, V_{ox} is negative if the SiO$_2$ energy band tilts up toward the gate as it does in Fig. 5–5, and positive if it tilts downward toward the gate.

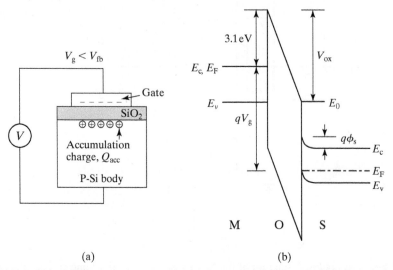

FIGURE 5–5 This MOS capacitor is biased into surface accumulation ($p_s > p_0 = N_a$). (a) Types of charge present. ⊕ represents holes and – represents negative charge. (b) Energy band diagram.

---

[2] In this case, ψ_g happens to be equal to χ_{Si}. In general, ψ_g is defined as the difference between E_0 and E_F.

Because E_v is closer to E_F at the surface than in the bulk, the surface hole concentration, p_s, is larger than the bulk hole concentration, $p_0 = N_a$. Specifically,

$$p_s = N_a e^{-q\phi_s/kT} \qquad (5.2.1)$$

Since ϕ_s may be −100 or −200 mV, $p_s \gg N_a$. That is to say, there are a large number of holes at or near the surface. They form an **accumulation layer** and these holes are called the **accumulation-layer holes**, and their charge the **accumulation charge, Q_{acc}**. This condition is known as **surface accumulation**. If the substrate were N type, the accumulation layer would hold electrons.

A relationship that we will use again and again is

$$\boxed{V_g = V_{fb} + \phi_s + V_{ox}} \qquad (5.2.2)$$

At flat band, $V_g = V_{fb}$, $\phi_s = V_{ox} = 0$ and Eq. (5.2.2) is satisfied. If $V_g \neq V_{fb}$, the difference must be picked up by ϕ_s and V_{ox}. In the case of surface accumulation, ϕ_s may be ignored in a first-order model since it is quite small and Eq. (5.2.2) becomes

$$V_{ox} = V_g - V_{fb} \qquad (5.2.3)$$

Using Gauss's Law, $\qquad \mathscr{E}_{ox} = -\dfrac{Q_{acc}}{\varepsilon_{ox}}$

$$V_{ox} = \mathscr{E}_{ox} T_{ox} = -\dfrac{Q_{acc}}{C_{ox}} \qquad (5.2.4)$$

where C_{ox} is the oxide capacitance per unit area (F/cm^2) and Q_{acc} is the accumulation charge (C/cm^2). Equation (5.2.4) is the usual capacitor relationship, $V = Q/C$ (or $Q = C\text{-}V$) except for the negative sign. In $V = Q/C$, the capacitor voltage and charge are both taken from the same electrode. In the MOS capacitor theory, the voltage is the gate voltage, but the charge is the substrate charge because interesting things happen in the substrate. This unusual choice leads to the negative sign in Eq. (5.2.4). Equations (5.2.4) and (5.2.3) tell us

$$Q_{acc} = -C_{ox}(V_g - V_{fb}) \qquad (5.2.5)$$

Therefore, the MOS capacitor in accumulation behaves like a capacitor with $Q = C\text{-}V$ (or $-C\text{-}V$ as explained earlier) but with a shift in V by V_{fb}. The shift is easily understandable because $Q_{acc} = 0$ when $V_g = V_{fb}$. In general, Eq. (5.2.4) should read

$$\boxed{V_{ox} = -Q_{sub}/C_{ox}} \qquad (5.2.6)$$

where Q_{sub} is all the charge that may be present in the substrate, including Q_{acc}.

5.3 • SURFACE DEPLETION •

How would Fig. 5–4 change if a more positive V_g than V_{fb} is applied? The band diagram on the gate side will be pulled downward as shown in Fig. 5–6b. Clearly, *there is now a depletion region at the surface* because E_F is far from both E_c and E_v

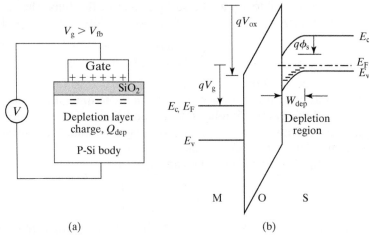

FIGURE 5–6 This MOS capacitor is biased into surface depletion. (a) Types of charge present; (b) energy band diagram.

and electron and hole densities are both small. This condition is called **surface depletion**. The depletion region has a width, W_{dep}. Equation (5.2.6) becomes

$$V_{ox} = -\frac{Q_{sub}}{C_{ox}} = -\frac{Q_{dep}}{C_{ox}} = \frac{qN_a W_{dep}}{C_{ox}} = \frac{\sqrt{qN_a 2\varepsilon_s \phi_s}}{C_{ox}} \quad (5.3.1)$$

$$\phi_s = \frac{qN_a W_{dep}^2}{2\varepsilon_s} \quad (5.3.2)$$

Q_{dep} is negative because the acceptor ions (after accepting the extra electrons) are negatively charged. In Eqs. (5.3.1) and (5.3.2), we used $W_{dep} = \sqrt{(2\varepsilon_s \phi_s)/(qN_a)}$ [Eq. (4.2.10)]. Combining Eqs. (5.3.1), (5.3.2), and (5.2.2),

$$V_g = V_{fb} + \phi_s + V_{ox} = V_{fb} + \frac{qN_a W_{dep}^2}{2\varepsilon_s} + \frac{qN_a W_{dep}}{C_{ox}} \quad (5.3.3)$$

This equation can be solved to yield W_{dep} as a function of V_g. With W_{dep} determined, V_{ox} [Eq. (5.3.1)] and ϕ_s [Eq. (5.3.2)] become known.

5.4 • THRESHOLD CONDITION AND THRESHOLD VOLTAGE •

Let's make V_g in Fig. 5–6 increasingly more positive. This bends the energy band down further. At some V_g, E_F will be close enough to E_c at the Si–SiO$_2$ interface that the surface is no longer in depletion but at the **threshold** of **inversion**. The term inversion means that the surface is inverted from P type to N type, or electron rich. Threshold is often defined as the condition when the surface electron concentration, n_s, is equal to the bulk doping concentration, N_a. That means $(E_c - E_F)_{surface} = (E_F - E_v)_{bulk}$, or $A = B$ in Fig. 5–7.[3] That, in turn, means

---

[3] Assuming $N_c = N_v$, we conclude that $A = B$ when $n_s = N_a$.

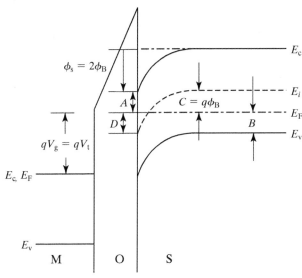

FIGURE 5–7 The threshold condition is reached when $n_s = N_a$, or equivalently, $A = B$, or $\phi_s = \phi_{st} = 2\phi_B$. Note that positive ϕ_{st} corresponds to downward band bending.

$C = D$. E_i is a curve drawn at **midgap**, which is half way between E_c and E_v. Let the surface potential (band bending) at the threshold condition be ϕ_{st}. It is equal to $(C + D)/q = 2C/q = 2\phi_B$.

Using Eqs. (1.8.12) and (1.8.8) and assuming $N_c = N_v$,

$$q\phi_B \equiv \frac{E_g}{2} - (E_F - E_v)|_{bulk}$$
$$= kT\ln\frac{N_v}{n_i} - kT\ln\frac{N_v}{N_a} = kT\ln\frac{N_a}{n_i} \quad (5.4.1)$$

ϕ_s at the threshold condition is

$$\boxed{\phi_{st} = 2\phi_B = 2\frac{kT}{q}\ln\frac{N_a}{n_i}} \quad (5.4.2)$$

The V_g at the threshold condition is called the **threshold voltage**, V_t. Substituting Eqs. (5.4.2) and (5.3.1) into Eq. (5.2.2),

$$\boxed{V_t = V_{fb} + 2\phi_B + \frac{\sqrt{qN_a 2\varepsilon_s 2\phi_B}}{C_{ox}}} \quad (5.4.3)$$

The threshold voltage as a function of T_{ox} and body doping using Eq. (5.4.3) is plotted in Fig. 5–8. In this figure, the gate dielectric is assumed to be SiO_2 with dielectric constant $\varepsilon_{ox} = 3.9$.

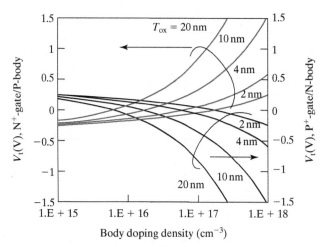

FIGURE 5–8 Theoretical threshold voltage vs. body doping concentration using Eq. (5.4.3). See Section 5.5.1 for a discussion of the gate doping type.

● **N-Type Body** ●

For an N-type body, Eq. (5.4.3) becomes

$$V_t = V_{fb} + \phi_{st} - \frac{\sqrt{2qN_d\varepsilon_s|\phi_{st}|}}{C_{ox}} \quad (5.4.4)$$

$$\phi_{st} = -2\phi_B \quad (5.4.5)$$

$$\phi_B = \frac{kT}{q}\ln\frac{N_d}{n_i} \quad (5.4.6)$$

Exercise: Draw the band diagram of an N-body MOS capacitor at threshold and show that the second term (ϕ_{st}) and the third term (V_{ox}) in Eq. (5.4.4) are negative.

5.5 ● STRONG INVERSION BEYOND THRESHOLD ●

Figure 5–9b shows the energy diagram at strong inversion, $V_g > V_t$. As shown in Fig. 5–9a, there is now an **inversion layer**, which is filled with **inversion electrons.** The **inversion charge density** is represented with Q_{inv} (C/cm^2). ϕ_s does not increase much further beyond $2\phi_B$ since even a 0.1 V further increase in ϕ_s would induce a much larger surface electron density and therefore a larger V_{ox} that would soak up the V_g in Eq. (5.2.2). If ϕ_s does not increase, neither will the depletion region width. Approximately speaking, W_{dep} has reached its maximum value

$$W_{dmax} = \sqrt{\frac{2\varepsilon_s 2\phi_B}{qN_a}} \quad (5.5.1)$$

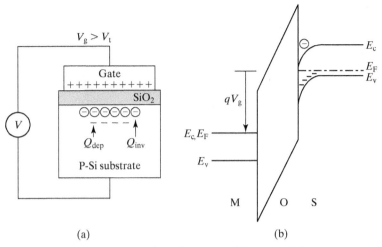

FIGURE 5–9 An MOS capacitor biased into inversion. (a) Types of charge present; (b) energy band diagram with arrow indicating the sense of positive V_g.

$$V_g = V_{fb} + 2\phi_B - \frac{Q_{dep}}{C_{ox}} - \frac{Q_{inv}}{C_{ox}} = V_{fb} + 2\phi_B + \frac{\sqrt{qN_a 2\varepsilon_s 2\phi_B}}{C_{ox}} - \frac{Q_{inv}}{C_{ox}}$$

$$= V_t - \frac{Q_{inv}}{C_{ox}} \qquad (5.5.2)$$

Equations (5.2.2) and (5.2.6) are used in deriving Eq. (5.5.2).

$$\therefore \quad \boxed{Q_{inv} = -C_{ox}(V_g - V_t)} \qquad (5.5.3)$$

Equation (5.5.3) confirms that the MOS capacitor in strong inversion behaves like a capacitor except for a voltage offset of V_t. At $V_g = V_t$, $Q_{inv} = 0$.

In this section, we have assumed that electrons will appear in the inversion layer whenever the closeness between E_c and E_F suggests their presence. However, there are few electrons in the P-type body, and it can take minutes for thermal generation to generate the necessary electrons to form the inversion layer. The MOS transistor structure shown in Fig. 5–2 solves this problem. The inversion electrons are supplied by the N^+ junctions, as shown in Fig. 5–10a. The inversion layer may be visualized as a very thin N layer (hence the term *inversion* of the surface conductivity type) as shown in Fig. 5–10b. The MOS transistor as shown in Figs. 5–2 and 5–10 is a more versatile structure for studying the MOS system than the MOS capacitor.

5.5.1 Choice of V_t and Gate Doping Type

The p-body transistor shown in Fig. 5–10 operates in an integrated circuit (IC) with V_g swinging between zero and a positive power supply voltage. To make circuit design easier, it is routine to set V_t at a small positive value, e.g., 0.4 V, so that, at $V_g = 0$, the transistor does not have an inversion layer and current does not flow between the two N^+ regions. A transistor that does not conduct current at $V_g = 0$ is called an **enhancement-type device**. This V_t value can be obtained with an N^+ gate and convenient body doping density as shown in Fig. 5–8. If the p-body device is paired with a P^+ gate,

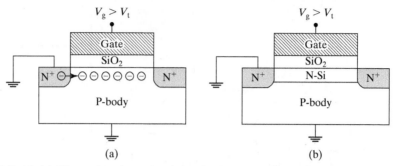

FIGURE 5–10 (a) The surface inversion behavior is best studied with a PN junction butting the MOS capacitor to supply the inversion charge. (b) The inversion layer may be thought of as a thin N-type layer.

V_t would be too large (over 1 V) and necessitate a larger power supply voltage. This would lead to larger power consumption and heat generation (see Section 6.7.3).

Similarly, an N-type body is routinely paired with a P$^+$ gate. In summary, P body is almost always paired with N$^+$ gate to achieve a small positive threshold voltage, and N body is normally paired with P$^+$ gate to achieve a small negative threshold voltage. The other body-gate combinations are almost never encountered.

• **Review: Basic MOS Capacitor Theory** •

Let us review the concepts, nomenclatures, common approximations, and simple relationships associated with the MOS capacitor theory. We will do so using a series of figures, starting with Fig. 5–11. The surface potential, ϕ_s, is zero at V_{fb} and approximately zero in the accumulation region. As V_g increases from V_{fb} into the depletion regime, ϕ_s increases from zero toward $2\phi_B$. When ϕ_s reaches $2\phi_B$, the surface electron concentration becomes so large that the surface is considered **inverted**. The V_g at that point is called V_t, the **threshold voltage**.

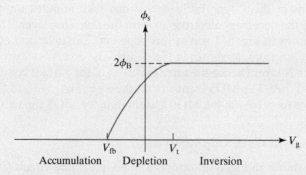

FIGURE 5–11 Surface potential saturates at $2\phi_B$ when V_g is larger than V_t.

Figure 5–12 uses W_{dep} to review the MOS capacitor. There is no depletion region when the MOS interface is in accumulation. W_{dep} in the PN junction and in the MOS capacitor is proportional to the square root of the band bending (ϕ_s in the MOS case). W_{dep} saturates at W_{dmax} when $V_g \geq V_t$, because ϕ_s saturates at $2\phi_B$.

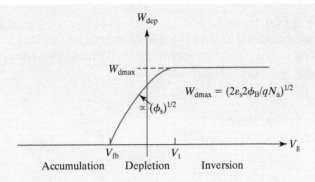

FIGURE 5–12 Depletion-region width in the body of an MOS capacitor.

Figure 5–13 reviews the three charge components in the substrate. The depletion charge Q_{dep} is constant in the inversion region because W_{dep} is a constant there. $Q_{inv} = -C_{ox}(V_g - V_t)$ appears in the inversion region. Q_{acc} shows up in the accumulation

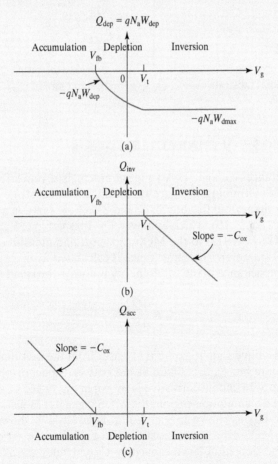

FIGURE 5–13 Components of charge (C/cm^2) in the MOS capacitor substrate: (a) depletion-layer charge; (b) inversion-layer charge; and (c) accumulation-layer charge.

region. In both (b) and (c), the slope is $-C_{ox}$. Figure 5–14 shows the total substrate charge, Q_{sub}. Q_{sub} in the accumulation region is made of accumulation charge. Q_{sub} is made of Q_{dep} in the depletion region. In the inversion region, there are two components, Q_{dep} that is a constant and Q_{inv} that is equal to $-C_{ox}(V_g - V_t)$.

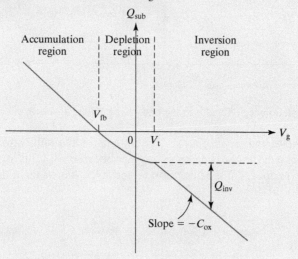

FIGURE 5–14 The total substrate charge, Q_{sub} (C/cm^2), is the sum of Q_{acc}, Q_{dep}, and Q_{inv}.

5.6 • MOS C–V CHARACTERISTICS •

The capacitance–voltage (C–V) measurement is a powerful and commonly used method of determining the gate oxide thickness, substrate doping concentration, threshold voltage, and flat-band voltage. The C–V curve is usually measured with a C–V meter (Fig. 5–15), which applies a DC bias voltage, V_g, and a small sinusoidal signal (1 kHz–10 MHz) to the MOS capacitor and measures the capacitive current with an AC ammeter. The capacitance is calculated from $i_{cap}/v_{ac} = \omega C$.

The capacitance in the MOS theory is always the **small-signal capacitance**

$$C \equiv \frac{dQ_g}{dV_g} = -\frac{dQ_{sub}}{dV_g} \qquad (5.6.1)$$

The negative sign in Eq. (5.6.1) arises from the fact that V_g is taken at the top capacitor plate but Q_{sub} is taken at the bottom capacitor plate (the body). Q_{sub} is given in Fig. 5–14 and its derivative is shown in Fig. 5–16.

In the accumulation region, the MOS capacitor is just a simple capacitor with capacitance C_{ox} as shown in Fig. 5–17a. Figure 5–17b shows that in the depletion region, the MOS capacitor consists of two capacitors in series: the oxide capacitor, C_{ox}, and the depletion-layer capacitor, C_{dep}. Under the AC small-signal voltage, W_{dep} expands and contracts slightly at the AC frequency. Therefore, the AC charge appears at the bottom of the depletion layer as shown in Fig. 5–17b.

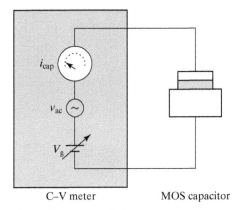

FIGURE 5–15 Setup for the C–V measurement.

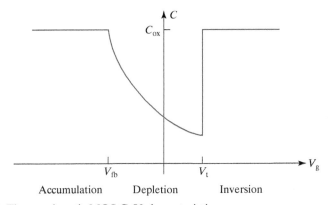

FIGURE 5–16 The quasi-static MOS C–V characteristics.

$$C_{dep} = \frac{\varepsilon_s}{W_{dep}} \tag{5.6.2}$$

$$\frac{1}{C} = \frac{1}{C_{ox}} + \frac{1}{C_{dep}} \tag{5.6.3}$$

$$\frac{1}{C} = \sqrt{\frac{1}{C_{ox}^2} + \frac{2(V_g - V_{fb})}{qN_a\varepsilon_s}} \tag{5.6.4}$$

To derive Eq. (5.6.4), one needs to solve Eq. (5.3.3) for W_{dep} as a function of V_g. The derivation is left as an exercise for the reader in the problems section at the end of the chapter. As V_g increases beyond V_{fb}, W_{dep} expands, and therefore C decreases as shown in Fig. 5–16.

Figure 5–17c shows that an inversion layer exists at the Si–SiO$_2$ interface. In response to the AC signal, Q_{inv} increases and decreases at the AC frequency. The inversion layer plays the role of the bottom electrode of the capacitor. Therefore, C reverts to C_{ox} in the inversion region as shown in Fig. 5–16. This C–V curve is called

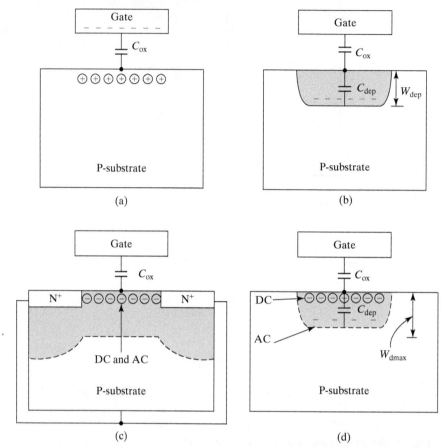

FIGURE 5–17 Illustration of the MOS capacitor in all bias regions with the depletion-layers shaded. (a) Accumulation region; (b) depletion region; (c) inversion region with efficient supply of inversion electrons from the N region corresponding to the transistor C–V or the quasi-static C–V; and (d) inversion region with no supply of inversion electrons (or weak supply by thermal generation) corresponding to the high-frequency capacitor C–V case.

the **quasi-static C–V** because Q_{inv} can respond to the AC signal as if the frequency were infinitely low (static case). That would require a ready source of electrons, which can be provided by the N region shown in Fig. 5–17c. PN junctions are always present in an MOS transistor. Therefore, the **MOS transistor C–V** characteristics at all frequencies follow the curve in Fig. 5–16, which is repeated as the upper curve in Fig. 5–18.

What if, as in Fig. 5–17d, the PN junctions are not present? The P-type substrate is an inefficient supplier of electrons. It produces electrons through thermal generation at a very slow rate (for the same reason the diode reverse leakage current is small.) Q_{inv} cannot respond to the AC signal and remains constant at its DC value. Instead, the AC signal causes ϕ_s to oscillate around $2\phi_B$

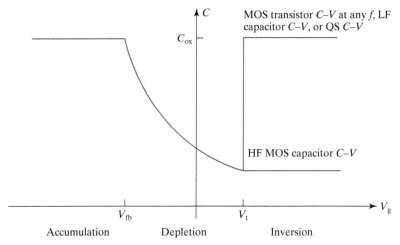

FIGURE 5–18 Two possible MOS C–V characteristics. The difference in the inversion region is explained in Fig. 5–17c and d.

and causes W_{dep} to expand and contract slightly around W_{dmax}. This change of W_{dep} can respond at very high frequencies because it only involves the movement of the abundant majority carriers. Consequently, the AC charge exists at the bottom of the depletion region. The result is a saturation of C at V_t as illustrated by the lower curve in Fig. 5–18. This curve is known as the capacitor C–V or the **high-frequency MOS capacitor C–V** (HF C–V). The name connotes that, in principle, at a sufficiently low frequency, even the MOS capacitor's C–V would follow the upper curve in Fig. 5–18. Following that reasoning, the upper curve is also known as the **low-frequency C–V** (LF C–V). In reality, even at a low frequency such as 1 kHz, the C–V of modern high-quality MOS capacitors does not follow the LF C–V curve. At yet lower frequencies, the C–V meter is ineffective (the capacitative current is too low) for studying the MOS capacitor. The term *low-frequency C–V* has a historical significance and is still used, but it no longer has a practical significance.

• **Measuring the Quasi-Static C–V Using an MOS Capacitor** •

There *is* a practical way to obtain the "low frequency" or quasi-static C–V (upper branch of Fig. 5–18) using an MOS capacitor without the PN junction. It involves applying a very slow linear-ramp voltage (<0.1V/s) to the gate and measuring I_g with a very sensitive DC ammeter during the ramp. C is calculated from $I_g = C \cdot dV_g/dt$. This technique provides sufficient time for Q_{inv} to respond to the slowly changing V_g. Plotting $I_g/(dV_g/dt)$ vs. V_g produces the QS C–V curve shown in Fig. 5–18. This technique becomes impracticable if the gate dielectric has too large a leakage current.

EXAMPLE 5-1 C–V of MOS Capacitor and Transistor

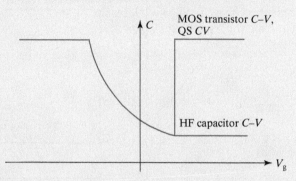

FIGURE 5–19 C–V curves of MOS capacitor and transistor.

For each of the following cases, does the QS C–V or the HF capacitor C–V apply?

(1) MOS transistor, 10 kHz.   (Answer: QS C–V).
(2) MOS transistor, 100 MHz.   (Answer: QS C–V).
(3) MOS capacitor, 100 MHz.   (Answer: HF capacitor C–V).
(4) MOS capacitor, 10 kHz.   (Answer: HF capacitor C–V).
(5) MOS capacitor, slow V_g ramp. (Answer: QS C–V).
(6) MOS transistor, slow V_g ramp. (Answer: QS C–V).

5.7 • OXIDE CHARGE—A MODIFICATION TO V_{fb} AND V_t[4] •

The basic MOS theory ignores the possible presence of electric charge in the gate dielectric. Assuming surface charge, Q_{ox} (C/cm^2), exists at the SiO$_2$–Si interface, the band diagram at the flat-band condition would be modified from Fig. 5–20a to 5–20b.

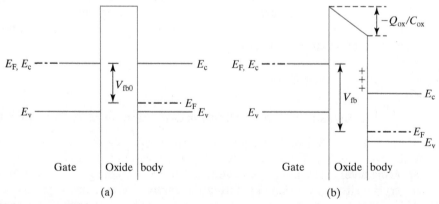

FIGURE 5–20 Flat-band condition (no band bending at body surface) (a) without any oxide charge; (b) with Q_{ox} at the oxide–substrate interface.

---
[4] This section may be omitted in an accelerated course.

The flat-band voltage in Fig. 5–20a is $\psi_g - \psi_s$ (Section 5.1). In Fig. 5–20b, the oxide charge (assumed to be located at the oxide–substrate interface for simplicity) induces an electric field in the oxide and an oxide voltage, $-Q_{ox}/C_{ox}$. Clearly, V_{fb} in part b is different from the V_{fb0} in part a. Specifically,

$$V_{fb} = V_{fb0} - Q_{ox}/C_{ox} = \psi_g - \psi_s - Q_{ox}/C_{ox} \qquad (5.7.1)$$

Because Q_{ox} changes V_{fb}, it also changes V_t through Eq. (5.4.3).

There are several types of **oxide charge**. Positive **fixed oxide charge** is attributed to silicon ions present at the Si–SiO$_2$ interface. **Mobile oxide charge** is believed to be mostly sodium ions. Mobile ions can be detected by observing V_{fb} and V_t shift under a gate bias at an elevated temperature (e.g., at 200 °C) due to the movement of the ions in the oxide. **Sodium contamination** must be eliminated from the water, chemicals, and containers used in an MOS fabrication line in order to prevent instabilities in V_{fb} and V_t. In addition, significant **interface traps** or **interface states** may be present and they can trap and release electrons and generate noise (see Section 6.15.3) and degrade the subthreshold current of MOSFET (see Section 7.2).

● **Reliability** ●

More interface states and fixed oxide charge appear after the oxide is subjected to high electric field for some time due to the breaking or rearrangement of chemical bonds. This raises a reliability concern because the threshold voltage and transistor current would change with usage and can potentially cause sensitive circuits to fail. Engineers ensure device reliability by controlling the stress field and improving the MOS interface quality and verifying or projecting the reliability with careful long-term testing.

EXAMPLE 5–2 Interpret the measured V_{fb} dependence on oxide thickness in Fig. 5–21 using Eq. (5.7.1). It is known that the gate electrode is N$^+$ poly-Si. What can you tell about the capacitors?

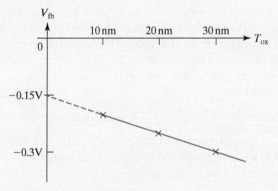

FIGURE 5–21 Measured V_{fb} of three capacitors with different oxide thicknesses.

SOLUTION:

$$V_{fb} = \psi_g - \psi_s - Q_{ox}T_{ox}/\varepsilon_{ox} \qquad (5.7.1)$$

Equation (5.7.1) suggests that V_{fb} at $T_{ox} = 0$ is $\psi_g - \psi_s$. Therefore, $\psi_g - \psi_s = -0.15$ V. This is illustrated in Fig. 5–22.

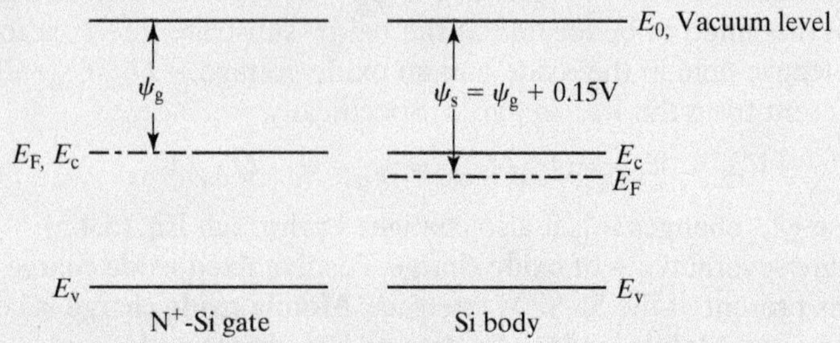

FIGURE 5–22 The relationship between ψ_g and ψ_s.

Because E_F is 0.15 V below E_c, we conclude that the substrate is N-type with

$$N_d = n = N_c e^{-0.15\,\text{eV}/kT} \approx 10^{17}\,\text{cm}^{-3}$$

Further, Eq. (5.7.1) suggests that

$$Q_{ox} = -\varepsilon_{ox} \times \text{slope of line in Fig. 5–21}$$

$$= -\varepsilon_{ox} \times \frac{-0.15\,\text{V}}{30\,\text{nm}} = \frac{3.9 \times 8.85 \times 10^{-14} \times 0.15\,\text{V}}{300 \times 10^{-8}} = 1.7 \times 10^{-8}\,\text{C/cm}^2$$

This corresponds to $1.7 \times 10^{-8}\,\text{cm}^2 \div q = 9 \times 10^{10}\,\text{cm}^2$ of positive charge at the interface. A high-quality MOS interface has about $10^{10}\,\text{cm}^2$ of charge. Both numbers are small fractions of the number of silicon atoms on a (100) crystal plane, $7 \times 10^{14}\,\text{cm}^{-2}$. In this sense, the SiO_2–Si interface is remarkably well-behaved and charge-free.

5.8 • POLY-SI GATE DEPLETION—EFFECTIVE INCREASE IN T_{OX} •

Consider an MOS capacitor with P$^+$ poly-Si gate and N body. The capacitor is biased into surface inversion. Figure 5–23a shows that the continuity of electric flux requires that the band bends in the gate. This indicates the presence of a thin depletion layer in the gate. Depending on the gate doping concentration and the oxide field, the **poly-Si gate depletion** layer thickness, W_{dpoly}, may be 1–2 nm. According to Gauss's Law,

$$W_{dpoly} = \varepsilon_{ox}\mathscr{E}_{ox}/qN_{poly} \tag{5.8.1}$$

Because a depletion layer is present in the gate, one may say that a poly-silicon-gate capacitor is added in series with the oxide capacitor as shown in Fig. 5–23b. The MOS capacitance in the inversion region becomes

$$C = \left(\frac{1}{C_{ox}} + \frac{1}{C_{poly}}\right)^{-1} = \left(\frac{T_{ox}}{\varepsilon_{ox}} + \frac{W_{dpoly}}{\varepsilon_s}\right)^{-1} = \frac{\varepsilon_{ox}}{T_{ox} + W_{dpoly}/3} \tag{5.8.2}$$

This **poly-depletion effect** effectively increases T_{ox} by $W_{dpoly}\varepsilon_{ox}/\varepsilon_s$ or $W_{dpoly}/3$, and can have a significant impact on the C–V curve if T_{ox} is thin. The gate capacitance drops as the capacitor is biased deeper into the inversion region due to increasing poly-depletion as shown in Fig. 5–26. The poly-depletion effect is

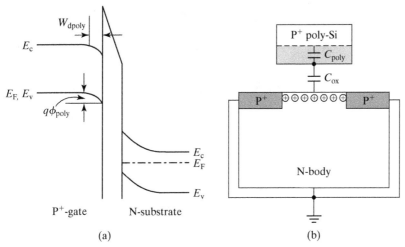

FIGURE 5–23 Poly-gate depletion effect illustrated with (a) the band diagram and (b) series capacitors representation. An N^+ poly-Si gate can also be depleted.

undesirable because a reduced C means reduced Q_{inv}, and reduced transistor current. The solution is to dope the poly-Si heavily. Unfortunately, very heavy doping may cause **dopant penetration** from the gate through the oxide into the substrate. **Poly-SiGe** gate can be doped to a higher concentration, thus improving gate depletion [1]. Poly-gate depletion is eliminated in advanced MOSFET technology by substitution of the poly-gate with a metal gate (see Section 7.4).

The effect of poly-gate depletion on Q_{inv} may be modeled in another way:

$$Q_{inv} = -C_{ox}(V_g - \phi_{poly} - V_t) \qquad (5.8.3)$$

Poly-gate depletion effectively reduces V_g by ϕ_{poly}. Even 0.1 V ϕ_{poly} would be highly undesirable when the power-supply voltage (the maximum V_g) is only around 1 V.

EXAMPLE 5-3 Poly-Si Gate Depletion

Assume that V_{ox}, the voltage across a 2 nm thin oxide is -1 V. The P^+ poly-gate doping is $N_{poly} = 8 \times 10^{19}$ cm^3 and substrate N_d is 10^{17}cm^3. Estimate (a) W_{dpoly}, (b) ϕ_{poly}, and (c) V_g.

SOLUTION:

a. Using Eq. (5.8.1),

$$W_{dpoly} = \varepsilon_{ox} \mathscr{E}_{ox}/qN_{poly} = \varepsilon_{ox} V_{ox}/T_{ox} q N_{poly}$$

$$= \frac{3.9 \times 8.85 \times 10^{-14} (\text{F/cm}) \cdot 1\text{V}}{2 \times 10^{-7}\text{cm} \cdot 1.6 \times 10^{-19}\text{C} \cdot 8 \times 10^{19}\text{cm}^{-3}}$$

$$= \frac{34.5 \times 10^{-14}\text{cm}}{256 \times 10^{-8}} = 0.13 \times 10^{-6}\text{cm} = 1.3\,\text{nm}$$

b. W_{dpoly} is related to ϕ_{poly} by the depletion-region model

$$W_{dpoly} = \sqrt{\frac{2\varepsilon_s \phi_{poly}}{qN_{poly}}}$$

$$\phi_{poly} = qN_{poly}W_{dpoly}^2/2\varepsilon_s$$

$$= \frac{1.6 \times 10^{-19}\text{C} \cdot 8 \times 10^{19}\text{cm}^{-3} \cdot (1.3 \times 10^{-7}\text{cm})^2}{2 \times 12 \times 8.85 \times 10^{-14}\text{F/cm}}$$

$$= \frac{2.3 \times 10^{-13}\text{V}}{2.1 \times 10^{-12}} = 0.11 \text{ V}$$

c. Equation (5.2.2) with a ϕ_{poly} term added is

$$V_g = V_{fb} + \phi_{st} + V_{ox} + \phi_{poly}$$

$$V_{fb} = \psi_g - \psi_s = \frac{E_g}{q} - \frac{kT}{q}\ln\frac{N_c}{N_d} = 1.1 - 0.15 \text{ V} = 0.95 \text{ V}$$

$$V_g = 0.95 - 0.8 - 1 - 0.11 \text{ V} = -0.96 \text{ V}$$

Using Eq. (5.4.5), $\phi_{st} = -2\phi_B = -2\frac{kT}{q}\ln\frac{N_d}{n_i} = -0.8$

Draw an energy band diagram to confirm the signs of terms in the last equation. The loss of 0.11 V to poly-depletion is a large loss relative to the 0.96 V applied voltage.

5.9 • INVERSION AND ACCUMULATION CHARGE-LAYER THICKNESSES AND QUANTUM MECHANICAL EFFECT •

So far, we have implicitly assumed that the inversion charge is a sheet charge at the Si–SiO$_2$ interface (i.e., the inversion layer is infinitely thin). In reality, the inversion-charge profile is determined by the solution of the Schrödinger equation and Poisson's equation [2]. For this reason, the present topic is often referred to as the **quantum mechanical effect** in an MOS device. An example of the charge profile is shown in Fig. 5–24. The average location or centroid of the inversion charge below the Si–SiO$_2$ interface is called the **inversion-layer thickness**, T_{inv}. Figure 5–25 shows T_{inv} as a function of V_g. When V_g is large, T_{inv} is around 1.5 nm. When V_g is low, T_{inv} can be 3 nm. It is shown in Eq. (6.3.6) that

$$\text{average field in the inversion layer} = \frac{V_g + V_t}{6T_{ox}} \quad (5.9.1)$$

It is reasonable that T_{inv} is a function of the average field, and therefore a function of $(V_g + V_t)/T_{ox}$ as shown in Fig. 5–25. The electron inversion layer is thinner than the hole inversion layer because the electron effective mass is smaller. It is valid to think that the bottom electrode of the MOS capacitor is not exactly at the Si–SiO$_2$ interface

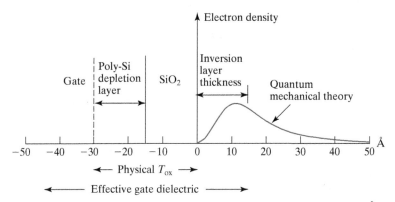

FIGURE 5–24 Average location of the inversion-layer electrons is about 15 Å below the Si–SiO$_2$ interface. Poly-Si gate depletion is also shown.

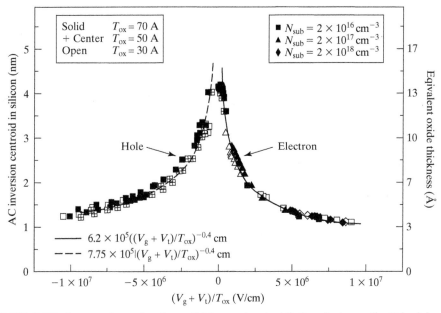

FIGURE 5–25 Average inversion-layer thickness (centroid) for electrons (in P body) and holes (in N body). (From [3]. © 1999 IEEE.)

but rather effectively located below the interface by T_{inv}. In other words, T_{ox} is effectively increased by $T_{inv}/3$, where 3 is the ratio of $\varepsilon_s/\varepsilon_{ox}$. The accumulation layer has a similar thickness. The effect on the C–V characteristics (shown in Fig. 5–26) is to depress the C–V curve at the onset of inversion and accumulation. Figure 5–27 explains the transition of the C–V curve in Fig. 5–26 from the depletion to the inversion region. Figure 5–27a is the general case. In the depletion region, C_{inv} is negligible (there is no inversion charge) and C_{poly} can be neglected because $W_{dpoly} \ll W_{dep}$. Therefore, Fig. 5–27 reduces to the basic series combination of C_{ox} and C_{dep} of Fig. 5–27b. As V_g increases toward V_t, C_{inv} increases as the inversion charge begins to appear, and the total capacitance rises above the basic C–V as shown in Fig. 5–27c and

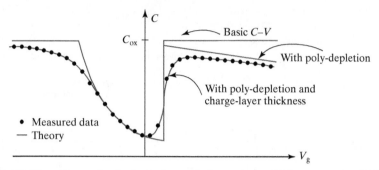

FIGURE 5–26 The effects of poly-depletion and charge-layer thickness on the C–V curve of an N⁺ poly-gate, P-substrate device.

Fig. 5–26. The capacitance rises smoothly toward C_{ox} because the inversion charge is not located exactly at the silicon–oxide interface, but at some depth that varies with V_g as shown in Fig. 5–25. At larger V_g, C_{poly} cannot be assumed to be infinity (W_{dpoly} increases), and C drops in Fig. 5–26.

T_{inv} and W_{dpoly} used to be negligible when T_{ox} was large (>10 nm). For thinner oxides, they are not. Because it is difficult to separate T_{ox} from T_{inv} and W_{dpoly} by measurement, an **electrical oxide thickness**, T_{oxe}, is often used to characterize the total effective oxide thickness. T_{oxe} is deduced from the inversion-region capacitance measured at $V_g = V_{dd}$. One may think of T_{oxe} as an **effective oxide thickness**, corresponding to an **effective gate capacitance**, C_{oxe}. T_{oxe} is the sum of three thicknesses,

$$T_{oxe} = T_{ox} + W_{dpoly}/3 + T_{inv}/3 \qquad (5.9.2)$$

where 3 is the ratio of $\varepsilon_s/\varepsilon_{ox}$, which translates W_{dpoly} and T_{inv} into equivalent oxide thicknesses. The total inversion charge per area, Q_{inv}, is

$$\boxed{\begin{aligned} Q_{inv} &= -C_{oxe}(V_g - V_t) \\ &= \frac{\varepsilon_{ox}}{T_{oxe}}(V_g - V_t) \end{aligned}} \qquad (5.9.3)$$

Typically, T_{oxe} is larger than T_{ox} by 6–10 Å.

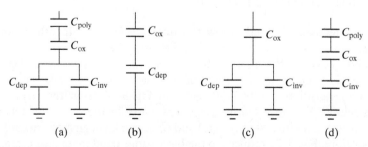

FIGURE 5–27 Equivalent circuit for understanding the C–V curve in the depletion region and the inversion region. (a) General case for both depletion and inversion regions; (b) in the depletion regions; (c) $V_g \approx V_t$; and (d) strong inversion.

In addition, there is another quantum effect that increases the threshold voltage [4]. At high substrate doping concentration, the high electric field in the substrate at the oxide interface in Fig. 5–7 causes the energy levels to be quantized and effectively increases E_g and decreases n_i in Eq. (5.4.1). This requires the band to bend down more before reaching threshold, i.e., causes ϕ_{st} in Eq. (5.4.2) to increase. The net effect is that the threshold voltage is increased by 100mV or so depending on the doping concentration due to this **quantum effect on threshold voltage**.

5.10 • CCD IMAGER AND CMOS IMAGER •

An **imager** is a sensing device that converts an optical image into an electronic signal. CCD imager and CMOS imager are used in digital cameras and camcorders. CCD imagers have higher performance but are more expensive. CMOS imagers are newer and less expensive. They are presented in the next two sub-sections.

5.10.1 CCD Imager

CCD stands for **charge-coupled device** [5]. The heart of a CCD imager is a large number of MOS capacitors densely packed in a two-dimensional array.

Let us first consider how a single MOS capacitor reacts to light. Figure 5–28a shows an MOS capacitor biased into **deep-depletion**. A voltage, $V_g > V_t$, has been suddenly applied to the gate. Because thermal generation is a slow process, there

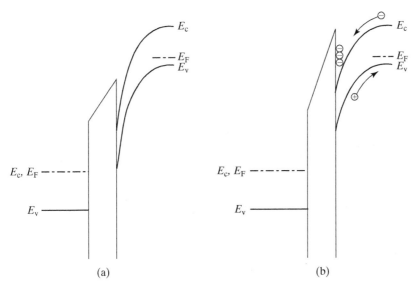

FIGURE 5–28 Deep depletion. (a) Immediately after a gate voltage $V_g >$, V_t is applied, there are no electrons at the surface. (b) After exposure to light, photo-generated electrons have been collected at the surface. The number of electrons is proportional to the light intensity.

are no electrons (no inversion layer) at the surface during at least the first fraction of a second. As a result, the band bends beyond $2\phi_B$ and the depletion region extends beyond W_{dmax}. This condition is called deep depletion. If light shines on the MOS capacitor in this condition for ten milliseconds, some photo-generated electrons will be collected at the interface as shown in Fig. 5–28b. The photo-generated holes flow into the substrate and are removed through the substrate contact. The number of electrons collected is proportional to the light intensity. This is the first function of a CCD array—to convert an image (two-dimensional pattern of light intensity) into packets of electrons stored in a two-dimensional array of MOS capacitors.

● **Deep-Depletion C–V** ●

If an MOS capacitor is biased into deep depletion by rapidly sweeping the gate bias, W_{dep} may exceed W_{dmax}. As a result, the capacitance continues to fall even at $V_g > V_t$ as shown in Fig. 5–29. Deep-depletion C–V again illustrates the impossibility of establishing the inversion layer rapidly in an MOS capacitor (without a PN junction supplying the inversion charge).

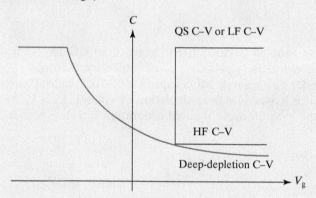

FIGURE 5–29 Deep-depletion C–V.

The second function of a CCD array is to transfer the collected charge packets to the edge of the array, where they can be read by a charge sensing circuit in a serial manner. To illustrate this charge transfer function, let us examine the one-dimensional array in Fig. 5–30, representing a small portion of a single row in the two-dimensional array. Every three MOS capacitors or elements constitute one sensor pixel. In Fig. 5–30a, exposure to a lens-projected image has produced some electrons in the element on the right, even more in the element on the left and yet more in the middle element in proportion to the image light intensity around those three locations. Electrons are collected only under these three elements, not the ones flanking them, because these three are biased to deeper band bendings (more positive ϕ_s) than their neighbor elements and any electrons that might show up in the neighbors would flow to these three more positive locations. Under the bias condition of Fig. 5–30b, V_2 creates the deepest depletion. After the gate biases are switched from (a) to (b), the charge packets will move to the elements connected to V_2 (i.e., shifted to the right by one element). The choice of $V_1 > V_3$ ensures that no

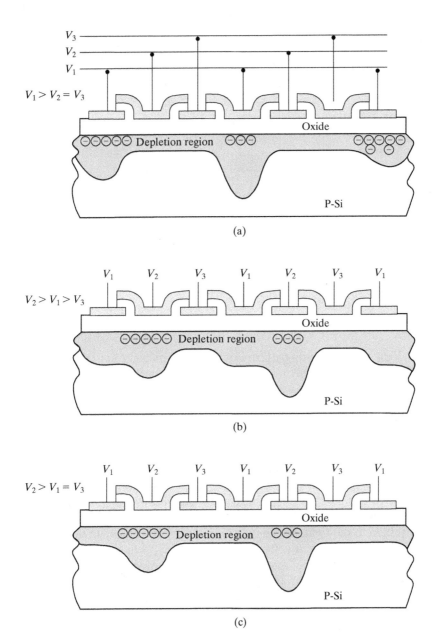

FIGURE 5–30 How CCD shifts the charge packets. The array is biased in the sequence (a), (b), (c), (a), (b), (c), (a) … . The drawing in (c) is identical to (a) but with all the charge packets shifted to the right by one capacitor element.

electrons are transferred to the left. Finally, in step (c), V_1 is reduced to the same value as V_3, thus making (c) identical to (a), except for the shift of the electron packets to the right, setting the stage for the next transfer operation. In this manner, the electron packets are shifted to the right element by element. Waiting at the right edge of the array is a charge-sensing circuit that generates a serial voltage signal that faithfully represents the image light pattern. In summary, a CCD imager first

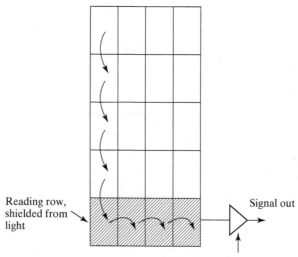

FIGURE 5–31 Architecture of a two-dimensional CCD imager. The arrows show the path of the charge-packet movement.

converts light patterns into patterns of electron packets and then transfers the charge packets one element at a time to the edge of the array, where they are converted into a serial electrical signal by a charge-sensing circuit. For example, the three charge packets in Fig. 5–30 would generate a small signal pulse, followed by a large pulse, and then a medium pulse.

Figure 5–31 depicts a two-dimensional CCD imager containing four rows and four columns of 16 MOS capacitors plus a reading row at the bottom. The reading row is shielded from the light by a metal film. The two-dimensional charge packets are read row by row. First, the charge packets in the 16 elements are shifted downward by one row. This action transfers the charge packets in the lowest sensing row (the fourth row from the top) into the reading row. Next, the charge packets in the reading row only are shifted to the right. To the right side of the row is a circuit that converts each arriving charge packet into a voltage pulse. After the packets of the fourth row have been read in this way, the remaining three rows of charge packets are shifted downward by one row again. Now the reading row begins to shift the new row of charge packets to the converter circuit. During the shifting-and-reading operation, the CCD array is blocked from light with a mechanical shutter. Otherwise, the image would be smeared. For example, the charge packets in the top row would be exposed to the light patterns of the other rows during the shifting and reading.

5.10.2 CMOS Imager

CMOS imagers do not shift the charge packets from row to row. They do not need mechanical shutters, use less power, and are cheaper than CCD imagers. For these reasons, CMOS imagers made mobile phone cameras practical and are widely used in low-cost digital cameras. In a CMOS imager, the charge collected in an array element is converted into voltage by a circuit integrated in that array element as

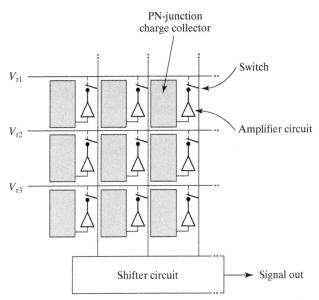

FIGURE 5–32 Architecture of a CMOS imager. Each array element has its own charge-to-voltage converter represented by the triangle. Actual imagers may support hundreds to over a thousand rows and columns of pixels.

shown in Fig. 5–32. An open-circuited N^+P junction collects the light-generated charge. The P substrate is grounded. Electrons generated by light near the PN junction diffuse to the junction and get collected and stored in the thin N^+ region. Since the PN junction is a capacitor, the stored electrons change the capacitor voltage, i.e., the N^+-region voltage. This voltage is amplified in the pixel as shown in Fig. 5–32.

Each pixel also contains a switch made of an MOS transistor and controlled by the voltages V_{r1}, V_{r2}, or V_{r3} that is carried by long horizontal metal lines. In order to read the top row of pixels, V_{r1} is raised to turn on (close) all the switches in the top row. This brings the signals from all the top-row pixels to the shifter circuit below by vertical-running metal lines.

CMOS imagers became attractive only after transistor size reduction made the circuitry in each array element, employing half a dozen or more transistors, small in comparison with the element area. CMOS imagers are so named because their circuitry and the N^+P junctions are fabricated with CMOS circuit (see Section 6.2) technology. CMOS IC technology is the mainstream manufacturing technology and its high volume has driven the wafer cost of CMOS imagers below that of CCD imagers. Because they share the same CMOS technology, CMOS imagers can be integrated with signal processing and control circuitries to further reduce system costs. A CMOS imager's image uniformity and contrast ratio are not as good as those of a CCD. The size constraint of the sensing circuits forces the CMOS imager to use very simple circuits and it is difficult to avoid variations among the very large number of sensing circuits. In contrast, a CCD imager employs a small number of sophisticated sensing circuits.

● **Color Imagers** ●

A color imager must produce three separate signals for the red, green, and blue light in the image. A color pixel usually contains four sensor array elements. The upper-left element senses red light. The upper-right senses green and the lower-right senses blue. The lower-left element senses green again because the human eye is more sensitive to the green light than red and blue. The color designation is accomplished by coating the elements with red, green, or blue filter films. These films containing color dyes may be deposited by a spin-on process and patterned with photolithography similar to photoresists (Section 3.3).

5.11 ● CHAPTER SUMMARY ●

The three regions **(accumulation, depletion, and inversion)** and the two transition points **(flat-band and threshold)** are reviewed in Fig. 5–33 for the two prevalent MOS device types. Upward arrows indicate negative V_g and downward arrows, positive V_g. Please review this figure carefully.

The flat-band voltage is

$$V_{fb} = \psi_g - \psi_s - Q_{ox}/C_{ox} \qquad (5.7.1)$$

ψ_g and ψ_s are the gate and substrate work functions. Q_{ox} is a sheet charge that may be present at the SiO_2–Si interface. The gate voltage in excess of V_{fb} is divided between the substrate and the oxide and the poly-gate depletion layer.

$$V_g = V_{fb} + \phi_s + V_{ox} + \phi_{poly} \qquad (5.2.2 \ \& \ Sec. \ 5.8)$$

$$= V_{fb} + \phi_s - Q_{sub}/C_{ox} + \phi_{poly} \qquad (5.2.6)$$

ϕ_s is the surface potential, or the substrate band bending. V_{ox} is the oxide voltage. Q_{sub} (C/cm^2) is all the accumulation, inversion, and depletion-layer charge. At the **threshold** of inversion, ϕ_s is

$$\phi_{st} = \pm 2\phi_B \qquad (5.4.2 \ \& \ 5.4.5)$$

$$\phi_B = \frac{kT}{q} \ln \frac{N_{sub}}{n_i} \qquad (5.4.2 \ \& \ 5.4.6)$$

$$V_t = V_{fb} + \phi_{st} \pm \frac{\sqrt{qN_{sub}2\varepsilon_s|\phi_{st}|}}{C_{ox}} \qquad (5.11.5 \ \& \ 5.4.4)$$

In the last three equations, the positive signs are for a P substrate (band bending downward) and the negative signs are for an N substrate (band bending upward).

There are two types of C–V curves as shown in Fig. 5–34. The quasi-static (QS) C–V curve, also known as the LF C–V, is applicable when the inversion charge can rapidly follow the change in V_g. It is the MOS transistor C–V at all frequencies because the short-circuited PN junction is a source of Q_{inv}. The lower C–V curve, the capacitor (HF) C–V, is applicable when Q_{inv} cannot follow the change in AC V_g. A third C–V curve, the deep-depletion C–V (Fig. 5–29), applies when Q_{inv} cannot even follow the rapid change in the bias V_g.

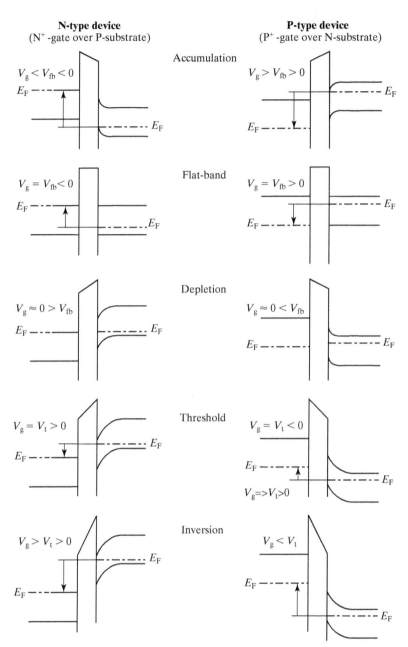

FIGURE 5–33 Energy band diagrams of the two dominant types of MOS capacitors. An N-type device is so named because it has N-type inversion charge that increases with a more positive V_g, and a P-type device has P-type inversion charge increasing with a more negative V_g.

The finite thickness of the inversion and accumulation layers, T_{inv} and T_{acc}, effectively increases T_{ox} by $T_{inv}/3$ and $T_{acc}/3$. The electrical oxide thicknesses is

$$T_{oxe} = T_{ox} + W_{dpoly}/3 + T_{ch}/3 \quad (5.9.2)$$

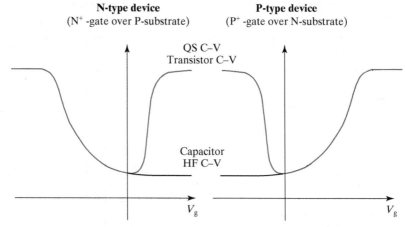

FIGURE 5–34 N-type and P-type MOS capacitors.

The number 3 is the ratio of silicon permittivity (11.9) to SiO$_2$ permittivity (3.9). T_{oxe} is usually determined from the inversion-region capacitance measured at $V_g = V_{dd}$. Quantization of states in the inversion layer causes the threshold voltage to increase beyond the prediction of the basic threshold voltage theory.

A **CCD (charge–coupled device)** is an imaging device based on an array of MOS capacitors operating under the **deep-depletion** condition, starved of inversion charge. Photo-generated carriers are collected in the surface potential wells, and the collected charge packets are transferred in a serial manner to the charge-sensing circuit located at the edge of the array. CCD imagers have been replaced by **CMOS imagers** where cost, size, and power consumption are more important than the best image quality. CMOS imagers integrate a charge-to-voltage conversion circuit in each sensing array element. In both types of imagers, color sensing is achieved with separate sensing elements for red, green, and blue in each pixel.

● PROBLEMS ●

● Energy Band Diagram ●

5.1 Sketch the energy band diagrams of an MOS capacitor with N-type silicon substrate and N$^+$ poly-Si gate at flatband, in accumulation, in depletion, at threshold, and in inversion.

5.2 Sketch the energy band diagrams (i) at thermal equilibrium and (ii) at flat band for the following MOS systems. Use a work function value that you find from any source.

  (a) Tungsten, W, gate with 1 Ωcm N-type silicon substrate.
  (b) Tungsten, W, gate with 1 Ωcm P-type silicon substrate.
  (c) Heavily doped P$^+$ polycrystalline silicon gate with 1 Ωcm N-type silicon substrate.
  (d) Heavily doped N$^+$-polycrystalline silicon gate with 1 Ωcm P-type silicon substrate.

● **MOS System: Inversion, Threshold, Depletion, and Accumulation** ●

5.3 The body of an MOS capacitor is N type. Match the "charge" diagrams (1) through (5) in Fig. 5–35 to (a) flat band, (b) accumulation, (c) depletion, (d) threshold, and (e) inversion.

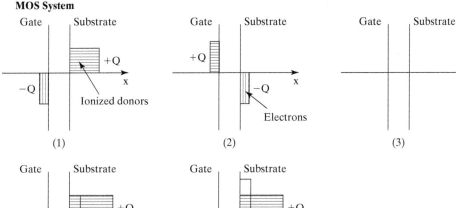

FIGURE 5–35

5.4 Consider an ideal MOS capacitor fabricated on a P-type silicon with a doping of $N_a = 5 \times 10^{16} \text{cm}^{-3}$ with an oxide thickness of 2 nm and an N^+ poly-gate.
  (a) What is the flat-band voltage, V_{fb}, of this capacitor?
  (b) Calculate the maximum depletion region width, W_{dmax}.
  (c) Find the threshold voltage, V_t, of this device.
  (d) If the gate is changed to P^+ poly, what would the threshold voltage be now?

5.5 Figure 5–36 shows the total charge per unit area in the P-type Si as a function of V_g for an MOS capacitor at 300 K.
  (a) What is the oxide thickness?
  (b) What is the doping concentration in Si?
  (c) Find the voltage drop in oxide (V_{ox}) when $V_g - V_{fb} = -1$ V.
  (d) Find the band bending in Si when $V_g - V_{fb} = 0.5$ V.

5.6 Make a series of qualitative sketches paralleling Figs. 5–11 to 5–14 (ϕ_s, W_{dep}, and charge as function of V_g) for an MOS capacitor having an N-type substrate and P^+poly gate. (Hint: At $V_g = V_t$, ϕ_s is negative. You may assume that V_t is negative.)

5.7 (a) Solve Eq. (5.3.1) for ϕ_s as a function of V_g.
  (b) Find an expression for V_{ox} as a function of V_g.
  (c) Make a rough sketch of ϕ_s vs. V_g and V_{ox} vs. V_g for -3 V $< V_g <$ 2 V, $V_{fb} = -0.9$ V, $N_a = 10^{17}\text{cm}^{-3}$, and $T_{ox} = 3$ nm.
  (d) Find W_{dep} as a function of V_g.

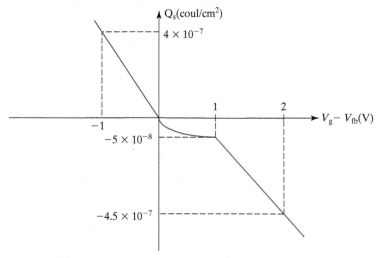

FIGURE 5–36

5.8 Consider an MOS capacitor fabricated on P-type Si substrate with a doping of 5×10^{16} cm^{-3} with oxide thickness of 10nm and N$^+$ poly-gate.

(a) Find C_{ox}, V_{fb}, and V_t.
(b) Find the accumulation charge (C/cm^2) at $V_g = V_{fb} - 1$ V.
(c) Find the depletion and inversion charge at $V_g = 2$ V.
(d) Plot the total substrate charge as a function of V_g for V_g from –2 to 2 V.

5.9 If we decrease the substrate doping concentration, how will the following parameters be affected? (Please indicate your answer by putting a mark, X, in the correct column.) Write down any relevant equation and explain briefly how you obtain the answer (a few words or one sentence). Assume the gate material is N$^+$poly and the body is P type.

Parameters	Increase	Decrease	Unchanged
A Accumulation region capacitance			
B Flat-band voltage, V_{fb}			
C Depletion-region capacitance			
D Threshold voltage, V_t			
E Inversion region capacitance			

5.10 From the high-frequency C–V measurements on an MOS capacitor with P-Si substrate performed at 300 K, the following characteristics were deduced:

Oxide thickness = 30 nm
Substrate doping = 10^{16} cm^{-3}
Flat-band voltage = –2 V

Construct the C–V curve, labeling everything, including the values of the oxide capacitance, flat-band voltage, and threshold voltage. Assuming an Al gate with 4.1 V work function, compute the effective oxide charge.

● Field Threshold Voltage ●

5.11 Metal interconnect lines in IC circuits form parasitic MOS capacitors as illustrated in Fig. 5–37. Generally, one wants to prevent the underlying Si substrate from becoming inverted. Otherwise, parasitic transistors may be formed and create undesirable current paths between the N^+ diffusions.

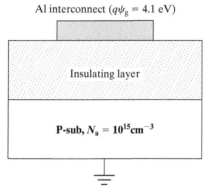

FIGURE 5–37

(a) Find V_{fb} of this parasitic MOS capacitor.

(b) If the interconnect voltage can be as high as 5 V, what is the maximum capacitance (F/cm^2) of the insulating layer that can be tolerated without forming an inversion layer?

(c) If the insulating layer thickness must be 1 μm for fabrication considerations, what should the dielectric constant $K = \varepsilon/\varepsilon_0$ of the insulating material be to make $V_t = 5$ V?

(d) Is the answer in (c) the minimum or maximum allowable K to prevent inversion?

(e) At $V_g = V_t + 2$ V ($V_t = 5$ V), what is the area charge density (C/cm^2) in the inversion layer?

(f) At $V_g = V_t = 5$ V, what is the high-frequency MOS capacitance (F/cm^2)?

(g) At $V_g = V_t + 2$ V ($V_t = 5$ V), what voltage is dropped across the insulating layer?

● Oxide Charge ●

5.12 Consider the C–V curve of an MOS capacitor in Fig. 5–38 (the solid line). The capacitor area is 6,400 μm^2. $C_0 = 45$ pF and $C_1 = 5.6$ pF.

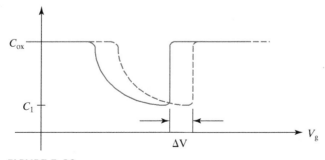

FIGURE 5–38

If, due to the oxide fixed charge, the C–V curve is shifted from the solid line to the dashed line with $\Delta V = 0.05$ V, what is the charge polarity and the area density (C/cm^2) of the oxide fixed charge?

5.13 Why is oxide charge undesirable? How do mobile charges get introduced into the oxide? How can this problem be overcome?

● **C–V Characteristics** ●

5.14 Derive $C(V_g)$ in Eq. (5.6.4). [Hint: Solve Eq. (5.3.3) for W_{dep}.]

5.15 Answer the following questions based on the C–V curve for an MOS capacitor shown in Fig. 5–39. The area of the capacitor is 10^4 μm^2.

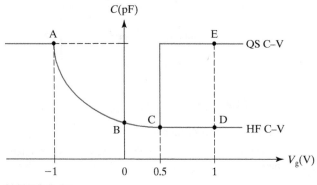

FIGURE 5–39

(a) Is the substrate doping N type or P type?
(b) What is the thickness of the oxide in the MOS capacitor?
(c) What is the doping concentration of the substrate, N_{sub}?
(d) What is the value of the capacitance at position C on the C–V curve shown above?
(e) Sketch the energy band diagram of the MOS structure at positions A, B, C, D, and E on the C–V curve.
(f) At location B on the C–V curve, what is the band bending, ϕ_s?

5.16 The C–V characteristics of MOS capacitors A (solid line) and B (dashed line), both having the same area, are shown in Fig. 5–40.

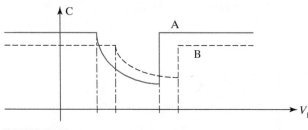

FIGURE 5–40

(a) Are the substrate P type or N type? How do you know this?

(b) Circle A or B to select the capacitor having larger

  X_{ox}: A B
  V_{fb}: A B
  X_{dmax}: A B
  N_{sub}: A B
  V_t: A B

5.17 Compare the maximum capacitance that can be achieved in an area $100 \times 100 \ \mu m^2$ by using either an MOS capacitor or a reverse-biased P^+N junction diode. Assume an oxide breakdown field of 8×10^6 V/cm, a 5V operating voltage, and a safety factor of two (i.e., design the MOS oxide for 10 V breakdown). The P^+N junction is built by diffusing boron into N-type silicon doped to 10^{16} cm^{-3}.

5.18 Consider the silicon–oxide–silicon structure shown in Fig. 5–41. Both silicon regions are N type with uniform doping of $N_d = 10^{16}$ cm^{-3}.

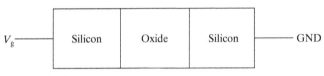

FIGURE 5–41

(a) What would be the flat-band voltage for this structure? Draw the energy band diagram for the structure for (i) $V_g = 0$, (ii) $V_g < 0$ and large, and (iii) $V_g > 0$ and large.

(b) Sketch the expected shape of the high-frequency C–V characteristics for the structure. What are the values of the capacitance for large positive and large negative V_g?

(c) If silicon on the left-hand side in the figure above is P-type doped with $N_a = 10^{16}$ cm^{-3}, sketch the C–V characteristics for the new structure.

5.19 Fill in the following table with appropriate mathematical expressions using the basic MOS C–V theory.

Bias condition	Surface potential	MOS capacitance (LF)	MOS capacitance (HF)	MOSFET capacitance
Accumulation				
Flat band				
Just below threshold				
Inversion				

5.20 The oxide thickness (T_{ox}) and the doping concentration (N_a or N_d) of the silicon substrate can be determined using the high-frequency C–V data shown in Fig. 5–42 for an MOS structure.

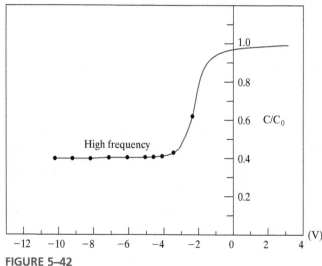

FIGURE 5–42

(a) Identify the regions of accumulation, depletion, and inversion in the substrate corresponding to this C–V curve. What is the doping type of the semiconductor?

(b) If the maximum capacitance of the structure C_0 (which is equal to $C_{ox} \times$ Area) is 82 pF and the gate area is 4.75×10^{-3} cm^2, what is the value of T_{ox}?

(c) Determine the concentration in the silicon substrate. Assume a uniform doping concentration.

(d) Assuming that the gate is P$^+$type, what is Q_{ox}?

● Poly-Gate Depletion ●

5.21 (a) Derive Eq. (5.8.1).

(b) Derive an expression for the voltage drop in the poly-depletion region, i.e., the band bending in the poly-Si gate, ϕ_s. Assume that the electric field inside the oxide, $\mathscr{E}_{ox}$, is known.

(c) Continue from (b) and express ϕ_{poly} in terms of V_g, not $\mathscr{E}_{ox}$. Assume surface inversion, i.e., $V_g > V_t$. Other usual MOS parameters such as V_{fb}, T_{ox}, and ϕ_B may also appear. Hint: $V_g = V_{fb} + V_{ox} + 2\phi_B + \phi_{poly}$.

(d) Using the result of (c), find an expression for W_{dpoly} in terms of V_g, not $\mathscr{E}_{ox}$. For part (e), (f), and (g), assume $T_{ox} = 2$ nm, $N_a = 10^{17}$cm^{-3}, $N_d = 6 \times 10^{19}$ cm^{-3} (for N$^+$poly-gate), and $V_g = 1.5$ V.

(e) Evaluate ϕ_{poly} and W_{dpoly}.

(f) Calculate V_t using Eq. (5.4.3). (The poly-depletion effect maybe ignored in V_t calculation because $\mathscr{E}_{ox}$ is very low at $V_g = V_t$.) Then, using ϕ_{poly} from part (e) in Eq. (5.8.3), find Q_{inv}.

(g) Calculate $Q_{inv} = C_{oxe}(V_g - V_t)$ with C_{oxe} given by Eq. (5.8.2).

Discussion:
Equation (5.8.2) is correct for the small signal capacitance

$$C(V_g) = dQ(V_g)/dV_g \Rightarrow Q(V_g) = \int C(V_g)dV_g$$

Here, part (g) does not yield the correct Q_{inv} because it assumes a constant C_{oxe}. C_{oxe} varies with V_g due to the poly-depletion effect even for $V_g > V_t$. The answer for part (f) is the correct value for Q_{inv}.

5.22 Draw an energy band diagram for Example 5–3 in Section 5.8. You need to decide whether V_g and V_{ox} are positive or negative. (Hint: The problem is about gate depletion.)

5.23 There is a voltage drop in the gate depletion region (V_{poly}). Express the following items using V_{poly}, the gate doping concentration N_{poly}, and the oxide capacitance C_{ox} as given variables.

(a) What is the charge density Q_{poly} in the gate depletion region?

(b) What is C_{poly}? ($C_{poly} = \varepsilon_s / W_{dpoly}$)

(c) What is the total MOS capacitance in the inversion region when poly depletion is included?

● **Threshold Voltage Expression** ●

5.24 After studying the derivation of Eq. (5.4.3), write down the steps of derivation on your own.

● **REFERENCES** ●

1. Lee, W. C., T-J. King, and C. Hu. "Observation of Reduced Boron Penetration and Gate Depletion for Poly-SiGe Gated PMOS Devices." *IEEE Electron Device Letters*. 20 (1) (1999), 9–11.
2. Stern, F. "Quantum Properties of Surface Space-Charge Layers." *CDC Critical Review Solid State Science*. 4 (1974), 499.
3. Yang, K., Y-C. King, and C. Hu. "Quantum Effect in Oxide Thickness Determination from Capacitance Measurement." *Technical Digest of Symposium on VLSI Technology*, 1999, 77–78.
4. Taur, Y., and T. H. Ning. *Fundamentals of Modern VLSI Devices*. Cambridge, UK: Cambridge University Press, 1998.
5. Tompsett, M.F. Video Signal Generation, in *Electronic Imaging*, T. P. McLean, ed. New York: Academic, 1979, 55.

● **GENERAL REFERENCES** ●

1. Muller, R. S., T. I. Kamins, and M. Chen. *Device Electronics for Integrated Circuits*, 3rd ed. New York: John Wiley & Sons, 2003.
2. Pierret, R. F. *Semiconductor Device Fundamentals*. Reading, MA: Addison-Wesley, 1996.

6

MOS Transistor

CHAPTER OBJECTIVES

This chapter provides a comprehensive introduction to the modern MOSFETs in their on state. (The off state theory is the subject of the next chapter.) It covers the topics of surface mobility, body effect, a simple IV theory, and a more complete theory applicable to both long- and short-channel MOSFETs. It introduces the general concept of CMOS circuit speed and power consumption, voltage gain, high-frequency operation, and topics important to analog circuit designs such as voltage gain and noise. The chapter ends with discussions of DRAM, SRAM, and flash nonvolatile memory cells.

The **MOSFET** is by far the most prevalent semiconductor device in ICs. It is the basic building block of digital, analog, and memory circuits. Its small size allows the making of inexpensive and dense circuits such as giga-bit (Gb) memory chips. Its low power and high speed make possible chips for gigahertz (GHz) computer processors and radio-frequency (RF) cellular phones.

6.1 ● INTRODUCTION TO THE MOSFET ●

Figure 6–1 shows the basic structure of a MOSFET. The two PN junctions are the **source** and the **drain** that supplies the electrons or holes to the transistor and drains them away respectively. The name **field-effect transistor** or **FET** refers to the fact that the gate turns the transistor (inversion layer) on and off with an electric *field* through the oxide. A **transistor** is a device that presents a high input resistance to the signal source, drawing little input power, and a low resistance to the output circuit, capable of supplying a large current to drive the circuit load. The hatched regions in Fig. 6–1a are the **shallow-trench-isolation** oxide region. The silicon surfaces under the thick isolation oxide have very high threshold voltages and prevent current flows between the N^+ (and P^+) diffusion regions along inadvertent surface inversion paths in an IC chip.

Figure 6–1 also shows the MOSFET IV characteristics. Depending on the gate voltage, the MOSFET can be off (conducting only a very small **off-state leakage current, I_{off}**) or on (conducting a large **on-state current, I_{on}**).

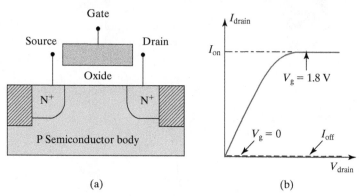

FIGURE 6–1 (a) Basic MOSFET structure and (b) IV characteristics.

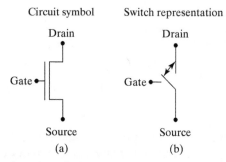

FIGURE 6–2 Two ways of representing a MOSFET: (a) a circuit symbol and (b) as an on/off switch.

At the most basic level, a MOSFET may be thought of as an on–off switch as shown in Fig. 6–2(b). The gate voltage determines whether a current flows between the drain and source or not. The circuit symbol shown in Fig. 6–2a connotes the much more complex characteristics of the MOSFET.

● **Early Patents on the FET** ●

The transistor and IC technologies owe their success mainly to the effort and ingenuity of a large number of technologists since the mid-1900s. Two early FET patents are excerpted here. These earliest patents are presented for historical interest only. Many more conceptual and engineering innovations and efforts were required to make MOSFETs what they are today.

J. E. Lilienfeld's 1930 U.S. patent is considered the first teaching of the FET. In Fig. 6–3, 10 is a glass substrate while 13 is the gate electrode (in today's terminology) and "consists of an … aluminum foil… ." 11 and 12 are metal contacts to the source and drain. 15 is a thin film of semiconductor (copper sulfide). Lilienfeld taught the following novel method of making a small (short) gate, the modern photolithography technique being yet unavailable to him. The glass substrate is broken into two pieces and then reassembled (glued back) with a thin aluminum foil inserted between the two pieces. The edge of the Al foil is used as the gate. The semiconductor film is deposited over the glass substrate and the gate, and source and drain contacts are provided. There is no oxide between the gate electrode and the semiconductor. The insulator in this FET would be the depletion layer at the metal–semiconductor junction (see Section 6.3.2).

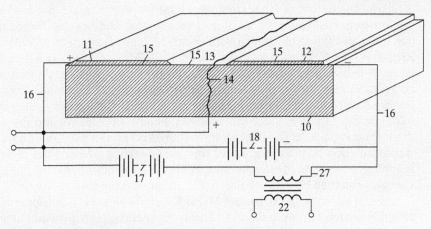

FIGURE 6–3 "A perspective view, on a greatly enlarged scale and partly in section, of the novel apparatus as embodied by way of example in an amplifier." (From [1].)

In a 1935 British patent, Oskar Heil gave a lucid description of a MOSFET. Referring to Fig. 6–4, "1 and 2 are metal electrodes between which is a thin layer 3 of semiconductor. A battery 4 sends a current through the thin layer of semiconductor and this current is measured by the ammeter 5. If, now, an electrode 6 in electro-static association with the layer 3 is charged positively or negatively in relation to the said layer 3, the electrical resistance of this layer is found to vary and the current strength as measured by the ammeter 5 also to vary."

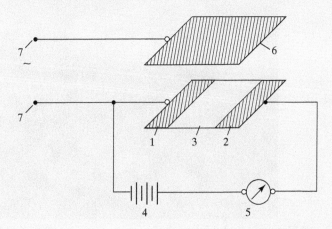

FIGURE 6–4 This 1935 drawing is a good illustration of a MOSFET even by today's standards. (From [2].)

6.2 • COMPLEMENTARY MOS (CMOS) TECHNOLOGY •

Modern MOSFET technology has advanced continually since its beginning in the 1950s. Figure 6–5 is a transmission electron microscope view of a part of a MOSFET. It shows the poly-Si gate and the single-crystalline Si body with visible individual Si atoms and a 1.2 nm amorphous SiO_2 film between them. 1.2 nm is the size of four SiO_2 molecules.

The basic steps of fabricating the MOSFET shown in Fig. 6–1 is to first make **shallow-trench-isolation** by etching a trench that defines the boundary of the transistor and filling the trench with chemical vapor deposition (CVD) oxide (see Section 3.7.2). Next, planarize the wafer with CMP (see Section 3.8), grow a thin layer of oxide (gate oxide) over the exposed silicon surface, deposit a layer of polycrystalline silicon as the gate material (Section 3.7.2), use optical lithography to pattern a piece of photoresist, and use the photoresist as a mask to etch the poly-Si to define the gate in Fig. 6–1 (Section 3.4). Finally, implant As into the source and drain (Section 3.5.1). The implantation is masked by the gate on one side and the trench isolation on the other. Rapid thermal annealing (see text box in Section 3.6) is applied to activate the dopant and repair the implantation damage to the crystal. Contacts can then be made to the source, drain, and the gate.

Figure 6–6a is an **N-channel MOSFET**, or **N-MOSFET** or simply **NFET**. It is called N-channel because the conduction channel (i.e., the inversion layer) is electron rich or N-type as shown in Fig. 6–6b. Figure 6–6c and d illustrate a **P-channel MOSFET**, or **P-MOSFET**, or **PFET**. In both cases, V_g and V_d swing between 0 V and V_{dd}, the power-supply voltage. The body of an NFET is connected to the lowest voltage in the circuit, 0 V, as shown in (b). Consequently, the PN junctions are always reverse-biased or unbiased and do not conduct forward diode current. When V_g is equal to V_{dd} as shown in (b), an inversion layer is present and the

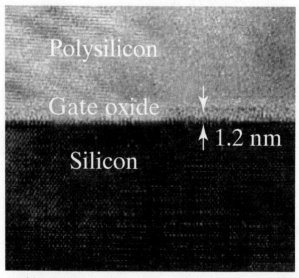

FIGURE 6–5 Gate oxides as thin as 1.2 nm can be manufactured reproducibly. Individual Si atoms are visible in the substrate and in the polycrystalline gate. (From [3]. © 1999 IEEE.)

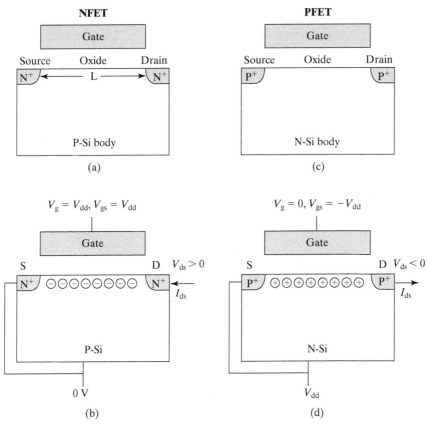

FIGURE 6–6 Schematic drawing of an N-channel MOSFET in the off state (a) and the on state (b). (c) and (d) show a P-channel MOSFET in the off and the on states.

NFET is turned on. With its body and source connected to V_{dd}, the PFET shown in (d) responds to V_g in exactly the opposite manner. When $V_g = V_{dd}$, the NFET is on and the PFET is off. When $V_g = 0$, the PFET is on and the NFET is off.

The complementary nature of NFETs and PFETs makes it possible to design low-power circuits called **CMOS** or **complementary MOS** circuits as illustrated in Fig. 6–7a. The circuit symbol of PFET has a circle attached to the gate. The example is an inverter. It charges and discharges the output node with its load capacitance, C, to either V_{dd} or 0 under the command of V_g. When $V_g = V_{dd}$, the NFET is on and the PFET is off (think of them as simple on–off switches), and the output node is pulled down to the ground ($V_{out} = 0$). When $V_g = 0$, the NFET is off and the PFET is on; the output node is pulled up to V_{dd}. In either static case, one of the two transistors is off and there is no current flow from V_{dd} through the two transistors directly to the ground. Therefore, CMOS circuits consume much less power than other types of circuits. Figure 6–7b illustrates how NFET and PFET can be fabricated on the same chip. Portions of the P-type substrate are converted into N-type wells by donor implantation and diffusion. Contacts to the P substrate and N well are included in the figure. Figure 6–7c illustrates the basic **layout** of a CMOS

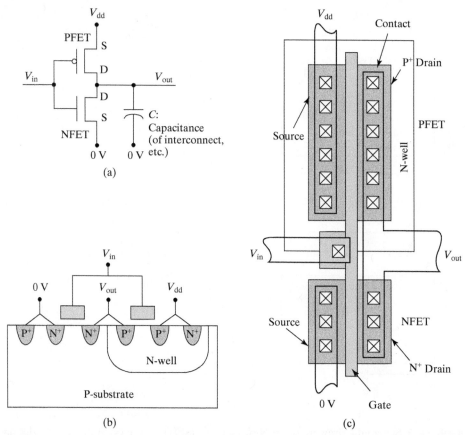

FIGURE 6–7 Three views of a CMOS inverter. (a) A CMOS inverter consists of a PFET **pull-up device** and an NFET **pull-down device**. (b) Integration of NFET and PFET on the same chip. For simplicity, trench isolation (see Fig. 6–1), which fills all the surface area except for the diffusion regions and the channel regions, is not shown. (c) Layout of a CMOS inverter.

inverter. It is a view of the circuit from above the Si wafer and may be thought of as a composite drawing of several photomasks used to fabricate the inverter. V_{in}, V_{out}, V_{dd}, and ground voltage are carried by metal lines. The poly-Si gate is the vertical bar connected to V_{in}. The metal to semiconductor contacts are usually made in multiple identical holes because it is more difficult to fabricate contact holes of varying sizes and shapes.

6.3 • SURFACE MOBILITIES AND HIGH-MOBILITY FETS •

It is highly desirable to have a large transistor current so that the MOSFET can charge and discharge the circuit capacitances (C in Fig. 6–7a) quickly and achieve a high circuit speed. An important factor that determines the MOSFET current is the electron or hole mobility in the surface inversion layer.

6.3.1 Surface Mobilities

When a small V_{ds} is applied, the drain to source current, I_{ds},[1] in Fig. 6–6b is

$$I_{ds} = W \cdot Q_{inv} \cdot v = W Q_{inv} \mu_{ns} \mathscr{E} = W Q_{inv} \mu_{ns} V_{ds}/L$$
$$= W C_{oxe}(V_{gs} - V_t) \mu_{ns} V_{ds}/L \quad (6.3.1)$$

W is the **channel width,** i.e., the channel dimension perpendicular to the page in Fig. 6–6 and the vertical dimension of the channel in Fig. 6–7c. Q_{inv} (C/cm^2) is the inversion charge density [Eq. (5.5.3)]. $\mathscr{E}$ is the channel electric field, and L is the **channel length.** μ_{ns} is the electron **surface mobility**, or the **effective mobility**. In MOSFETs, μ_{ns} and μ_{ps} (hole surface mobility) are several times smaller than the bulk mobilities presented in Section 2.2. In Eq. (6.3.1), all quantities besides μ_{ns} are known in Eq. (6.3.1) or can be measured, and therefore μ_{ns} can be determined.

μ_{ns} is a function of the average of the electric fields at the bottom and the top of the inversion charge layer, $\mathscr{E}_b$ and $\mathscr{E}_t$ in Fig. 6–8 [4]. From Gauss's Law, using the depletion layer as the Gaussian box

$$\mathscr{E}_b = -Q_{dep}/\varepsilon_s \quad (6.3.2)$$

From Eq. (5.4.4)

$$V_t = V_{fb} + \phi_{st} - Q_{dep}/C_{oxe} \quad (6.3.3)$$

Therefore,

$$\mathscr{E}_b = \frac{C_{oxe}}{\varepsilon_s}(V_t - V_{fb} - \phi_{st}) \quad (6.3.4)$$

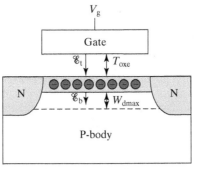

FIGURE 6–8 Surface mobility is a function of the average of the electric fields at the bottom and the top of the inversion charge layer, $\mathscr{E}_b$ and $\mathscr{E}_t$.

---

[1] We will follow the convention that positive I_{ds} refers to the normal direction of channel current from V_{dd} to ground, i.e., drain to source in NFET and source to drain in PFET. Therefore, I_{ds} is always positive.

Apply Gauss's Law to a box that encloses the depletion layer and the inversion layer.

$$\mathscr{E}_t = -(Q_{dep} + Q_{inv})/\varepsilon_s$$

$$= \mathscr{E}_b - Q_{inv}/\varepsilon_s = \mathscr{E}_b + \frac{C_{oxe}}{\varepsilon_s}(V_{gs} - V_t)$$

$$= \frac{C_{oxe}}{\varepsilon_s}(V_{gs} - V_{fb} - \phi_{st}) \tag{6.3.5}$$

$$\frac{1}{2}(\mathscr{E}_b + \mathscr{E}_t) = \frac{C_{oxe}}{2\varepsilon_s}(V_{gs} + V_t - 2V_{fb} - 2\phi_{st})$$

$$\approx \frac{C_{oxe}}{2\varepsilon_s}(V_{gs} + V_t + 0.2 \text{ V})$$

$$= \frac{\varepsilon_{ox}}{2\varepsilon_s T_{oxe}}(V_{gs} + V_t + 0.2 \text{ V})$$

$$= \frac{V_{gs} + V_t + 0.2 \text{ V}}{6 T_{oxe}} \quad \text{for N}^+ \text{ poly-gate NFET} \tag{6.3.6}$$

μ_{ns} has been found to be a function of the average of $\mathscr{E}_b$ and $\mathscr{E}_t$. (This conclusion is sometimes presented with the equivalent statement that μ_{ns} is a function of $Q_{dep} + Q_{inv}/2$.) The measured μ_{ns} is plotted in Fig. 6–9 and can be fitted with [4]:[2]

$$\mu_{ns} = \frac{540 \text{ cm}^2/\text{Vs}}{1 + \left(\frac{V_{gs} + V_t + 0.2 \text{ V}}{5.4 T_{oxe}}\right)^{1.85}} \tag{6.3.7}$$

Empirically, the hole surface mobility is a function of $(\mathscr{E}_t + 1.5\mathscr{E}_b)/2$ [5].

$$\mu_{ps} = \frac{185 \text{ cm}^2/\text{Vs}}{1 - \left(\frac{V_{gs} + 1.5 V_t - 0.25 \text{ V}}{3.38 T_{oxe}}\right)} \tag{6.3.8}$$

T_{oxe} is defined in Eq. (5.9.2). Normally, V_{gs} and V_t are negative for a PFET, i.e., in Eq. (6.3.8). This mobility model accounts for the effects of the major variables on the surface mobility. When device variables V_{gs}, V_t, and T_{oxe} are properly considered, all silicon MOSFETs exhibit essentially the same surface mobility as illustrated in Fig 6–9. This is said to be Si's **universal effective mobility**. The surface mobility is lower than the bulk mobility because of **surface roughness scattering** [5, 6]. It makes the mobilities

---

[2] Equation (6.3.7) is for the common case of NMOSFET with N$^+$ poly-Si gate. In general, the 0.2 V term should be replaced with $-2(V_{fb} + \phi_{st})$. See Eq. (5.4.2) for ϕ_{st}. Eq. (6.3.8) is for the common case of PMOSFET with P$^+$ poly-Si gate. In general, the -0.25 V term should be replaced with $2.5(V_{fb} + \phi_{st})$.

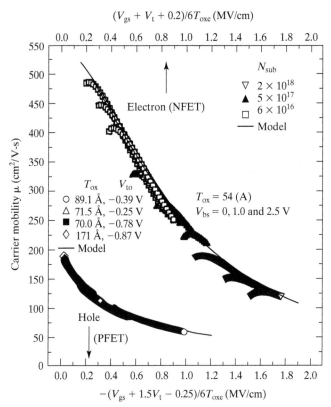

FIGURE 6–9 Electron and hole surface mobilities are determined by V_{gs}, V_t, and T_{oxe}. T_{oxe} is the SiO_2 equivalent electrical oxide thickness. (From [4]. © 1996 IEEE.)

● Effect of Wafer Surface Orientation and Drift Direction ●

The surface mobility is a function of the surface orientation and the drift direction. The standard CMOS technology employs the [100] surface silicon wafers, and the transistors are laid out so that the electrons and holes flow along the identical (0 ±1 ±1) directions on the wafer surface. (See Section 1.1 for explanation of the notation). One of the reasons for the choice is that this combination provides the highest μ_{ns}, though not the highest μ_{ps}. The mobility data in Fig. 6–9 are for this standard choice. The wafer orientation and current direction also determine how μ_{ns} and μ_{ps} respond to mechanical stress (see Section 7.1.2). These orientation effects can be explained by the solution of the Schrödinger's wave equation.

decrease as the field in the inversion layer ($\mathcal{E}_b$, $\mathcal{E}_t$) becomes stronger and the charge carriers are confined closer to the Si–SiO$_2$ interface.

μ_{ns} and μ_{ps} still roughly follow the $T^{-3/2}$ temperature dependence that is characteristic of phonon scattering (see Eq. 2.2.5). In Fig. 6–9, the surface mobility around $V_g \approx V_t$, especially in the heavily doped semiconductor (2×10^{18} cm^{-3}), is lower than the universal mobility. Dopant ion scattering is the culprit. At higher V_g, dopant ion scattering effect is screened out by the inversion layer carriers (see Section 2.2.2).

EXAMPLE 6–1

What is the surface mobility at $V_{gs} = 1V$ in an N-channel MOSFET with $V_t = 0.3$ V and $T_{oxe} = 2$ nm?

SOLUTION:

$$(V_{gs} + V_t + 0.2)/6T_{oxe} = (1.5\text{V}/12 \times 10^{-7}\text{cm} = 1.25\,\text{MV/cm})$$

A megavolt (10^6 V) is 1 MV. From Fig. 6–9, $\mu_{ns} \approx 190$ cm^2/V·s. To the dismay of MOSFET engineers, this is several times smaller than μ_n, the bulk mobility. μ_{ps} for a PMOSFET of similar design is only 60 cm^2/V·s.

6.3.2 GaAs MESFET

Higher carrier mobility allows the carriers to travel faster and the transistors to operate at higher speeds. High-speed devices not only improve the throughput of electronic equipment but also open up new applications such as inexpensive microwave communication. The most obvious way to improve speed is to use a semiconductor having higher mobility than silicon such as germanium, Ge (see Table 2–1) or strained Si (see Section 7.1.2). Single-crystalline Ge and SiGe alloy films can be grown epitaxially over Si substrates. The extension of Si technology to include Ge or SiGe transistor is a promising way to improve the device speed.

Table 2–1 indicates that GaAs and some other compound semiconductors have much higher electron mobilities than Si. For some applications, only N-channel FETs are needed and the hole mobility is of no importance. Unfortunately, it is very difficult to produce high-quality MOS transistors in these materials. There are too many charge traps at the semiconductor/dielectric interface for MOSFET application. Fortunately, a Schottky junction can serve as the control gate of a GaAs FET in place of an MOS gate. The device, called **MESFET** for **metal–semiconductor field-effect transistor**, is shown in Fig. 6–10. Because GaAs has a large E_g and small n_i, undoped GaAs has a very high resistivity and can be considered an insulator. The metal gate may be made of Au, for example. A large Schottky barrier height is desirable for minimizing the input gate current, i.e., the Schottky diode current.

When a reverse-bias voltage or a small forward voltage (small enough to keep the gate diode current acceptable) is applied to the gate, the depletion region under the gate expands or contracts. This modulates the thickness of the conductive channel, the part that is not depleted. This change, in turn, modulates the channel

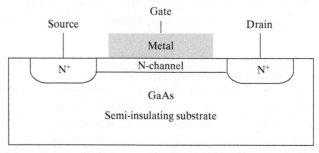

FIGURE 6–10 Schematic of a Schottky gate FET called **MESFET**.

current I_{ds}. Because I_{ds} does not flow in a surface inversion layer, the electron mobility is not degraded by surface scattering. This fact further enhances GaAs MESFET's speed advantage.

If the N-channel thickness is larger than the depletion-layer width at $V_g = 0$, the MESFET is conductive at $V_g = 0$ and requires a (reverse bias) gate voltage to turn it off. It is called a **depletion-mode transistor**. If the N-channel is thinner than the depletion-layer width at $V_g = 0$, a (forward) gate voltage is needed to turn the transistor on. This is known as an **enhancement-mode transistor**. Modern Si MOSFETs are all enhancement-mode transistors, which make circuit design much easier. GaAs FETs of both depletion-mode and enhancement-mode types are used. The depletion-type device is easier to make.

6.3.3 HEMT

The dopants in the channel in Fig. 6–10 significantly reduce the electron mobility through impurity scattering (see Section 2.2.2). If the channel is undoped, the mobility can be much higher. A MOSFET does not rely on doping to provide the conduction channel. Can GaAs FET do the same? The answer is yes. A MOS-like structure can be made by growing a thin epitaxial layer of GaAlAs over the undoped GaAs substrate as shown in Fig. 6–11a. Under the gate the GaAlAs film is

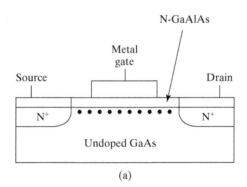

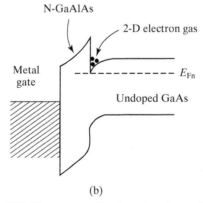

FIGURE 6–11 (a) The basic HEMT structure. The large band gap GaAlAs functions like the SiO_2 in a MOSFET. The conduction channel is in the undoped GaAs. (b) The energy diagram confirms the similarity to a MOSFET.

depleted. GaAlAs has a larger band gap than GaAs and Fig. 6–11b shows that it functions like the oxide in a MOSFET (see Fig. 5–9) in that it creates an energy well and a thin layer of electrons at the GaAs–GaAlAs interface. The curvature in the GaAlAs band diagram is due to the presence of the dopant ions as in the depletion layer of a PN junction. E_F is the Fermi level of the N+ source and it (with E_c) determines the electron concentration in the conduction channel. The channel electrons come from the N+ source. Because the epitaxial interface of the two semiconductors is smoother than the Si–SiO$_2$ interface, this device does not suffer from mobility degradation by surface scattering as MOSFET does. This device is called **HEMT** or **high electron-mobility transistor**, or **MODFET** for **modulation-doped FET**. It is used in microwave communication, satellite TV receivers, etc.

6.3.4 JFET

If the Schottky junction in Fig. 6–10 is replaced with a P+N junction, the new structure is called a **JFET** or **junction field-effect transistor**. The P+ gate is of course connected to a metal for circuit connections. As in a MESFET, a reverse bias would expand the depletion layer and constrict the conduction channel. In this manner, the JFET current can be controlled with the gate voltage. Before the advent of MOSFET, ICs were built

● **How to Measure the V_t of a MOSFET** ●

V_t is rarely determined from the CV data. Instead it can be more easily measured from the I_{ds} – V_{gs} plot shown in Fig. 6–12.

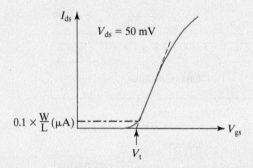

FIGURE 6–12 V_t can be measured by extrapolating the I_{ds} vs. V_{gs} curve to $I_{ds} = 0$. Alternatively, it can be defined as the V_{gs}, at which I_{ds} is a small fixed amount.

I_{ds} measured at a small V_{ds} such as 50 mV is plotted against V_{gs}. At $V_{gs} > V_t$, I_{ds} increases linearly with ($V_{gs} - V_t$) according to Eq. (6.3.1), if μ_{ns} were a constant. Because μ_{ns} decreases with increasing V_{gs} (see Section 6.3), the curve is sublinear. It is a common practice to extrapolate the curve at the point of maximum slope and take the intercept with the x-axis as V_t.

An increasingly popular alternative is to define V_t as the V_{gs} at which I_{ds} is equal to a small value such as

$$I_{ds} = 0.1 \, \mu A \times \frac{W}{L}$$

Also see Fig. 7–2 d.

with bipolar transistors, which have forward-biased diodes at the input and draw significant input current (see Chapter 8). The high input currents and capacitances were quite undesirable for some circuits. JFET provided a low input current and capacitance device because its input is a reverse-biased diode. JFET can be fabricated with bipolar transistors and coexist in the same IC chip.

6.4 • MOSFET V_t, BODY EFFECT, AND STEEP RETROGRADE DOPING •

The inversion layer of a MOSFET can be thought of as a resistive N-type film (1–2 nm thin) that connects the source and the drain as shown in Fig. 6–13. This film, at potential V_s, forms a capacitor with the gate, the oxide being the capacitor

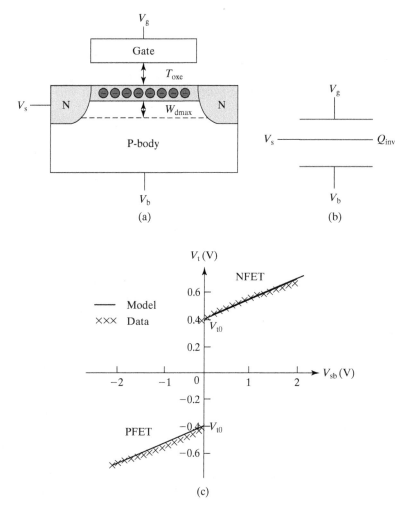

FIGURE 6–13 (a, b) The inversion layer can be viewed as a conductive film that is coupled to V_g through the oxide capacitance and coupled to V_b through the depletion-layer capacitance. The drain is open-circuited. (c) V_t is an approximately linear function of the body to source bias voltage. The polarity of the body bias is normally that which would reverse bias the body-source junction.

dielectric. It also forms a second capacitor with the body and the capacitor dielectric is the depletion layer. The depletion-layer capacitance is

$$C_{dep} = \frac{\varepsilon_s}{W_{dmax}} \qquad (6.4.1)$$

In Chapter 5, with $V_b = V_s$, we concluded that the gate voltage induces a charge in the invesion layer,

$$Q_{inv} = -C_{oxe}(V_{gs} - V_t) \qquad (6.4.2)$$

Let us now assume that there is also a voltage between the source and the body, V_{sb}. Since the body and the channel are coupled by C_{dep}, V_{sb} induces a charge in the inversion layer, $C_{dep}V_{sb}$. Therefore

$$Q_{inv} = -C_{oxe}(V_{gs} - V_t) + C_{dep}V_{sb} \qquad (6.4.3)$$

$$= -C_{oxe}\left(V_{gs} - \left(V_t + \frac{C_{dep}}{C_{oxe}}V_{sb}\right)\right) \qquad (6.4.4)$$

Equation (6.4.4) can be rewritten in the simple form of Eq. (6.4.2) if we adopt a modification to V_t. (What we have called V_t up to this point will henceforth be called V_{t0}.)

$$Q_{inv} = -C_{oxe}(V_{gs} - V_t(V_{sb})) \qquad (6.4.5)$$

$$\boxed{V_t(V_{sb}) = V_{t0} + \frac{C_{dep}}{C_{oxe}}V_{sb} = V_{t0} + \alpha V_{sb}} \qquad (6.4.6)$$

$$\alpha = C_{dep}/C_{oxe} = 3T_{oxe}/W_{dmax} \qquad (6.4.7)$$

The factor 3 is the ratio of the relative dielectric constants of silicon (11.9) and SiO_2 (3.9). Figure 6–13c illustrates the conclusion that V_t is a function of V_{sb}. *When the source-body junction is reverse-biased, the NFET V_t becomes more positive and the PFET V_t becomes more negative.* Normally, the source-body junctions are never forward biased so that there is no forward diode current.

The fact that V_t is a function of the body bias is called the **body effect**. When multiple NFETs (or PFETs) are connected in series in a circuit, they share a common body (the silicon substrate) but their sources do not have the same voltage. Clearly some transistors' source–body junctions are reversed biased. This raises their V_t and reduces I_{ds} and the circuit speed. Circuits therefore perform best when V_t is as insensitive to V_{sb} as possible, i.e., the body effect should be minimized. This can be accomplished by minimizing the T_{ox}/W_{dmax} ratio. (We will see again and again that a thin oxide is desirable.) α in Eq. (6.4.6) can be extracted from the slope of the curve in Fig. 6–13c and is called the **body-effect coefficient**.

Modern transistors employ steep **retrograde body doping profiles** (light doping in a thin surface layer and very heavy doping underneath) illustrated in Fig. 6–14. Steep retrograde doping allows transistor shrinking to smaller sizes for cost reduction and reduces impurity scattering. Section 7.5 explains why. The depletion-layer thickness is basically the thickness of the lightly doped region. As V_{sb} increases, the depletion layer does not change significantly. Therefore C_{dep} and

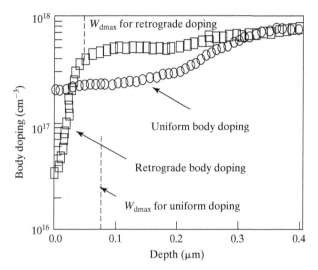

FIGURE 6–14 Comparison of a steep retrograde doping profile and a uniform doping profile.

α are basically constants. As a result, modern transistors exhibit a more or less linear relationship between V_t and V_{sb}. A linear relationship means that W_{dmax} and therefore the C_{dep}/C_{oxe} ratio are independent of the body bias.

In earlier generations of MOSFETs, the body doping density is more or less uniform (see the lower curve in Fig. 6–14) and W_{dmax} varies with V_{sb}. In that case, the theory for the body effect is more complicated. V_t can be obtained by replacing the $2\phi_B$ term (band bending in the body) in Eq. (5.4.3) with $2\phi_B + V_{sb}$.[3]

$$V_t = V_{t0} + \frac{\sqrt{qN_a 2\varepsilon_s}}{C_{oxe}}(\sqrt{2\phi_B + V_{sb}} - \sqrt{2\phi_B})$$

$$\equiv V_{t0} + \gamma(\sqrt{2\phi_B + V_{sb}} - \sqrt{2\phi_B}) \quad (6.4.8)$$

γ is called the **body-effect parameter**. Equation (6.4.8) predicts that V_t is a sublinear function of V_{sb}. A hint of the sublinearity is observable in the data in Fig. 6–13c. Equation (6.4.8) is sometimes linearized by Taylor expansion so that V_t is expressed as a linear function of V_{sb} in the form of Eq. (6.4.6).

6.5 • Q_{INV} IN MOSFET •

Let us consider Fig. 6–15 with $V_d > V_s$. The channel voltage, V_c, is now a function of x. $V_c = V_s$ at $x = 0$ and $V_c = V_d$ at $x = L$. Compare a point in the middle of the channel where $V_c > V_s$ with a point at the source-end of the channel, where

---

[3] When the source–body junction is reverse biased, there are two quasi-Fermi levels, E_{Fn} and E_{Fp} (similar to Fig. 4–7c with the P-region being the MOSFET body and the N-region being the source), which are separated by qV_{sb}. The inversion layer does not appear when E_c at the interface is close to E_{Fp} (E_F in Fig. 5–7). It appears when E_F is close to E_{Fn} (qV_{sb} below E_F in Fig. 5–7). This requires the band bending to be $2\phi_B + V_{sb}$, not $2\phi_B$.

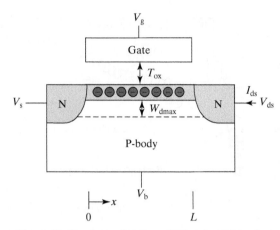

FIGURE 6–15 When $V_{ds} \neq 0$, the channel voltage V_c is a function of x.

$V_c = V_s$. Because the voltage in the middle of the channel is higher at $V_c(x)$, there is less voltage across the oxide capacitor (and across the depletion layer capacitor). Therefore, there will be fewer electrons on the capacitor electrode (the inversion layer). Specifically, the V_{gs} term in Eq. (6.4.5) should be replaced by $V_{gc}(x)$ or $V_{gs} - V_{cs}(x)$ and V_{sb} by $V_{sb} + V_{cs}(x)$.

$$Q_{inv}(x) = -C_{oxe}(V_{gs} - V_{cs} - V_{t0} - \alpha(V_{sb} + V_{cs}))$$

$$= -C_{oxe}(V_{gs} - V_{cs} - (V_{t0} + \alpha V_{sb}) - \alpha V_{cs})$$

$$= -C_{oxe}(V_{gs} - mV_{cs} - V_t) \qquad (6.5.1)$$

$$m \equiv 1 + \alpha = 1 + C_{dep}/C_{oxe} = 1 + 3T_{oxe}/W_{dmax} \qquad (6.5.2)$$

m is typically around 1.2. It is acceptable and easier at the beginning to simply assume $m = 1$. However, including m in the equations significantly improves their accuracies for later reference. The body is sometimes called the **back gate** since it clearly has a similar though weaker effect on the channel charge. The back-gate effect on Q_{inv} is often called the **bulk-charge effect**. m is called the **bulk-charge factor**. Clearly the bulk-charge effect is closely linked to the body-effect of Section 6.4.

6.6 • BASIC MOSFET IV MODEL •

Using Eq. (6.5.1) and dropping the negative sign for simplicity (I_{ds} in Fig. 6–15 is understood to flow from the high-voltage terminal to the low-voltage terminal).

$$I_{ds} = W \cdot Q_{inv}(x) \cdot v = W \cdot Q_{inv}\mu_{ns}\mathscr{E}$$

$$= WC_{oxe}(V_{gs} - mV_{cs} - V_t)\mu_{ns}dV_{cs}/dx \qquad (6.6.1)$$

6.6 • Basic MOSFET IV Model

$$\int_0^L I_{ds} dx = WC_{oxe}\mu_{ns}\int_0^{V_{ds}}(V_{gs} - mV_{cs} - V_t)dV_{cs} \qquad (6.6.2)$$

$$I_{ds}L = WC_{oxe}\mu_{ns}\left(V_{gs} - V_t - \frac{m}{2}V_{ds}\right)V_{ds} \qquad (6.6.3)$$

$$\boxed{I_{ds} = \frac{W}{L}C_{oxe}\mu_{ns}\left(V_{gs} - V_t - \frac{m}{2}V_{ds}\right)V_{ds}} \qquad (6.6.4)$$

Equation (6.6.4) shows that I_{ds} is proportional to W (channel width), μ_{ns}, V_{ds}/L (the average field in the channel), and $C_{ox}(V_g - V_t - mV_{ds}/2)$, which may be interpreted as the average Q_{inv} in the channel. When V_{ds} is very small, the $mV_{ds}/2$ term is negligible and $I_{ds} \propto V_{ds}$, i.e., the transistor behaves as a resistor. As V_{ds} increases, the average Q_{inv} decreases and dI_{ds}/dV_{ds} decreases. By differentiating Eq. (6.6.4) with respect to V_{ds}, it can be shown that dI_{ds}/dV_{ds} becomes zero at a certain V_{ds}.

$$\frac{dI_{ds}}{dV_{ds}} = 0 = \frac{W}{L}C_{ox}\mu_{ns}(V_{gs} - V_t - mV_{ds}) \quad \text{at} \quad V_{ds} = V_{dsat}$$

$$\boxed{V_{dsat} = \frac{V_{gs} - V_t}{m}} \qquad (6.6.5)$$

V_{dsat} is called the **drain saturation voltage**, beyond which the drain current is saturated as shown in Fig. 6–16. For each V_g, there is a different V_{dsat}. The part of the IV curves with $V_{ds} \ll V_{dsat}$ is the **linear region**, and the part with $V_{ds} > V_{dsat}$ is the **saturation region**. Analog designers often refer to the regions as the **Ohmic region** and the **active region**.

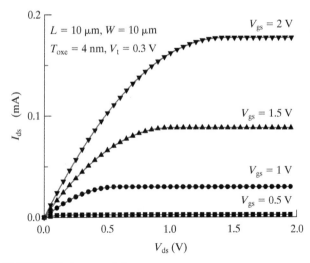

FIGURE 6–16 MOSFET IV characteristics.

The saturation current can be obtained by substituting V_{dsat} [Eq. (6.6.5)] for V_{ds} in Eq. (6.6.4).

$$I_{dsat} = \frac{W}{2\,mL} C_{oxe} \mu_{ns} (V_{gs} - V_t)^2 \qquad (6.6.6)$$

What happens at $V_d = V_{dsat}$ and why does I_{ds} stay constant beyond V_{dsat}? The first question can be answered by substituting V_{dsat} [Eq. (6.6.5)] for V_{cs} in Eq. (6.5.1). Q_{inv} at the drain end of the channel, when $V_{ds} = V_{dsat}$, is zero! This disappearance of the inversion layer is called channel **pinch-off**. Figure 6–17 plots V_{cs}, Q_{inv}, and I_{ds} at $V_{ds} = V_{dsat}$ and $V_{ds} > V_{dsat}$. In these two cases, $V_{cs}(x)$, $Q_{inv}(x)$ and therefore I_{ds} are the same. This explains why I_{ds} does not change with V_{ds} beyond V_{dsat}. The only difference is that, at $V_{ds} > V_{dsat}$, there exists a short, high-field **pinch-off region** where $Q_{inv} = 0$ and across which the voltage $V_{ds} - V_{dsat}$ is dropped. Section 6.9.1 will present an improvement to the concept of pinch-off such that Q_{inv} does not drop to zero. For now, the concept of pinch-off is useful for introducing the phenomenon of current saturation.

How can a current flow through the pinch-off region, which is similar to a depletion region? The fact is that a depletion region does not stop current flow as long as there is a supply of the right carriers. For example, in solar cells and photodiodes, current can flow through the depletion region of PN junctions. Similarly, when the electrons reach the pinch-off region of a MOSFET, they are swept down the steep potential drop in Fig. 6–17h. Therefore, the pinch-off region does not present a barrier to current flow. Furthermore, Fig. 6–17d and h show that the electron flow rates (current) are equal in the two cases because they have the same drift field and Q_{inv} in the channel. In other words, the current is independent of V_{ds} beyond V_{dsat}. The situation is like a mountain stream feeding into a waterfall. The slope of the river bed (dE_c/dx) and the amount of water in the stream determine the water flow rate in the stream, which in turn determines the flow rate down the waterfall. The height of the waterfall ($V_{ds} - V_{dsat}$), whether 1 or 100 m, has no influence over the flow rate.

• **Channel Voltage Profile** •

First consider the case of $V_{ds} = V_{dsat}$. Substituting the upper limits of integration in Eq. (6.6.2), L and V_{ds}, with x and V_{cs} and using $I_{ds} = I_{dsat}$ = Eq. (6.6.6), you can show that (see Problem 6.9 at the end of the chapter).

$$V_{cs} = \frac{V_{gs} - V_t}{m}\left(1 - \sqrt{1 - \frac{x}{L}}\right) \qquad (6.6.7)$$

As expected, $V_{cs} = 0$ at $x = 0$ and $V_{cs} = V_{dsat} = (V_g - V_t)/m$ at $x = L$. From this, you can show that $WQ_{inv}\mu_s\mathscr{E}$ or $WC_{ox}(V_{gs} - mV_{cs} - V_t)\mu_s dV_{cs}/dx$ is independent of x and yields the I_{dsat} expressed in Eq. (6.6.6). Equation (6.6.7) is plotted in Fig. 6–17a.

See Fig. 6–17e for the $V_{ds} > V_{dsat}$ case. V_{cs} still follows Eq. (6.6.7) from the source to the beginning of the pinch-off region. $V_{ds} - V_{dsat}$ is dropped in a narrow pinch-off region next to the drain.

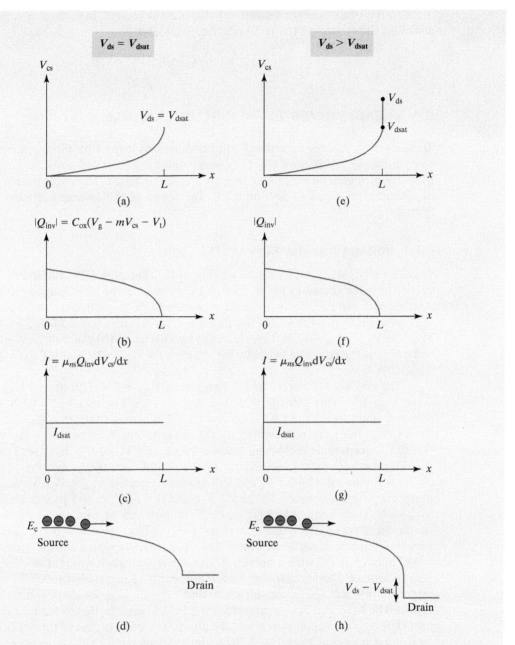

FIGURE 6–17 (a)–(d) $V_{ds} = V_{dsat}$ and (e)–(h) $V_{ds} > V_{dsat}$. Current does not change when V_{ds} increases beyond V_{dsat}. (d) and (h) are $E_c(x)$ from the energy band diagrams.

Transconductance, defined as

$$g_m \equiv dI_{ds}/dV_{gs}\big|_{V_{ds}} \quad (6.6.8)$$

is a measure of a transistor's sensitivity to the input voltage. In general, a large g_m is desirable. Substituting Eq. (6.6.6) into Eq. (6.6.8), we find

$$g_{msat} = \frac{W}{mL} C_{oxe} \mu_{ns}(V_{gs} - V_t) \qquad (6.6.9)$$

6.7 • CMOS INVERTER—A CIRCUIT EXAMPLE •

Transistors' influences on circuits will be illustrated using CMOS inverters, which were introduced in Section 6.2. They consume little power and have the important property of regenerating or cleaning up the digital signal. The latter property will be discussed in detail in Section 6.7.1. The speed of the inverters is analyzed in Section 6.7.2.

6.7.1 Voltage Transfer Curve (VTC)

Consider the CMOS inverter shown in Fig. 6–18a. The NFET IV characteristics are similar to those shown in Fig. 6–16 and are plotted on the right half of Fig. 6–18b. Assume that the PFET has identical (symmetric) IV as plotted on the left half of the figure. From (a), the V_{ds} of the PFET and NFET are related to V_{out} by $V_{dsN} = V_{out}$ and $V_{dsP} = V_{out} - 2$ V. Therefore, the two halves of (b) can be replotted in (c) using V_{out} as the common variable. For example, at $V_{out} = 2$V in (c), $V_{dsN} = 2$V and $V_{dsP} = 0$ V.

The two $V_{in} = 0$ curves in (c) intersect at $V_{out} = 2$ V. This means $V_{out} = 2$ V when $V_{in} = 0$ V. This point is recorded in Fig. 6–19. The two $V_{in} = 0.5$ V curves intersect at around $V_{out} = 1.9$ V. The two $V_{in} = 1$ V curves intersect at $V_{out} = 1$ V. All the V_{in}/V_{out} pairs are represented by the curve in Fig. 6–19, which is the **voltage transfer characteristic** or **voltage transfer curve** or **VTC** of the inverter. The VTC provides the important noise margin of the digital circuits. V_{in} may be anywhere between 0 V and the NFET V_t and still produce a perfect $V_{out} = V_{dd}$. Similarly, V_{in} may be anywhere between 2V and 2 V plus the PFET V_t and produce a perfect $V_{out} = 0$ V. Therefore, perfect "0" and "1" outputs can be produced by somewhat corrupted inputs. This regenerative property allows complex logic circuits to function properly in the face of inductive and capacitive noises and IR drops in the signal lines. A VTC with a narrow and steep middle region would maximize the noise tolerance. Device characteristics that contribute to a desirable VTC include a large g_m, low leakage in the off state, and a small $\partial I_{ds}/\partial V_{ds}$ in the saturation region. The latter two device properties will be discussed further in the next chapter.

For optimal circuit operation, the sharp transition region of the VTC should be located at or near $V_{in} = V_{dd}/2$. To achieve this symmetry, the *IV* curves of NFET and PFET Fig. 6–18b need to be closely matched (symmetric). This is accomplished by choosing a larger W for the PFET than the NFET. The W_P/W_N ratio is usually around two to compensate for the fact that μ_{ps} is smaller than μ_{ns}.

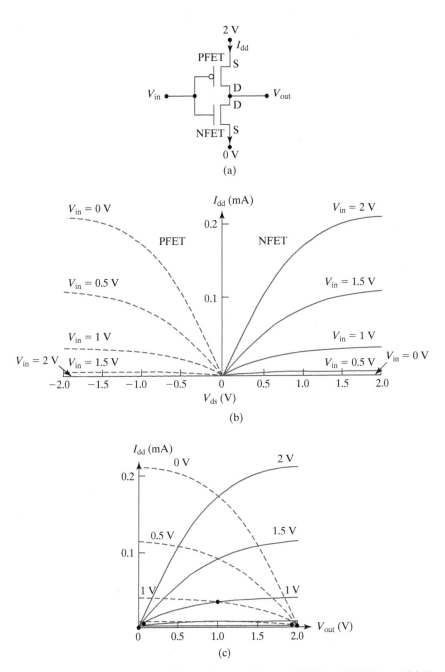

FIGURE 6–18 (a) CMOS inverter; (b) IV characteristics of NFET and PFET; and (c) $V_{out} = V_{dsN} = 2\,V + V_{dsP}$ according to (a).

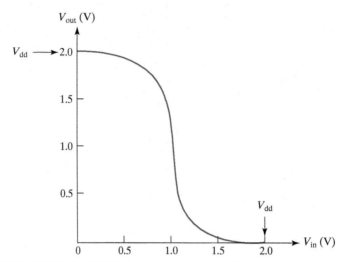

FIGURE 6–19 The VTC of a CMOS inverter.

6.7.2 Inverter Speed—The Importance of I_{on}

Propagation delay is the time delay for a signal to propagate from one gate to the next in a chain of identical gates as shown in Fig. 6–20.

τ_d is the average of the delays of pull-down (rising V_1 pulling down the output, V_2) and pull-up (falling V_2 pulling up the output, V_3). The propagation delay of an inverter may be expressed as [7]

$$\tau_d \approx \frac{CV_{dd}}{4}\left(\frac{1}{I_{onN}} + \frac{1}{I_{onP}}\right) \qquad (6.7.1)$$

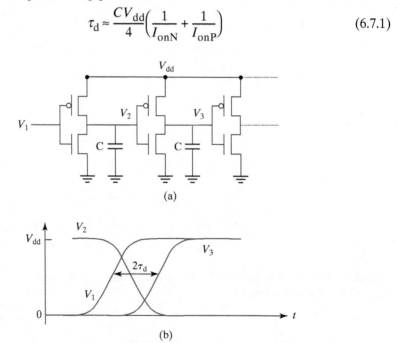

FIGURE 6–20 (a) A CMOS inverter chain. A circle on the gate indicates a PFET. (b) Propagation delay, τ_d, defined.

where I_{onN} is taken at $V_{gs} = V_{dd}$ and I_{onP} taken at $V_{gs} = -V_{dd}$. They are called the **on-state current**, of the NFET and the PFET

$$I_{on} \equiv I_{dsat}\big|_{\text{maximum }|V_{gs}|} \quad (6.7.2)$$

Equation (6.7.1) has a simple explanation

$$\tau_d = \frac{1}{2}(\text{pull-down delay} + \text{pull-up delay}) \quad (6.7.3)$$

$$\text{pull-down delay} \approx \frac{CV_{dd}}{2I_{onN}}$$

$$\text{pull-up delay} \approx \frac{CV_{dd}}{2I_{onP}} \quad (6.7.4)$$

The delay is the time for the on-state transistor supplying a current, I_{on}, to change the output by $V_{dd}/2$ (not V_{dd}). $V_{dd}/2$ is plausible in view of Fig. 6–17. The charge drained from (or supplied to) C by the FET during the delay is $CV_{dd}/2$. Therefore, the delay is $Q/I = CV_{dd}/2I_{on}$. One may interpret the delay as RC with $V_{dd}/2I_{on}$ as the switching resistance of the transistor. In order to maximize circuit speed it is clearly important to maximize I_{on}. We will further improve the I_{on} model in the next two sections.

The capacitance C represents the sum of all the capacitances that are connected to the output node of the inverter. They are the input capacitance of the next inverter in the chain, all the parasitic capacitances of the drain, and the capacitance of the metal interconnect that feeds the output voltage to the next inverter. In a large circuit, some interconnect metal lines can be quite long and their capacitances slow down the circuit significantly. This is ameliorated with the low-k dielectric technology described in Section 3.8 and circuit design techniques such as using a transistor with large W (a large I_{on}) to drive a longer interconnect and using repeaters.

Although the inverter is a very simple circuit, it is the basis of other more complex logic gates and memory cells. For example, Fig. 6–21 shows a NAND gate with two inputs. It is an inverter circuit with two series transistors in the pull-down path and two parallel transistors in the pull-up path.

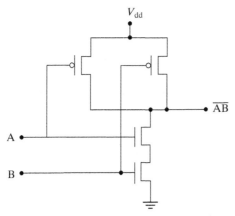

FIGURE 6–21 Inverters are the foundation of more complex circuits such as this two-input NAND gate.

> ● **Ring Oscillators** ●
>
> τ_d of a logic gate can be conveniently measured by connecting the end of a chain of identical logic gates (see Fig. 6–20a, for example) to the beginning of the chain to form a **ring oscillator**. The signal of any of the drain nodes in the ring oscillates with a period equal to τ_d times the number of gates in the ring. By using a large number of gates in the ring, the oscillation frequency can be conveniently low for easy measurement. Dividing the measured period of oscillation by the number of gates yields τ_d.
>
> The number of gates in a ring oscillator must be an odd number such as 91. If the number is an even number such as 92, the circuit will not oscillate. Instead, it will be static at one of two stable states.

6.7.3 Power Consumption

An important goal of device design is to minimize circuit power consumption. In each switching cycle, a charge CV_{dd} is transferred from the power supply to the load, C. The charge taken from the power supply in each second, $kCV_{dd}f$, is the average current provided by the power supply. Here, f is the clock frequency and $k(<1)$ is an **activity factor** that represents the fact that a particular gate in a given circuit is not switched every clock cycle all the time. Therefore

$$P_{\text{dynamic}} = V_{dd} \times \text{average current} = kCV_{dd}^2 f \qquad (6.7.6)$$

This **dynamic power** dominates the power consumption when the inverter is switched frequently. *Power consumption can be reduced by lowering V_{dd} and by minimizing all capacitances in the circuit as well as by reducing k.* It is interesting to note that *making I_{on} large by using a small L or improving the carrier mobility does not increase P_{dynamic}*.

It is desirable for a transistor to provide a large I_{on} (to reduce circuit switching delay) at a low V_{dd} (to reduce circuit power consumption). Reducing the transistor L and W, other parameters being equal, would lower C through reduction in the gate capacitance and the source–drain junction capacitance. Furthermore, smaller transistors make the chip smaller and therefore reduce the interconnect capacitance, too. Both device size reduction and V_{dd} reduction have been powerful means of lowering the power consumption per circuit function.

Another component of power consumption is the **static power**, or **leakage power** or **stand-by power** that is consumed when the inverter is static.

$$P_{\text{static}} = V_{dd} I_{\text{off}} \qquad (6.7.7)$$

I_{off} is the off-state leakage current when the transistor is supposed to be off. In an ideal transistor, I_{off} would be zero. It is difficult to keep I_{off} low in very high speed IC technologies as explained in detail in Chapter 7. The total power consumption is

$$P_{\text{static}} = P_{\text{static}} + P_{\text{dynamic}} \qquad (6.7.8)$$

6.8 • VELOCITY SATURATION •

A major weakness of the basic MOSFET IV model is that a finite current flows through the pinch-off region, where $Q_{inv} = 0$. This requires the carrier velocity to be infinite, a physical impossibility. We will now remove this shortcoming.

When the electric field is low, the carrier drift velocity, v, is $\mu\mathcal{E}$. As $\mathcal{E}$ increases, the kinetic energy of the carriers rises. When the energy of a carrier exceeds the optical phonon energy,[4] it generates an optical phonon and loses much of its velocity. Consequently, the kinetic energy and therefore the drift velocity cannot exceed a certain value. The limiting velocity is called the **saturation velocity**. The v–$\mathcal{E}$ relationship is shown in Fig. 6–22.

The flattening of the v–$\mathcal{E}$ curve is called **velocity saturation** and can be approximated with

$$v = \frac{\mu_{ns}\mathcal{E}}{1 + \mathcal{E}/\mathcal{E}_{sat}} \quad (6.8.1)$$

where μ_{ns} is the electron surface mobility and $\mathcal{E}_{sat}$ is the field at which velocity saturation becomes significant or dominant. When $\mathcal{E} \ll \mathcal{E}_{sat}$, Eq. (6.8.1) reduces to $v = \mu\mathcal{E}$. When $\mathcal{E} \gg \mathcal{E}_{sat}$, v is a constant regardless of how large $\mathcal{E}$ is. *Velocity saturation has a large and deleterious effect on the I_{on} of MOSFETs.*

• Velocity Overshoot •

Figure 6–22b shows the v–$\mathcal{E}$ characteristics of inversion-layer electrons at 85 K [8]. This is offered as clearer evidence that velocity saturates at high field than the room-temperature data (Fig. 6–22a). Because the velocity saturation phenomenon is clearer, we can see an important detail—v_{sat} is larger in transistors with very small channel lengths.

In the basic velocity-saturation model, v_{sat} is independent of the channel length. However, this figure shows that v_{sat} becomes larger when L is very small. When the channel length is sufficiently small, electrons may pass through the channel in too short a time for all the energetic carriers to lose energy by emitting optical phonons. As a result, the carriers can attain somewhat higher velocities in very small devices. This phenomenon is called **velocity overshoot**.

Velocity overshoot frees the extremely short transistors from the limit of velocity saturation. Unfortunately, another velocity limit (see Section 6.12) sets in before velocity overshoot offers a lot of relief.

---

[4] Optical phonon is a type of phonons (atom vibration) that has much higher energy than the acoustic phonons that are partially responsible for the low-field mobility (see Section 2.2.2). The optical phonons involve large displacements of neighboring atoms. These displacements create electrical dipole field that interact very strongly with electrons and holes. An electron or a hole that has enough energy to generate an optical phonon will do so readily and lose its kinetic energy in the process.

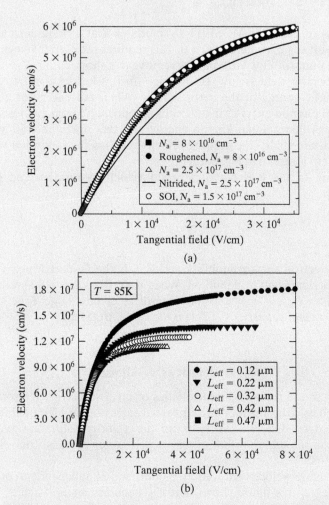

FIGURE 6–22 (a) The inversion-layer electron velocity saturates at high field regardless of the body doping concentration and surface treatment. (b) Velocity saturation is more prominent at low temperature. Velocity overshoot is also evident. (From [8]. © 1997 IEEE.)

6.9 • MOSFET IV MODEL WITH VELOCITY SATURATION

The basic MOSFET IV theory presented in Section 6.6 assumes a constant mobility. It provides an excellent introduction to the theory of MOSFET. The present section refines the theory by including the important velocity saturation effect. If we apply Eq. (6.8.1) to Eq. (6.6.1), using an NMOSFET for example

$$I_{ds} = WC_{oxe}(V_{gs} - mV_{cs} - V_t)\frac{\mu_{ns}dV_{cs}/dx}{1 + \frac{dV_{cs}}{dx}/\mathscr{E}_{sat}} \quad (6.9.1)$$

$$\int_0^L I_{ds} dx = \int_0^{V_{ds}} [WC_{oxe}\mu_{ns}(V_{gs} - mV_{cs} - V_t) - I_{ds}/\mathscr{E}_{sat}] dV_{cs} \qquad (6.9.2)$$

$$I_{ds} = \frac{\frac{W}{L} C_{oxe}\mu_{ns}\left(V_{gs} - V_t - \frac{m}{2} V_{ds}\right) V_{ds}}{1 + \frac{V_{ds}}{\mathscr{E}_{sat} L}} \qquad (6.9.3)$$

When L is large, Eq. (6.9.3) reduces to Eq. (6.6.4). Therefore the latter is known as the **long-channel IV model**.

$$\boxed{I_{ds} = \frac{\text{long-channel } I_{ds} \text{ (Eq. (6.6.4))}}{1 + V_{ds}/\mathscr{E}_{sat} L}} \qquad (6.9.4)$$

The effect of velocity saturation is to reduce I_{ds} by a factor of $1 + V_{ds}/\mathscr{E}_{sat}L$. This factor reduces to one (i.e., velocity saturation becomes negligible) when V_{ds} is small or L is large. This factor may be interpreted as $1 + \mathscr{E}_{ave}/\mathscr{E}_{sat}$, where $\mathscr{E}_{ave} \equiv V_{ds}/L$ is the average channel field. The saturation voltage, V_{dsat}, can be found by solving $dI_{ds}/dV_{ds} = 0$:

$$V_{dsat} = \frac{2(V_{gs} - V_t)/m}{1 + \sqrt{1 + 2(V_{gs} - V_t)/m\mathscr{E}_{sat}L}} \qquad (6.9.5)$$

Equation (6.9.5) is rather inconvenient to use. A simpler and even more accurate V_{dsat} model may be derived from a piece-wise model that actually fits the v–$\mathscr{E}$ data better than Eq. (6.8.1)[9]. It assumes that

$$v = \frac{\mu_{ns}\mathscr{E}}{1 + \mathscr{E}/\mathscr{E}_{sat}} \qquad \text{for } \mathscr{E} \leq \mathscr{E}_{sat} \qquad (6.9.6)$$

$$v = v_{sat} \qquad \text{for } \mathscr{E} \geq \mathscr{E}_{sat} \qquad (6.9.7)$$

Equating Eqs. (6.9.6) and (6.9.7) at $\mathscr{E} = \mathscr{E}_{sat}$ yields

$$\mathscr{E}_{sat} = 2v_{sat}/\mu_{ns} \qquad (6.9.8)$$

Equation (6.9.6) leads to Eq. (6.9.3), which is valid when the carrier speed is less than v_{sat}, i.e., $V_{ds} \leq V_{dsat}$. Equation (6.9.7) leads to the following equation describing the current at the drain end of the channel at the onset of velocity saturation (i.e., at $V_d = V_{dsat}$):

$$I_{ds} = WQ_{inv} v$$
$$= WC_{oxe}(V_g - V_t - mV_{dsat})v_{sat} \qquad (6.9.9)$$

Equating Eqs. (6.9.3) and (6.9.9) leads to

$$\boxed{\frac{1}{V_{dsat}} = \frac{m}{V_{gs} - V_t} + \frac{1}{\mathscr{E}_{sat}L}} \qquad (6.9.10)$$

V_{dsat} in Eq. (6.9.6) is an average of $\mathscr{E}_{sat}L$ and the long-channel V_{dsat}, $(V_{gs} - V_t)/m$ [Eq. (6.6.5)]. It is smaller than the latter. Note that $\mathscr{E}_{sat}$ is defined with Eq. (6.9.8).[5] It is known that v_{sat} is 8×10^6 cm/s for electrons and 6×10^6 cm/s for holes.

EXAMPLE 6–2 Drain Saturation Voltage

At $V_{gs} = 1.8$ V, what is the V_{dsat} of an NMOSFET with $T_{oxe} = 3$ nm, $V_t = 0.25$ V, and $W_{dmax} = 45$ nm for (a) $L = 10$ μm, (b) $L = 1$ μm, (c) $L = 0.1$ μm, and (d) $L = 0.05$ μm?

SOLUTION:

From Fig. 6–9 or Eq. (6.3.7), μ_n is 200 cm²/V/s. Using Eq. (6.9.8)

$$\mathscr{E}_{sat} = 2v_{sat}/\mu_{ns} = 2 \times 8 \times 10^6 \text{cm/s} \div 200 \text{ cm}^2/\text{Vs} = 8 \times 10^4 \text{V/cm}$$

Using Eq. (6.5.2)

$$m = 1 + 3T_{oxe}/W_{dmax} = 1 + 9\,\text{nm}/45\,\text{nm} = 1.2$$

Using Eq. (6.9.10)

$$V_{dsat} = \left(\frac{m}{V_{gs} - V_t} + \frac{1}{\mathscr{E}_{sat}L}\right)^{-1}$$

a. $L = 10$ μm,

$$V_{dsat} = \left(\frac{1.2}{1.55\,\text{V}} + \frac{1}{8 \times 10^4 \text{V/cm} \cdot L}\right)^{-1} = \left(\frac{1}{1.3\,\text{V}} + \frac{1}{80\,\text{V}}\right)^{-1} = 1.3\,\text{V}$$

b. $L = 1$ μm,

$$V_{dsat} = \left(\frac{1}{1.3\,\text{V}} + \frac{1}{8\,\text{V}}\right)^{-1} = 1.1\,\text{V}$$

c. $L = 0.1$ μm

$$V_{dsat} = \left(\frac{1}{1.3\,\text{V}} + \frac{1}{0.8\,\text{V}}\right)^{-1} = 0.5\,\text{V}$$

d. $L = 0.05$ μm

$$V_{dsat} = \left(\frac{1}{1.3\,\text{V}} + \frac{1}{0.4\,\text{V}}\right)^{-1} = 0.3\,\text{V}$$

Clearly, short-channel V_{dsat} is much smaller than long-channel V_{dsat}, $V_g - V_t$.

Substituting Eq. (6.9.10) for V_{ds} in Eq. (6.9.3)

$$I_{dsat} = \frac{W}{2mL}C_{oxe}\mu_{ns}\frac{(V_{gs} - V_t)^2}{1 + \dfrac{V_{gs} - V_t}{m\mathscr{E}_{sat}L}} = \frac{\text{long channel } I_{dsat}\,(\text{Eq. (6.6.6)})}{1 + \dfrac{V_{gs} - V_t}{m\mathscr{E}_{sat}L}} \qquad (6.9.11)$$

---

[5] You may find this $\mathscr{E}_{sat}$ definition to be inconsistent with Eq. (6.8.1). Equations (6.9.6)–(6.9.8) match the sharp curvature and the asymptotic values of the velocity-field data better than Eq. (6.8.6) [9].

Two special cases of Eqs. (6.9.10) and (6.9.11) are discussed below.

1. Long-channel or low V_{gs} case, $\mathscr{E}_{sat}L \gg V_{gs} - V_t$

$$V_{dsat} = (V_{gs} - V_t)/m \quad (6.9.12a)$$

$$I_{dsat} = \frac{W}{2mL}C_{oxe}\mu_{ns}(V_{gs} - V_t)^2 \quad (6.9.12b)$$

These are identical to Eqs. (6.6.5) and (6.6.6). The long-channel model is valid when L is large.

● How Large Must L Be to Be "Long Channel"? ●

The condition $\mathscr{E}_{sat}L \gg V_{gs} - V_t$ can be satisfied when L is large or when V_{gs} is close to V_t. The latter case is frequently encountered in analog circuits where the gate is biased close to V_t to reduce power consumption. Assuming $\mathscr{E}_{sat} = 6 \times 10^4$ V/cm and $V_{gs} V_t = 2$ V (for digital circuits), a 0.2 μm channel length would not satisfy the condition of $\mathscr{E}_{sat}L \gg V_{gs} - V_t$. Therefore, it exhibits significant short-channel behaviors. But, read on. If $V_{gs} - V_t = 0.1$ V (for low-power analog circuits), even a 0.1 μm channel length would satisfy the inequality and the transistor would exhibit some *long-channel* characteristics, i.e., $I_{dsat} \propto (V_{gs} - V_t)^2/L$ and $V_{dsat} = (V_{gs} - V_t)/m$. For applications to this low-power analog circuit, the "long-channel" equations such as Eq. (6.6.6) may be used even if L is 0.05 μm.

There are other short-channel behaviors that are observable even at small $V_{gs} - V_t$, e.g., a larger leakage current and a larger slope in the $I_d - V_d$ plot at $V_{ds} > V_{dsat}$. These other behaviors are sensitive to transistor design parameters such as T_{oxe} as explained in the next chapter.

2. Very short-channel case, $\mathscr{E}_{sat}L \ll V_{gs} - V_t$

$$V_{dsat} \approx \mathscr{E}_{sat}L < \frac{(V_g - V_t)}{m} \quad (6.9.13)$$

$$I_{dsat} \approx W v_{sat} C_{oxe}(V_{gs} - V_t - m\mathscr{E}_{sat}L) \quad (6.9.14)$$

I_{dsat} is proportional to $V_{gs} - V_t$ rather than $(V_{gs} - V_t)^2$ and is less sensitive to L than the long-channel I_{dsat} ($\propto 1/L$). Equation (6.9.14), derived from Eq. (6.9.11) by Taylor expansion, is quite easy to understand. I_{dsat} is proportional to W. Carriers travel at the saturation velocity at the drain end of the channel where $Q_{inv} = C_{oxe}(V_{gs} - V_t - mV_{dsat})$ and V_{dsat} is $\mathscr{E}_{sat}L$.

Figure 6–23a and b compare the measured IV characteristics of two NFETs with $L = 0.15$ μm and $L = 2$ μm. The shorter channel device shows an approximately linear relationship between I_{dsat} and V_{gs} in agreement with Eq. (6.9.14). V_{dsat} is significantly less than $(V_{gs} - V_t)/m$. (The behavior at $V_{ds} > V_{dsat}$ is explained in Sec. 7.9.) The 2 μm channel device shows a superlinear increase of I_{dsat} with increasing V_g in rough agreement with Eq. (6.9.12).

To raise I_{dsat}, we must increase $C_{oxe}(V_{gs} - V_t)$, i.e., reduce T_{oxe}, minimize V_t, and use high V_{gs}. The limit of T_{oxe} is set by oxide tunneling leakage and reliability. The lower limit of V_t is set by MOSFET leakage in the off state. These will be discussed in the next chapter. The maximum V_{gs} is the power supply voltage, V_{dd}, which is limited by concerns over circuit power consumption and device reliability.

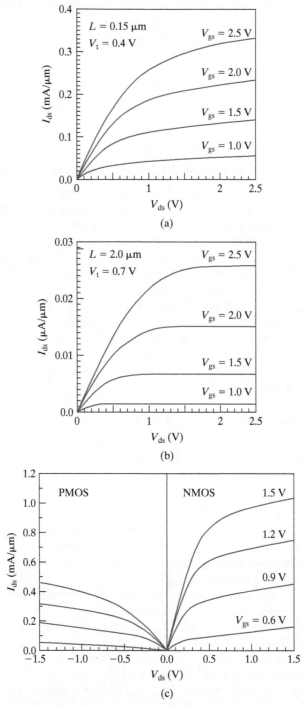

FIGURE 6–23 Measured IV characteristics. (a) A 0.15 μm channel device ($V_t = 0.4$ V) shows a linear relationship between I_{dsat} and V_{gs}. V_{dsat} is significantly less than $V_{gs} - V_t$. (b) A 2 μm device ($V_t = 0.7$ V) exhibits the $I_{dsat} \propto (V_{gs} - V_t)^2$ relationship. (c) IV characteristics of PFET and NFET with $T_{oxe} = 3$ nm and $L \approx 100$ nm.

Figure 6–23c shows that PFET and NFET have similar IV characteristics, e.g., both exhibit a linear I_{dsat}–V_g relationship. I_P is about half of I_N. The holes' mobility is three times smaller and their saturation velocity is 30% smaller than that of the electrons.

6.9.1 Velocity Saturation vs. Pinch-Off

The concept of pinch-off in Section 6.6 suggests that I_{ds} saturates when Q_{inv} becomes zero at the drain end of the channel. A more accurate description of the cause of current saturation is that the carrier velocity has reached v_{sat} at the drain. Instead of the pinch-off region, there is a **velocity saturation region** next to the drain where Q_{inv} is a constant (I_{dsat}/Wv_{sat}). The series of plots in Fig. 6–17 are still valid with one modification. In (b) and (f), $Q_{inv} = I_{dsat}/Wv_{sat}$ at L. In (f), of course, there is a very short region next to L, the velocity saturation region, where Q_{inv} remains constant. This region is not shown in Fig. 6–17 for simplicity.

6.10 ● PARASITIC SOURCE-DRAIN RESISTANCE ●

The main effect of the parasitic resistance shown in Fig. 6–24a is that V_{gs} in the I_{ds} equations is reduced by $R_s \cdot I_{ds}$. For example, Eq. (6.9.14) becomes

$$I_{dsat} = \frac{I_{dsat0}}{1 + R_s I_{dsat0}/(V_{gs} - V_t)} \qquad (6.10.1)$$

I_{dsat0} is the current in the absence of R_s. I_{dsat} may be significantly reduced by the parasitic resistance, and the impact is expected to rise in the future. The shallow diffusion region under the dielectric spacer is a contributor to the parasitic resistance. The shallow junction is needed to prevent excessive off-state leakage I_{ds} in short-channel transistors (see Section 7.6). The silicide (e.g., $TiSi_2$ or $NiSi_2$) reduces the **sheet resistivity**[6] of the N^+ (or P^+) source–drain regions by a factor of ten. It also reduces the

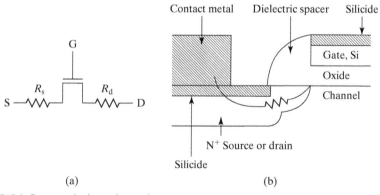

FIGURE 6–24 Source–drain series resistance.

---

[6] If the sheet resistivity of a film is 1 Ω per square, the resistance between two opposite edges of a square-shaped piece of this film (regardless of the size of the square) will be 1 Ω.

contact resistance between the silicide and the N^+ or P^+ Si. The contact resistance is another main source of resistance and more on this subject may be found in Section 4.21. The dielectric spacer is produced by coating the structure in Fig. 5–1 with a conformal film of dielectric followed by anisotropic dry etching to remove the dielectric from the horizontal surfaces. The silicides over the source/drain diffusion regions and over the gate are formed simultaneously by reaction between metal and silicon at a high temperature. The unreacted metal over the surface of the dielectric spacer is removed with acid. A second effect of the series resistance is an increase in V_{dsat}:

$$V_{dsat} = V_{sat0} + I_{dsat}(R_s + R_d) \qquad (6.10.2)$$

where V_{dsat0} is the V_{dsat} in the absence of R_s and R_d.

6.11 • EXTRACTION OF THE SERIES RESISTANCE AND THE EFFECTIVE CHANNEL LENGTH[7] •

Figure 6–25 illustrates the channel length and two other related quantities. A circuit designer specifies a channel length in the circuit layout, called the drawn gate length, L_{drawn}. This layout is transferred to a photomask, then to a photoresist pattern, and finally to the physical gate. The final physical **gate length**, L_g, may not be equal to L_{drawn} because each pattern transfer can introduce some dimensional change. However, engineers devote extraordinary efforts, e.g., by OPC (optical proximity correction) (see Section 3.3) to minimize the difference between L_{drawn} and L_g. As a result, one may assume L_{drawn} and L_g to be equal. L_g can be measured using scanning electron microscopy (SEM).

For device analysis and modeling, it is necessary to know the channel length, L, also called the effective channel length (L_{eff}) or the electrical channel length (L_e) to differentiate it from L_{drawn} and L_g. It is particularly useful to know the

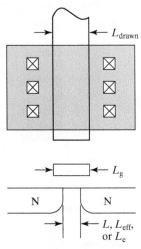

FIGURE 6–25 L_{drawn}, L_g, and L (also known as L_{eff} or L_e) are different in general.

---

[7] This section may be omitted in an accelerated course.

difference between L_{drawn} and L. This difference is called ΔL, which is assumed to be a constant, independent of L_{drawn}

$$L = L_{\text{drawn}} - \Delta L \quad (6.11.1)$$

Measuring ΔL in short transistors is quite difficult. There are several imperfect options. The following method is the oldest and still commonlly used. From Eq. (6.3.1),

$$V_{ds} = \frac{I_{ds}(L_{\text{drawn}} - \Delta L)}{WC_{\text{oxe}}(V_{gs} - V_t)\mu_{ns}} \quad (6.11.2)$$

When the series resistance, $R_{ds} \equiv R_d + R_s$, shown is Fig. 6–24a is included, Eq. (6.11.2) becomes

$$V_{ds} = I_{ds}R_{ds} + \frac{I_{ds}(L_{\text{drawn}} - \Delta L)}{WC_{\text{oxe}}(V_{gs} - V_t)\mu_{ns}} \quad (6.11.3)$$

$$\frac{V_{ds}}{I_{ds}}(= R_{ds} + \text{channel resistance}) = R_{ds} + \frac{L_{\text{drawn}} - \Delta L}{WC_{\text{oxe}}(V_{gs} - V_t)\mu_{ns}} \quad (6.11.4)$$

Figure 6–26 plots the measured V_{ds}/I_{ds} against L_{drawn} using three MOSFETs that are identical (fabricated on the same test chip) except for their L_{drawn}s. I_{ds} is measured at a small V_{ds} (≤ 50 mV) and at least two values of $V_{gs} - V_t$. V_{ds}/I_{ds} is a linear function of L_{drawn}. The two straight lines intersect at a point where V_{ds}/I_{ds} is independent of $V_{gs} - V_t$ according to Eq. (6.11.4), i.e., where $L_{\text{drawn}} = \Delta L$ and $V_{ds}/I_{ds} = R_{ds}$. Once ΔL is known, L can be calculated using Eq. (6.11.1).

Detailed measurements indicate that R_{ds} tends to decrease with increasing V_g. One reason is that the gate voltage induces more (accumulation) electrons in the source–drain diffusion region and therefore reduces R_{ds}. More puzzling is the observation that ΔL decreases (or L increases) with increasing V_g. The dependence of both ΔL and R_{ds} on V_g suggests the interpretation of channel length illustrated in Fig. 6–27 [10]. The sheet conductivities (inverse of sheet resistivity, introduced in Section 6.10) of the source–drain diffusion regions and the channel inversion layer (the horizontal lines) are plotted. The inversion-layer sheet conductivity increases with increasing V_g, of course. The channel length may be interpreted as the length of the part of the channel where the inversion-layer sheet conductivity is larger than the source/drain sheet conductivity. In other words, the channel is where the

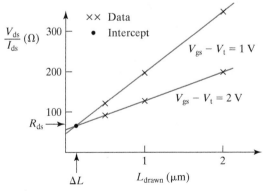

FIGURE 6–26 Method of extracting R_{ds} and ΔL.

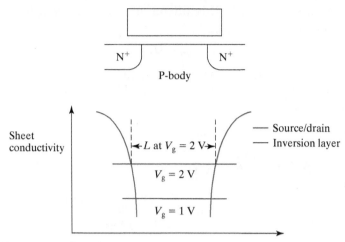

FIGURE 6–27 Interpretation of channel length and its dependence on V_g.

conductivity is determined by V_g, not by the source–drain doping profiles. Any resistance from outside the "channel" is attributed to R_{ds}. It is clear from Fig. 6–27 that the channel expands (i.e., L increases and R_{ds} decreases) with increasing V_g.

6.12 • VELOCITY OVERSHOOT AND SOURCE VELOCITY LIMIT[8] •

The concept of mobility is dubious when the channel length is comparable to or smaller than the mean free path (see Section 2.2.2). For this reason, Eq. (6.9.14) is particularly interesting because it does not contain mobility. The carrier velocity at the drain end of the channel is limited by the saturation velocity, which determines I_{dsat}. However, when the channel length is reduced much below 100 nm, the saturation velocity may be greatly raised by velocity overshoot as explained in Section 6.8. In that case, some other limit on I_{dsat} may set in.

The carrier velocity at the source becomes the limiting factor. There, the velocity is limited by the thermal velocity, with which the carriers enter the channel from the source. This is known as the **source injection velocity** limit.

The source is a reservoir of carriers moving at the thermal velocity. As the channel length approaches zero, all the carriers moving from the source into the channel are captured by the drain. No carriers flow from the drain to the source due to the voltage difference (or energy barrier) shown in Fig. 6–28.

$$I_{dsat} = WBv_{thx}Q_{inv} = WBv_{thx}C_{oxe}(V_{gs} - V_t) \quad (6.12.1)$$

Equation (6.12.1) is similar to Eq. (6.9.14) except that v_{sat} is replaced by v_{thx}, the x-direction component of the thermal velocity. Thorough analysis of v_{thx} shows that v_{thx} is about 1.6×10^7 cm/s for electrons and 1×10^7 cm/s for holes in silicon MOSFETs [11]. B is the fraction of carriers captured by the drain in a real transistor. The rest of the injected carriers are scattered back toward the source.

---

[8] This section may be omitted in an accelerated course.

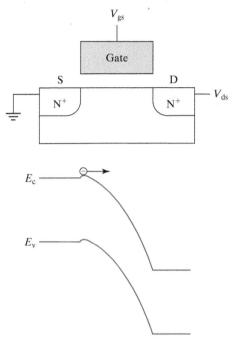

FIGURE 6–28 In the limit of no scattering in a very short channel, carriers are injected from the source into the channel at the thermal velocity and travel ballistically to the drain.

A particle simulation technique called the Monte Carlo simulation arrived at 0.5 as a typical value of B [11]. This makes Eq. (6.12.1) practically identical to Eq. (6.9.14) because v_{sat} is about 8×10^6 cm/s for electrons and 6×10^6 cm/s for holes. Both the drain-end velocity saturation limit and the source-end injection velocity limit predict similar I_{dsat}. B in Eq. (6.12.1) is expected to increase somewhat with decreasing L as v_{sat} in Eq. (6.9.14) is expected to do, too.

6.13 • OUTPUT CONDUCTANCE •

The saturation of I_{ds} (at $V_{ds} > V_{dsat}$) is rather clear in Fig. 6–23b. The saturation of I_{ds} in Fig. 6–23a is gradual and incomplete. The cause for the difference is that the channel length is long in the former case and short in the latter. The slope of the I–V curve is called the **output conductance**

$$g_{ds} = \frac{dI_{dsat}}{dV_{ds}} \tag{6.13.1}$$

A clear saturation of I_{ds}, i.e., a small g_{ds} is desirable. The reason can be explained with the simple amplifier circuit in Fig. 6–29. The bias voltages are chosen such that the transistor operates in the saturation region. A small-signal input, v_{in}, is applied.

$$i_{ds} = g_{msat} \cdot v_{gs} + g_{ds} \cdot v_{ds} \tag{6.13.2}$$
$$= g_{msat} \cdot v_{in} + g_{ds} \cdot v_{out}$$

$$v_{out} = -R \times i_{ds} \tag{6.13.3}$$

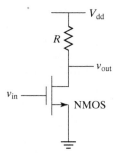

FIGURE 6–29 A simple MOSFET amplifier.

Eliminate i_{ds} from the last two equations and we obtain

$$v_{out} = \frac{-g_{msat}}{g_{ds} + 1/R} \times v_{in} \quad (6.13.4)$$

The magnitude of the output voltage, according to Eq. (6.13.4) is amplified from the input voltage by a **gain** factor of $\frac{g_{msat}}{g_{ds} + 1/R}$. The gain can be increased by using a large R. Even with R approaching infinity, the voltage gain cannot exceed

$$\text{Maximum Voltage Gain} = \frac{g_{msat}}{g_{ds}} \quad (6.13.5)$$

This is the **intrinsic voltage gain** of the transistor. If g_{ds} is large, the voltage gain will be small. As an extreme example, the maximum gain will be only 1 if g_{ds} is equal to g_{msat}. A large gain is obviously beneficial to analog circuit applications. A reasonable large gain is also needed to obtain a steep transition in the VTC, i.e., needed for digital circuit applications to enhance noise immunity. Therefore, g_{ds} must be kept much lower than g_{msat}.

The physical causes of the output conductance are the influence of V_{ds} on V_t and a phenomenon called channel length modulation. They are discussed in Section 7.9. The conclusions may be summed up this way. In order to achieve a small g_{ds} and a large voltage gain, L should be large and/or T_{ox}, W_{dep}, and X_j should be small.

6.14 ● HIGH-FREQUENCY PERFORMANCE ●

The high-frequency performance of the MOSFET shown in Fig. 6–30a is limited by the input RC time constant. C is the gate capacitance, $C_{ox}WL_g$. At high frequencies, the gate capacitive impedance, $1/2\pi fC$, decreases and the gate AC current increases. More of the gate signal voltage is dropped across R_{in}, and the output current is reduced. At some high frequency, the output current becomes equal to the input current. This unit current-gain frequency is called the **cutoff frequency**, f_T. In narrow-band analog circuits operating at a particular high frequency, the gate capacitance may be compensated with an on-chip inductor at that frequency to

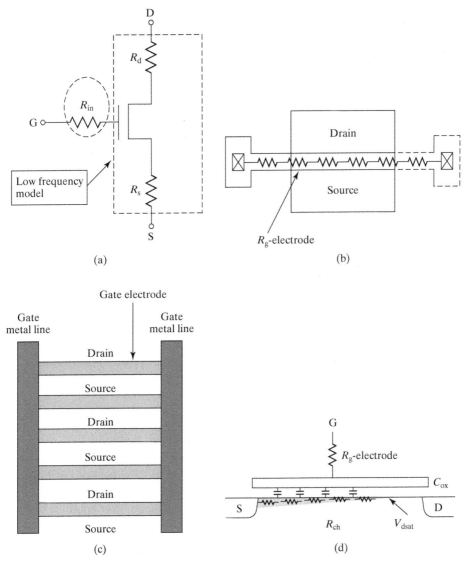

FIGURE 6–30 (a) The input resistance together with the input capacitance sets the high-frequency limit. (b) One component of R_{in} is the gate-electrode resistance. (c) The multi-finger layout dramatically reduces the gate-electrode resistance. (d) The more fundamental and important component of R_{in} is the channel resistance, which is also in series with the gate capacitor.

overcome the f_T limit. In that case, R_{in} still consumes power and at some frequency, typically somewhat higher than f_T, the power gain drops to unity. This frequency is called the **maximum oscillation frequency**, f_{max}. In either case, it is important to minimize R_{in}.

R_{in} consists of two components, the **gate-electrode resistance**, $R_{g\text{-electrode}}$, and the **intrinsic input resistance**, R_{ii}.

$$R_{in} = R_{g-electrode} + R_{ii} \qquad (6.14.1)$$

The gate-electrode resistance is straightforward as shown in Fig. 6–30b. A powerful way to reduce the gate-electrode resistance is **multi-finger layout** shown in Fig. 6–30c, which means designing a MOSFET with a large channel width, say 10 μm, as 10 MOSFETs connected in parallel each having a width of 1 μm. This reduces the gate-electrode resistance by a factor of 100 because each finger's resistance is ten times smaller and there are now ten finger resistors in parallel.

$$R_{g-\text{electrode}} = \rho W / 12 T_g L_g N_f^2 \quad (6.14.2)$$

ρ is the gate resistivity of the gate material, W, is the total channel width, T_g is the gate thickness, L_g is the gate length, and N_f is the number of fingers. The factor 12 comes from two sources. A factor of three comes from the fact that the gate current is distributed over the finger width and all the gate capacitor current does not flow through the entire finger resistor. The remaining factor of four arises from contacting the gate fingers at both the left and the right ends of the fingers as shown in Fig. 6–30c. Doing so effectively doubles the number of fingers and halves the finger width as if each finger is further divided into two at the middle of the finger. Using multifinger layout, the gate-electrode resistance can be quite low if the gate material is silicided poly-silicon. If the gate material is metal, this component of R_{in} becomes negligible.

The more important, fundamental, and interesting component is the intrinsic input resistance. The concept is illustrated in Fig. 6–30d. Even if $R_{g\text{-electrode}}$ is zero, there is still a resistor in series with the gate capacitor. The gate capacitor current flows through the channel resistance, R_{ch}, to the source, then through the input signal source (not shown) back to the gate to complete the current loop. R_{ii} is a resistance in the path of the gate current[12].

$$R_{ii} = \kappa \int dR_{\text{ch}} = \kappa \frac{V_{\text{ds}}}{I_{\text{ds}}} \quad (6.14.3)$$

κ is a number smaller than one [12] because due to the distributed nature of the RC network in Fig. 6–30d, the capacitance current does not flow through the entire channel resistance. V_{ds} Eq. (6.14.3) saturates at V_{dsat} when $V_{\text{ds}} > V_{\text{dsat}}$.

With each new generation of MOSFET technology, the gate length is reduced making R_{ii} smaller for a fixed W due to larger I_{ds} and smaller V_{dsat}. Furthermore, the input capacitance $C_{\text{ox}}WL_g$ is reduced somewhat when L_g is made smaller although C_{ox} is made larger (T_{oxe} thinner) at the same time. As a result, f_T and $f_{\max}$ have been improving linearly with the gate length. They are about 200 GHz in the 45 nm technology node, sufficient for a wide range of new applications.

6.15 • MOSFET NOISES •

Noise is whatever that corrupts the desired signal. One type of noise, the inductive and capacitive interferences or **cross talk** created by the interconnect network, may be called external noise. This kind of noise is important but can be reduced in principle by careful shielding and isolation by the circuit designers. The other noise category is called **device noise** that is inherent to the electronic devices. This kind of noise is due to the random behaviors of the electric carriers inside the device that create voltage and current fluctuations measurable at the terminals of the device.

This section is concerned with the device noise. *Noise, power consumption, speed, and circuit size (cost) are the major circuit-design constraints.*

6.15.1 Thermal Noise of a Resistor

If a resistor is connected to the input of an oscilloscope, the noise voltage across the resistor can be observed as shown in Fig. 6–31a. The origin of the noise is the random thermal motion of the charge carriers shown in Fig. 2–1, and the noise is called the **thermal noise**. The noise contains many frequency components. If one inserts a frequency filter with bandwidth Δf and measures the root-mean-square value of the noise in this frequency band, the results are

$$\overline{v_n^2} = S_{v_n} \Delta f = 4kT\Delta f R \qquad (6.15.1)$$

$$\overline{i_n^2} = S_{i_n} \Delta f = 4kT\Delta f / R \qquad (6.15.2)$$

where R is the resistance and Eq. (6.15.2) presents the noise current that would flow if the resistor's terminals were short-circuited. Clearly, the noise is proportional to Δf but is independent of f. This characteristic is called **white noise** and its **noise spectral density** is shown in Fig. 6–31b. S is called the noise power density.

6.15.2 MOSFET Thermal Noise

The intrinsic thermal noise of MOSFETs originates from the channel resistance. The channel may be divided into many segments as shown in Fig. 6–32 and each contributes some noise. The channel noise voltage can be expressed by Eq. (6.15.1).

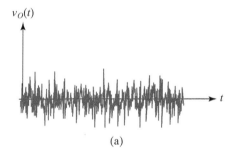

(a)

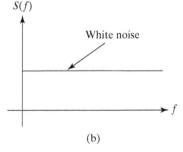

(b)

FIGURE 6–31 (a) The thermal noise voltage across a resistor and (b) the spectral density of white noise.

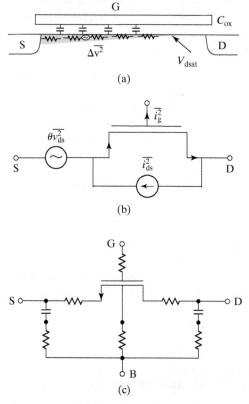

FIGURE 6–32 (a) Each segment of the channel may be considered a resistor that contributes thermal noise. (b) The noise current is added to the normal MOSFET current as a parallel current source. The noise voltage is multiplied by the transconductance into another component of noise current. (c) Parasitic resistances also contribute to the thermal noises.

However, there are several theories of what value should be assigned to R. A classical and popular theory interprets it as dV_{ds}/dI_{ds}, or $1/g_{ds}$ in the linear (small V_{ds}) region, as shown in Eqs. (6.15.3) and (6.15.4). γ is a function of V_{ds} and V_{gs}. At $V_{ds} > V_{dsat}$, γ saturates at 2/3. While this model works well at long-channel length, it underestimates the noise in short-channel MOSFETs. In circuit design practice, γ is chosen to fit noise measurements to improve the accuracy of the noise model.

$$\overline{v_{ds}^2} = 4\gamma kT\Delta f/g_{ds} \qquad (6.15.3)$$

$$\overline{i_{ds}^2} = 4\gamma kT\Delta f g_{ds} \qquad (6.15.4)$$

As in a resistor, this white noise of Eqs. (6.15.4) presents itself as a parallel current source added to the regular MOSFET current in Fig. 6–32b.

The channel noise voltage also induces a gate current through the gate capacitance. As a result, a portion of the channel noise current flows into the gate network. The gate noise current multiplied by the impedance of the gate input network and the transconductance produces a second noise current at the output. The complete model of the MOSFET noise therefore includes a partially correlated noise source appearing at the gate terminal. This effect can be approximately modeled by lumping the channel noise voltage at the source. θ in Fig. 6–32b is a function of L and V_{gs} and accounts for the fact that the noise voltage is actually distributed throughout the channel rather than lumped at the source [12,13]. Due to the partial correlation between the gate noise and the channel noise, the channel and gate noises can partially cancel each other at the output of the device. By optimizing the gate network impedance, design engineers can minimize the output noise.

The gate electrode, source, drain, and substrate parasitic resistances shown in Fig. 6–31c also contribute thermal noises. These resistances are usually minimized through careful MOSFET layout. It is important to reduce the gate electrode resistance as its noise is amplified by g_{msat} into the I_{ds} noise. The gate resistance can be minimized with the same **multifinger layout** discussed in Section 6.14.

6.15.3 MOSFET Flicker Noise

Flicker noise, also known as **1/ƒ noise**, refers to a noise spectral density that is inversely proportional to the frequency as shown in Fig. 6–33a. The mechanism for flicker noise is the random capture and release of electrons by traps located in the gate dielectric. When a trap captures an electron from the inversion layer, there is one less electron to conduct current. Also the trap becomes charged and reduces the channel carrier mobility due to **Coulombic scattering** similar to the effect of an impurity ion (see Section 2.2.2). In other words, both the carrier number and the mobility fluctuate due to charge trapping and detrapping. In a MOFSET with very small W and L, there is often only a single operative trap at a given bias condition and I_{ds} fluctuates between a high and a low current level with certain average cycle period as shown in Fig. 6–33b. This noise is called the **random telegraph noise**. The two current states reflect the empty and filled states of the trap. In a larger area ($W \times L$) MOSFET, there are many traps. The traps located at or near the oxide–semiconductor interface can capture and release electrons with short time constants and they contribute mostly high-frequency noises. Traps located far from the interface have long time constants and contribute mostly low-frequency noises. It can be shown that adding these contributions up with the assumption of a uniform distribution of traps in the oxide leads to the 1/ƒ noise spectrum [14].

$$\overline{i_{ds}^2} = \frac{KF \cdot W}{fL^2 C_{ox}} \left(\frac{I_{ds}}{W}\right)^{AF} \cdot kT\Delta f \qquad (6.15.7)$$

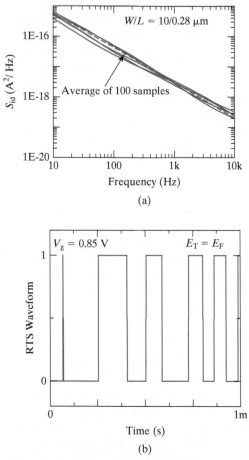

FIGURE 6–33 (a) Flicker noise is also known as $1/f$ noise because the noise power density is proportional to 1/frequency. (b) In a MOSFET with very small W and L, there may be only one operative trap and I_{ds} fluctuates between two levels. This is the random telegraph noise.

The constant KF is proportional to the oxide trap density, which is technology specific. AF is between 1 and 2 depending on the importance of Coulombic scattering to carrier mobility and W, L, and C_{ox} are the width, length, and per-area oxide capacitance of the MOSFET. The flicker noise is the dominant noise at low frequency. At frequencies above 100 MHz, one can safely ignore the flicker noise as it is much smaller than the thermal noise. In non-linear or time-varying circuits such as oscillators and mixers, which operate periodically with a large-amplitude high-frequency signal, the flicker noise is shifted up or down in frequency to the beat (sum and difference) frequencies of the signal and the noise. This creates a noise in the oscillator output, for example. HEMT (see Section 6.3.3) and bipolar transistors (see Chapter 8) have significantly lower flicker noise than MOSFET because they do not employ the MOS structure.

6.15.4 Signal to Noise Ratio, Noise Factor, Noise Figure

The input to a device or a circuit is in general a combination of the desired signal and some noise. The ratio of the signal power to the noise power is called the **signal** to noise ratio or **SNR**. SNR is a measure of the detectability of the signal in the presence of noise. The device or circuit also has some internal noise that is added to the amplified input noise and forms the total noise at the output. As a result, the SNR at the output of a linear device or circuit is smaller than the SNR at the input. The ratio of the input SNR and output SNR is called the **noise factor**.

$$F = \frac{S_i/N_i}{S_0/N_0} \tag{6.15.8}$$

The **noise figure** is defined as ten times the base-10 logarithm of the noise factor.

$$N_F = 10 \times \log F \tag{6.15.9}$$

The unit of noise figure is decibel or dB. As discussed earlier (see Sec. 6.15.2), the noise can be minimized with an optimum gate network impedance. Achieving this $N_{F\text{-min}}$ is an important goal of low-noise circuit design.

● **Noise and Digital Circuits** ●

The above discussion of MOSFET noise is more relevant to analog circuits than digital circuits. For a linear circuit such as a linear amplifier that must faithfully preserve the input waveform while amplifying its magnitude, the SNR at the output is at best the same as at the input. A digital circuit such as an inverter can generate an output that is 0 or V_{dd} even when the input is somewhat lower than V_{dd} or higher than 0. It eliminates the small noise at the input with its nonlinear voltage-transfer characteristic (see Fig. 6–19). In other words, a digital circuit has no gain for the small-amplitude noise at the input and has gain only for the larger real digital signal.

You may have had the pleasant experience of getting a photocopy of a black and white document that is cleaner looking than the original. The light smudges or erased pencil writings on the original are absent in the copy. That photocopier is a nonlinear system as is the digital circuit. If a photocopier is called on to reproduce a gray tone photograph as a linear system, it cannot reduce the noise in the original photograph because the copier cannot tell whether a smudge in the original is noise or part of the photograph.

This signal sharpening property of the digital circuits makes it possible to pack the digital circuits densely with long signal wires running close to each other. The dense wiring creates large cross-talk noise that is typically much larger than the thermal noise and flicker noise. Engineers reduce the cross talk by electrically shielding the sensitive lines, using low-k dielectrics between the lines (to reduce capacitive coupling), and limiting the line lengths.

When the MOSFET becomes very small as in advanced flash memory cells (see Section 6.16.3), a single trap can produce enough random telegraph noise (see Fig. 6–33b) to cause difficulty reading the 1 and 0 stored in a cell. Although this happens to only a very small portion of the memory cells, it is a concern for high-density memory design [15].

6.16 ● SRAM, DRAM, NONVOLATILE (FLASH) MEMORY DEVICES ●

Most of the transistors produced every year are used in semiconductor memories. Memory devices are commonly embedded in digital integrated circuits (ICs). For example, memory can occupy most of the area of a computer processor chip. Memory devices are also available in stand-alone memory chips that only perform the memory function. There are three types of semiconductor memories—**static RAM** or **SRAM**, **dynamic RAM** or **DRAM**, and **nonvolatile memory** with **flash memory** being the most prevalent nonvolatile memory. RAM stands for **random-access memory** meaning every data byte is accessible any time unlike hard disk memory, which has to move the read head and the disk to fetch new data with a significant delay. "Nonvolatile" means that data will not be lost when the memory is disconnected from electrical power source. The three types coexist because each has its own advantages and limitations. Table 6–1 summarizes their main differences.

SRAM only requires the same transistors and fabrication processes of the basic CMOS technology. It is therefore the easiest to integrate or embed into CMOS circuits. A DRAM cell is many times smaller than an SRAM cell but requires some special fabrication steps. High-density stand-alone DRAM chips are produced at large specialized DRAM fabrication plants. Low cost DRAMs has helped to proliferate PCs. A flash memory cell employs one of a variety of physical mechanisms to perform nonvolatile storage and has even smaller size than DRAM. Flash memory has not replaced DRAM or SRAM because of its slower writing speed and limited write cycles. Flash memory is economical and compact and has enabled advanced portable applications such as cell phones, media players, and digital cameras. Less aggressive (larger cell size) versions of DRAM and flash memory can be embedded in CMOS logic chips with some modification of the CMOS process technology. **Embedded DRAM** can be more economical than embedded SRAM when the required number of memory bits is very large.

6.16.1 SRAM

A basic SRAM cell uses six transistors to store one bit of data. As shown in Fig. 6–34a, its core consists of two cross-coupled inverters. M_1 and M_3 make up the left inverter. M_2 and M_4 make up the right inverter. The output of the left inverter is connected to the input of the right inverter and vice versa. If the left-inverter output, which is the input of the right inverter is high (hi), the right-inverter output would be low. This low output in turn makes the left-inverter out high. The positive feedback ensures that this state is stored and stable. If we change the left-inverter output to low and the right-inverter output to high, that would be a second stable state. Therefore this cell has two stable states, which represent the "0" and "1" and can store one bit of data. Many identical SRAM cells are arranged in an XY array. Each row of cells is connected to one word line (WL) and each column of cells is connected to a pair of bit lines (BL and BLC).

Two pass transistors M_5 and M_6 connect the outputs of the inverters to the bit lines. In order to read the stored data (determine the inverter state), the selected cell's WL is raised to turn on the pass transistors. A sensitive **sense amplifier** circuit compares the voltages on BL and BLC to determine the stored state.

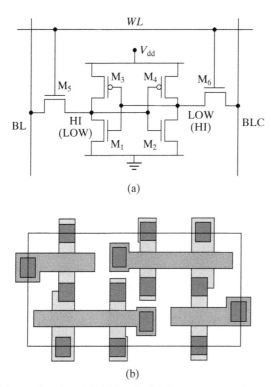

FIGURE 6–34 (a) Schematic of an SRAM cell. (b) Layout of a 32 nm technology SRAM. (From [16]. © 2007 IEEE.) The dark rectangles are the contacts. The four horizontal pieces are the gate electrodes and the two PFETs have larger Ws than the six NFETs. Metal interconnects (not shown) cross couple the two inverters.

TABLE 6–1 • The differences among three types of memories.

	Keep Data Without Power?	Cell Size and Cost/bit	Rewrite Cycles	Write-One-byte Speed	Compatible with Basic CMOS Manufacturing	Main Applications
SRAM	No	Large	Unlimited	Fast	Totally	Embedded in logic chips
DRAM	No	Small	Unlimited	Fast	Need modifications	Stand-alone chips and embedded
Flash memory	Yes	Smallest	Limited	Slow	Need extensive modifications	Nonvolatile storage stand-alone

In order to write the left-low state into the cell, for example, BL is set to low and BLC is set to high. Next, the word-line voltage is raised and the inverters will be forced into this (new) state.

SRAM cells provide the fastest operation among all memories. But since it requires six transistors to store one bit of data, the cost per bit is the largest. SRAM cells are often used as cache memory embedded in a processing unit where speed is critical. The steady increase in the clock speed of the processors requires the cache size to increase as well. Much effort is spent on size reduction, called *scaling*, for SRAM and for other types of memories. Figure 6–34b shows the layout of the six transistors of a 32 nm technology node SRAM cell [16].

6.16.2 DRAM

A DRAM cell contains only one transistor and one capacitor as shown in Fig. 6–35. Therefore it can provide a large number of bits per area and therefore lower cost per bit. Figure 6–35 is a portion of a schematic DRAM cell array. One end of the cell capacitor is grounded. The states "1" and "0" are represented by charging the cell capacitor to V_{dd} or zero. To write data into the upper-left cell, WL 1 is raised high to turn on the transistor (connecting the capacitor to bit line 1) and bit line 1 is set to V_{dd} to write "1" or 0 V to write "0." The cell to the right can be written at the same time by setting bit line 2 to the appropriate value (V_{dd} or 0 V).

Each bit line has its own (unavoidable) capacitance, $C_{\text{bit line}}$. In order to read the stored data from the upper-left cell, bit line 1 is precharged to $V_{dd}/2$ and then left floating. WL 1 voltage is raised to connect the cell capacitor in parallel with the larger $C_{\text{bit line}}$. Depending on the cell capacitor voltage (V_{dd} or 0), the cell capacitor either raises or lowers the bit line voltage by $C \cdot V_{dd}/2(C + C_{\text{bit line}})$, usually tens of milivolts. A sense amplifier circuit connected to the bit line monitors this voltage change to determine (read) the stored data. All cells connected to one WL are read at the same time. After each read operation the same data are automatically written back to the cell because the capacitor charge has been corrupted by the read.

The DRAM capacitor can only hold the data for a limited time because its charge gradually leaks through the capacitor dielectric, the PN junction (transistor S/D), and the transistor subthreshold leakage (see Section 7.2). To prevent data loss, the change must be refreshed (read and rewritten) many times each second.

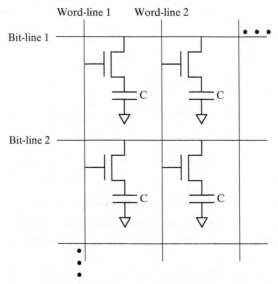

FIGURE 6–35 A schematic DRAM cell array. Each cell consists of a transistor and a capacitor.

The D in DRAM refers to this dynamic **refresh** action. Refresh consumes stand-by power. To increase the refresh interval, the cell capacitance should be large so that more charge is stored.

A large cell capacitance (not too much smaller than $C_{\text{bit line}}$) is also important for generating a large read signal for fast and reliable reading. However, it has become increasingly difficult to provide a large C while the cell area has been reduced to a few percent of 1 μm^2. Besides deploying very thin capacitor dielectrics, engineers have resorted to complex three-dimensional capacitor structures that provide capacitor areas larger than the cell area. Figure 6–36a shows a cup-shaped capacitor and Fig. 6–36b shows a scanning electron microscope view of the cross section of

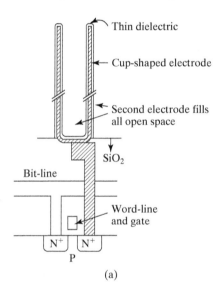

(a)

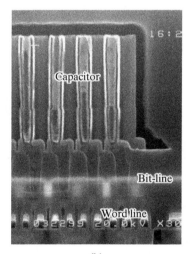

(b)

FIGURE 6–36 (a) Schematic drawing of a DRAM cell with a cup-shaped capacitor. (b) Cross-sectional image of DRAM cells. The capacitors are on top and the transistors are near the bottom. (From [17]. © 2002 IEEE.)

several DRAM cells. The four deep-cup shaped elements are four capacitors. Each capacitor has two electrodes. One electrode is cup-shaped and made of polysilicon or metal. It is connected at the bottom by a poly-Si post to the transistor below. Both the inside and the outside of the cup electrode are coated with a thin dielectric film. The other electrode is also made of poly-Si and it fills the inside of the cup as well as all the spaces between the cups. This second electrode is grounded (see Fig. 6–35). This complex structure provides the necessary large capacitor area.

A much simplified DRAM process technology can be integrated into logic CMOS technology at significant sacrifice of the cell area. Such an embedded DRAM technology is an attractive alternative to embedded SRAM when the number of bits required is large.

6.16.3 Nonvolatile (Flash) Memory

SRAMs and DRAMs lose their stored content if they are not connected to an electric power source. **Nonvolatile memory** or **NVM** is a memory device that keeps its content without power for many years. NVMs are used for program **code storage** in cell phones and most microprocessor based systems. They are also the preferred **data storage** medium (over hard disks and CDs) in portable applications for storing documents, photos, music, and movies because of their small size, low power consumption, and absence of moving parts. There are many variations of NVM devices [18], but the prevalent type is illustrated in Fig. 6–37a.

The structure may be understood as a MOSFET with one modification. The gate insulator is replaced with two insulators sandwiching a charge-storage layer. For example, the charge-storage layer can be silicon nitride or another insulator with a high density of electron traps. When the traps are empty or neutral, the transistor has a low V_t. When electrons are trapped in the insulator, the transistor has a high V_t as discussed in Section 5.7 and illustrated in Fig. 6–37b. The low and high V_t states represent the "0" and "1," respectively, and can be easily read with a sense circuit that checks the V_t. The charge storage layer may be a conductor, and in fact the most important and prevalent charge storage layer material is the familiar polycrystalline Si. NVM employing a poly-Si charge storage layer is called the **floating-gate memory** because the poly-Si layer is a transistor gate that is electrically floating.

Figure 6–37c shows how to put electrons into the charge-storage layer, i.e., how to write "1" into the NVM cell. About 20 V is applied to the gate and the high field causes electrons to tunnel (see Section 4.20) from the inversion layer into the charge storage layer. In Fig. 6–37d the cell is erased into "0" when the stored electrons tunnel into the substrate (the P-type accumulation layer).

Because the erase operation by tunneling is slow (taking milliseconds compared to nano-seconds for SRAM and DRAM), these NVMs are erased in blocks of kilobytes rather than byte by byte. Electrical erase by large memory blocks is called **flash erase** and this type of memory is called **flash memory**. Flash memory is the dominant type of NVM so that the two terms are often used interchangeably. Writing by tunneling is also slow so that it is also performed on hundreds of bytes at the same time.

There is another way of writing the cell in Fig. 6–37 (a and e). When the source is grounded and higher-than-normal voltages are applied to the gate and the drain, a high electric field exists in the pinch-off (or velocity-saturation) region near

6.16 • SRAM, DRAM, Nonvolatile (Flash) Memory Devices 243

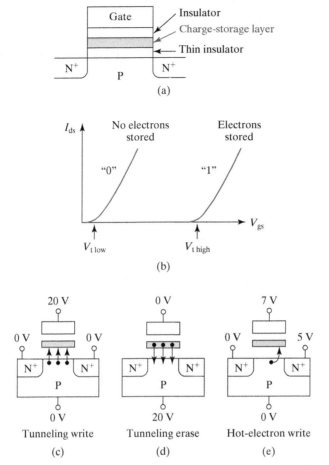

FIGURE 6–37 (a) A charge-storage NVM cell has a charge-storage layer in the gate dielectric stack; (b) V_t is modified by trapping electrons; (c) electron injection by tunneling; (d) electron removal by tunneling; and (e) electron injection by hot-electron injection.

the drain. A small fraction of electrons traveling through this region can gain enough energy to jump over the insulator energy barrier into the charge-storage layer. This method of writing is faster than tunneling but takes more current and power. The energetic electrons are called the **hot electrons** and this writing mechanism is called **hot carrier injection** or **HCI**.

● **Hot-Carrier-Injection Reliability of MOSFETs** ●

The high-quality gate oxide of the best CMOS transistors still contains charge traps. Even under normal CMOS circuit operation, a small number of hot carriers may be injected and trapped in the oxide. Over many years the trapped charge may change V_t and the I–V characteristics. Before releasing a CMOS technology for production, engineers must carry out accelerated tests of **hot-carrier reliability** and conduct careful analysis of the data to ensure that circuit performance will not change appreciably [19] over the product lifetime.

A limitation of the flash memory is that repeated write and erase cycling under high-electric field can break chemical bonds in the insulator and create leakage paths with diameters of a few atoms and at random locations. A single leakage path can discharge a floating gate and cause data loss. This sets an NVM endurance limit of less than 10^6 write/erase cycles. If the floating gate is replaced with a dielectric film containing many isolated electron traps or isolated nanocrystals of metal or semiconductor, one leakage path can only discharge a fraction of the stored electrons in the cell. Endurance may be improved. They are called **charge-trap NVM** and the **nano-crystal NVM**.

For several reasons, NVMs can store larger numbers of bits per centimeter square than DRAMs and SRAMs. First, the NVM cell (see Fig. 6–37a) is simple and small even in comparison with a DRAM cell. Second, it is possible to write and store more than two V_t values (see Fig. 6–37b) in a flash memory cell by controlling the number of stored electrons. Two V_ts can code one bit of data. Four V_ts can code two bits of data (00, 01, 10, and 11). This technique is called the **multilevel cell** technology. **NAND flash** memory gets even higher integration density (measured in bits/cm^2) by stringing dozens of flash memory cells in series. Imagine a long and narrow silicon strip area covered with the gate dielectric stack and flanked by shallow-trench-isolation oxide on its left and its right. Thirty-two parallel poly-Si gate lines, separated by minimum spacing, cross over the silicon strip. The spaces between the poly-Si gates are doped into N$^+$ regions by ion implantation. This creates 32 NFETs (NVM cells) connected in series. Doing so eliminates the need to make metal contacts to every cell because the N$^+$ source of one cell doubles as the drain of the next cell and so on. To illustrate the operation, let us consider only two cells in series. To read the data (the V_t) stored in the top cell, the gate voltage of the bottom cell is raised to higher than $V_{t\text{-high}}$. Similarly, reading the other cell as well as writing and erasing the cells can be performed by cleverly choosing the control voltages. It is call NAND flash because the string of transistors resembles a part of the NAND logic gate.

Charge storage is the most common but not the only mechanism for data storage. Figure 6–38a shows a **resistance-change NVM** or **RRAM** cell employing a programmable resistor. The resistor can be made of metal oxide or other inorganic or organic materials and programmed by electric field or current and sits over the transistor to save area. In one version, it is programmed by a heat pulse and made

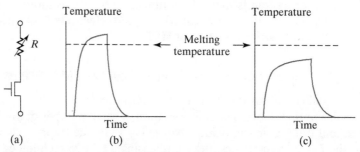

FIGURE 6–38 (a) Concept of a resistance-change memory such as a PCM. (b) Program the PCM into high-resistance state by rapid solidification, producing a highly resistive amorphous phase. (c) Program the PCM into low-resistance state by annealing, turning the amorphous material into a conductive crystalline phase.

of an alloy of Ge, Sb, and Te. If a current pulse is applied to heat the material above its melting temperature as shown in Fig. 6–38b, the subsequent rapid solidification creates an amorphous phase (see Fig. 3–15) of the material that is highly resistive. In Fig. 6–38c, another current pulse heats the resistor to a below-melting temperature, at which the amorphous material is annealed into a (poly)crystalline phase that has order-of-magnitude lower resistivity. The R_{low} and R_{high} states represent the "0" and "1." Reading is performed at a much lower current level with less heating. This memory is known as the **phase change memory** or **PCM**. PCM can be written and erased at SRAM speed and has much better endurance than the charge-storage memory.

In another technology, the resistor in Fig. 6–38a is an extremely thin filament of metal ions. The filament can be broken to create R_{high} by moving just a few metal ions with an electrical pulse. It can be restored with an electrical pulse of the opposite polarity. This memory concept is called **metal migration memory**.

6.17 • CHAPTER SUMMARY •

The basic CMOS technology is presented in Fig. 6–7. The CMOS inverter, as a representative digital gates, is analyzed in Section 6.7. The PFET pull-up device and the NFET pull-down device create a highly nonlinear **VTC**. This nonlinearity gives the inverter its ability to refresh digital signals and provides the much-needed noise margin in a noisy digital circuit. The inverter *propagation delay* is

$$\tau_d \approx \frac{CV_{dd}}{4}\left(\frac{1}{I_{\text{onN}}} + \frac{1}{I_{\text{onP}}}\right) \quad (6.7.1)$$

CMOS circuits' power consumption is

$$P = kCV_{dd}^2 f + V_{dd}I_{\text{off}} \quad (6.7.9)$$

where $k < 1$ accounts for the *activity of the circuit*. The first term is the *dynamic power* and the second, the *static power*.

It is highly desirable to have large $I_{\text{on}}s$ without using a large power supply voltage, V_{dd}. It is also desirable to reduce the total load capacitance, C (including the junction capacitance of the driver devices, the gate capacitance of the driven devices, and the interconnect capacitance). Both capacitance and cost reductions provide strong motivations for reducing the size of the transistors and therefore the size of the chip. In addition, speed has benefited from the relentless push for smaller L, thinner T_{ox}, and lower V_t; and power consumption has benefited greatly from the lowering of V_{dd}.

Electron and hole *surface mobilities*, μ_{ns} and μ_{ps}, are well-known functions of the average electric field in the inversion layer, which can be roughly expressed as $(V_{gs} + V_t)/6T_{\text{ox}}$. As this *effective vertical field* increases, the surface mobility decreases. At typical operating fields, surface mobilities are only fractions of the bulk mobilities. All of these are captured in Fig. 6–9.

GaAs has a high electron mobility but poor quality of dielectric–semiconductor interface. GaAs MESFET is an FET structure that does not require an MOS structure. Instead, the channel conductance is controlled by a Schottky contact gate.

HEMT uses an epitaxial high-band-gap semiconductor as an insulator in a MOSFET-like structure. The epitaxial interface is smooth. The electron mobility is very high and the device speed is very fast.

The V_t of a MOSFET can be easily measured from the I_{ds} vs. V_g plot. V_t increases with increasing body-to-source reverse bias, V_{sb}. This *body effect* is deleterious to circuit speed.

$$V_t(V_{sb}) = V_{t0} + \alpha V_{sb} \quad \text{for steep retograde body doping} \quad (6.4.6)$$

$$\alpha = 3T_{oxe}/W_{dmax} \quad (6.4.7)$$

$$V_t = V_{t0} + \frac{\sqrt{qN_a 2\varepsilon_s}}{C_{oxe}}(\sqrt{2\phi_B + V_{sb}} - \sqrt{2\phi_B}) \quad \text{for uniform body doping} \quad (6.4.8)$$

where V_{t0} is the threshold voltage in the absence of body bias.

The basic I_{ds} model is

$$I_{ds} = \frac{W}{L} C_{oxe} \mu_{ns} \left(V_{gs} - V_t - \frac{m}{2}V_{ds}\right) V_{ds} \quad (6.6.4)$$

$$m = 1 + 3T_{oxe}/W_{dmax} \approx 1.2 \quad (6.5.2)$$

The IV characteristics may be divided into the *linear region* and the *saturation region*. I_{ds} saturates at

$$V_{dsat} = \frac{V_{gs} - V_t}{m} \quad (6.6.5)$$

$$I_{dsat} = \frac{W}{2mL} C_{oxe} \mu_{ns} (V_{gs} - V_t)^2 \quad (6.6.6)$$

The *transconductance* of a MOSFET in the saturation region is

$$g_{msat} \equiv dI_{dsat}/dV_{gs} = \frac{W}{mL} C_{oxe} \mu_{ns}(V_{gs} - V_t) \quad (6.6.8), (6.6.9)$$

The above basic I_{ds} model can be significantly improved by considering velocity saturation. The result is

$$V_{dsat} = \left(\frac{m}{V_{gs} - V_t} + \frac{1}{\mathscr{E}_{sat}L}\right)^{-1} \quad (6.9.10)$$

$$\mathscr{E}_{sat} = 1.6 \times 10^7 \text{ cm/s} \div \mu_{ns} \quad \text{for electrons, and}$$

$$1.2 \times 10^7 \text{ cm/s} \div \mu_{ns} \quad \text{for holes.}$$

$$I_{dsat} = \frac{\text{long channel } I_{dsat} \text{ (Eq. (6.6.6))}}{1 + \frac{V_{gs} - V_t}{m\mathscr{E}_{sat}L}} \quad (6.9.11)$$

If $\mathscr{E}_{sat}L \gg V_{gs} - V_t$, Eqs. (6.9.10) and (6.9.11) reduce to the long-channel model, Eqs. (6.6.5) and (6.6.6). If $\mathscr{E}_{sat}L \ll V_{gs} - V_t$

$$V_{dsat} \approx \mathscr{E}_{sat}L < \text{long-channel } V_{dsat} \qquad (6.9.13)$$

$$I_{dsat} = W v_{sat} C_{oxe}(V_{gs} - V_t - \mathscr{E}_{sat}L) \qquad (6.9.14)$$

If L is reduced to tens of nanometers, *velocity overshoot* will raise $\mathscr{E}_{sat}$ and v_{sat} in the above equations somewhat. Eventually, the *carrier injection velocity* at the source will limit I_{dsat}. Interestingly, The present estimate of this limit is not significantly different from what Eq. (6.9.14) would predict.

The *intrinsic voltage gain* of a MOSFET is g_{msat}/g_{ds}. $g_{ds} = dI_{dsat}/dV_d$ is the output conductance. To achieve a small g_{ds} requires a large L and/or small T_{ox}, W_{dep}, and X_j (see Section 7.9).

For high-frequency applications, it is important to reduce the (poly-Si) gate electrode resistance by breaking a wide-W transistor into a large number of smaller-W transistors connected in parallel. Reducing the channel length can reduce the intrinsic input resistance as shown in Eq. (6.14.3).

MOSFET noise arises from the channel, gate, substrate thermal noise, and the flicker noise. While the thermal noise is a white noise, the flicker noise per bandwidth is proportional to $1/f$. The flicker ($1/f$) noise is reduced if the trap densities in the gate dielectric or the oxide–semiconductor interface are reduced.

A basic SRAM cell employs six MOSFETs. SRAM is commonly embedded in logic chips. DRAM cell consists of one transistor and one capacitor. Its size is very small. DRAM requires refreshing and a specialized technology, partly because of the complex capacitor structure that has a large surface area. The prevalent NVM is the flash memory. It uses even smaller Si area per bit than DRAM and can store data without power for many years. While floating-gate NAND is the dominant NVM, several new NVM concepts are under active investigation.

● PROBLEMS ●

● MOSFET AND MESFET V_t ●

6.1 An N-channel MOSFET with N$^+$-poly gate is fabricated on a 15 Ω cm P-type Si wafer with oxide fixed charge density $= q \times 8 \times 10^{10} \text{cm}^{-2}$, $W = 50$ μm, $L = 2$ μm, $T_{ox} = 5$ nm.

(a) Determine the flat-band voltage, V_{fb}.

(b) What is the threshold voltage, V_t?

(c) A circuit designer requested N-MOSFET with $V_t = 0.5$ V from a device engineer. It was not allowed to change the gate oxide thickness. If you are the device engineer, what can you do? Give specific answers including what type of equipment to use.

6.2 A GaAs MESFET has a 0.2 μm thick N-channel doped to $N_d = 10^{17} \text{cm}^{-3}$. Assume that ϕ_{Bn} of the Au–GaAs Schottky gate is 1 V. ε_s of GaAs is 13 times the vacuum dielectric constant. $V_d = V_s = 0$.

(a) What is W_{dep} at $V_g = 0$? (Hint: Please refer to Table 1–4 for the value of N_c of GaAs at room temperature.)

(b) At what V_g (including the sign) will W_{dep} be equal to the channel thickness? This is the cut-off gate voltage of the MESFET. The channel is shut off at this V_g.

(c) Can any gate voltage of the opposite sign to (b) be applied to the gate without producing expression gate current? What is its effect on W_{dep} and I_{ds}?

(d) What needs to be done to redesign this MESFET so that its channel is cut off at $V_g = 0$ and the channel only conducts current at V_g larger than a threshold voltage?

Discussion: The device in (d) is called an enhancement-mode transistor. The device of (b) is a depletion mode transistor.

6.3 An N-MOSFET and a P-MOSFET are fabricated with substrate doping concentration of $6 \times 10^{17} \text{cm}^{-3}$ (P-type substrate for N-MOSFET and N-type substrate for P-MOSFET). The gate oxide thickness is 5 nm. See Fig. 6–39.

(a) Find V_t of the N-MOSFET when N^+ poly-Si is used to fabricate the gate electrode.

(b) Find V_t of the P-MOSFET when N^+ poly-Si is used to fabricate the gate electrode.

(c) Find V_t of the P-MOSFET when P^+ poly-Si is used to fabricate the gate electrode.

(d) Assume that the only two voltages available on the chip are the supply voltage $V_{dd} = 2.5$ V and ground, 0 V. What voltages should be applied to each of the terminals (body, source, drain, and gate) to maximize the source-to-drain current of the N-MOSFET?

(e) Repeat part (d) for P-MOSFET.

(f) Which of the two transistors (b) or (c) is going to have a higher saturation current. Assuming that the supply voltage is 2.5 V, find the ratio of the saturation current of transistor (c) to that of transistor (b).

(g) What is the ratio of the saturation current of transistor (c) to that of transistor (a)? Use the mobility values from Fig. 6–9.

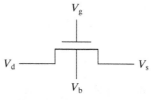

FIGURE 6–39

● **Basic MOSFET IV Characteristics** ●

6.4 CV and $I_d - V_g$ characteristics of a hypothetical MOSFET with channel length $L = 1$ μm are given in Fig. 6–40.

(a) Is the CV characteristic obtained at high frequency or low frequency? Or, is it impossible to determine? Explain.

(b) Is this a PMOSFET or an NMOSFET?

(c) Find the threshold voltage of this transistor.

(d) Determine the mobility of the carriers in the channel of the transistor.

(e) Plot $I_d - V_d$ curves at $V_g = 1$ V and $V_g = 2.5$ V.

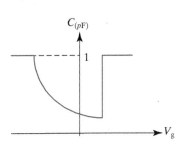
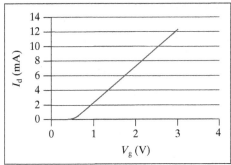

FIGURE 6–40

6.5 Figure 6–41 is the IV characteristics of an NMOSFET with $T_{ox} = 10$ nm, $W = 10$ μm, and $L = 2$ μm. (Assume $m = 1$ and do not consider velocity saturation.)

(a) Estimate V_t from the plot.

(b) Estimate μ_{ns} in the inversion layer.

(c) Add the I–V curve corresponding to $V_{gs} = 3$ V to the plot.

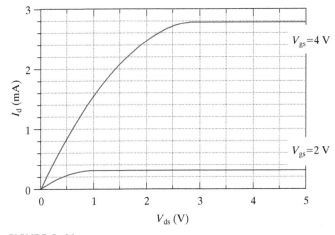

FIGURE 6–41

6.6 The MOSFET in the circuit shown in Fig. 6–42 is described by

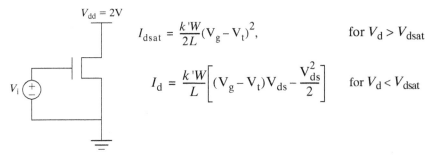

$$I_{dsat} = \frac{k'W}{2L}(V_g - V_t)^2, \quad \text{for } V_d > V_{dsat}$$

$$I_d = \frac{k'W}{L}\left[(V_g - V_t)V_{ds} - \frac{V_{ds}^2}{2}\right] \quad \text{for } V_d < V_{dsat}$$

FIGURE 6–42

where k' is $\mu_{ns}C_{ox}$ and obtained in practical case by measuring I_{dsat} at a given gate bias. When $k' = 25$ μA/V^2, $V_t = 0.5$ V, $W = 10$ μm, $L = 1$ μm, and V_i varied from 0 to 3 V,

(a) Make a careful plot of $\sqrt{I_{dsat}}$ as a function of V_i showing any break points on the curve.

(b) Make a plot of the MOSFET transconductance using a solid line.

(c) On the plot of part (b), use a dotted line to indicate a curve of the output conductance, dI_{ds}/dV_{ds}.

6.7 One I_{ds}–V_{ds} curve of an ideal MOSFET is shown in Fig. 6–43. Note that $I_{dsat} = 10^{-3}$ A and $V_{dsat} = 2$ V for the given characteristic. (You may or may not need the following information: $m = 1$, $L = 0.5$ μm, $W = 2.5$ μm, $T_{ox} = 10$ nm. Do not consider velocity saturation.)

(a) Given a V_t of 0.5 V, what is the gate voltage V_{gs} one must apply to obtain the I–V curve?

(b) What is the inversion-layer charge per unit area at the drain end of the channel when the MOSFET is biased at point (1) on the curve?

(c) Suppose the gate voltage is changed such that $V_{gs} - V_t = 3$ V. For the new condition, determine I_{ds} at $V_{ds} = 4$ V.

(d) If $V_d = V_s = V_b = 0$ V, sketch the general shape of the gate capacitance C vs. V_g to be expected from the MOSFET, when measured at 1 MHz. Do not calculate any capacitance but do label the $V_g = V_t$ point in the C–V curve.

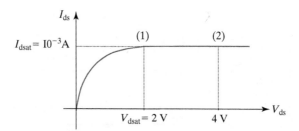

FIGURE 6–43

6.8 An ideal N-channel MOSFET has the following parameters: $W = 50$ μm, $L = 5$ μm, $T_{ox} = 0.05$ μm, $N_a = 10^{15}$ cm^{-3}, N$^+$ poly-Si gate, $\mu_{ns} = 800$ cm^2/V/s (and independent of V_g). Ignore the bulk charge effect and velocity saturation.
Determine:

(a) V_t

(b) I_{dsat} if $V_g = 2$ V

(c) dI_{ds}/dV_{ds} if $V_g = 2$ V and $V_d = 0$

(d) dI_{ds}/dV_{gs} if $V_g = 2$ V and $V_d = 2$ V.

● **Potential and Carrier Velocity in MOSFET Channel** ●

6.9 Derive the equation $V_c(x) = (V_g - V_t)[1\sqrt{1-x/L}\,]$ in Section 6.6. Assume $m = 1$. (Do not consider velocity saturation.)

6.10 This is an expanded version of Problem 6.9.

(a) Provide the derivation of Eq. (6.6.7).

(b) Find the expression for $Q_{inv}(x)$.

(c) Find the expression for $v(x) = \mu_n dV_{cs}/dx$.

(d) Show that $WQ_{inv}(x)v(x) = I_{dsat}$ expressed in Eq. (6.6.6).

(e) Make a qualitative sketch of $V_{cs}(x)$.

● **IV Characteristics of Novel MOSFET** ●

6.11 An NMOSFET has thinner T_{ox} at the center of the channel and thicker T_{ox} near the source and drain (Fig. 6–44). This could be approximately expressed as $T_{ox} = Ax^2 + B$. Assume that V_t is independent of x and $m = 1$. (Do not consider velocity saturation.)

(a) Derive an expression for I_d.

(b) Derive an expression for V_{dsat}.

(c) Does the assumption of nearly constant V_t suggest a large or small W_{dmax}?

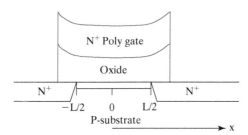

FIGURE 6–44

6.12 Suppose you have a MOSFET whose gate width changes as a function of distance along the channel as:

$$W(x) = W_0 + x$$

where $x = 0$ at the source and $x = L$ at the drain. Except for its gate width, assume that this MOSFET is like the typical MOSFET you studied in Chapter 6. (Do not consider velocity saturation.)

(a) Find an expression for I_d for this device. Ignore the bulk charge effect ($m = 1$).

(b) Derive an expression for I_{dsat} for this device.

● **CMOS** ●

6.13 MOS circuits perform best when the V_t of NMOS and the V_t of PMOS devices are about equal in magnitude and of opposite signs. To achieve this symmetry in V_t, PFET and NFET should have equal $N_{substrate}$, and symmetrical flat-band voltages, i.e., $V_{fb, PMOS} = -V_{fb, NMOS}$.

(a) Calculate the V_{fb} of NMOS and PMOS devices if the substrate doping is 5×10^{16} cm^{-3} and the gate is N$^+$. Are the flat-band voltages symmetrical?

(b) Assume the NMOS and PMOS devices now have a P$^+$ gate. Redo (a).

(c) If you were restricted to one type of gate material, what work function value would you choose to achieve the same $|V_t|$?

(d) If you were allowed to use both N$^+$ and P$^+$ gates, which type of gate would you use with your NMOS and which with your PMOS devices?

(Hint: Use the results of (a) and (b). Consider the need to achieve symmetrical V_t and the fact that large $|V_t|$ is bad for circuit speed.)

6.14

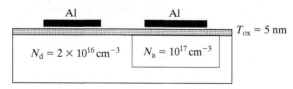

FIGURE 6–45

(a) Determine the flat-band voltage of the NMOS and PMOS capacitors fabricated on the same chip. (The devices are shown in Fig. 6–45.)

(b) Find the threshold voltages of these two devices.

(c) It is desirable to make the NMOS and PMOS threshold voltages equal in magnitude ($V_{tPMOS} = -V_{tNMOS}$). One can in principle implant dopant with ionized dopant charge $Q_{impl}(C/cm^2)$ at the Si–SiO$_2$ interface to change the threshold voltage. Assume that such an implant is applied to PMOS only. Find the value of Q_{impl} necessary to achieve $V_{tPMOS} = -V_{tNMOS}$.

6.15 Supply the missing steps between (a) Eqs. (6.7.1) and (6.7.3) and between (b) Eqs. (6.7.3) and (6.7.4).

6.16

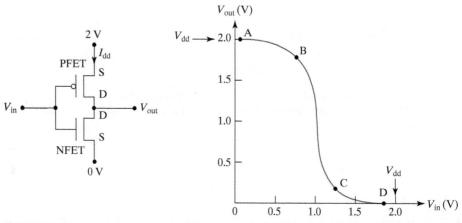

FIGURE 6–46

The voltage transfer curve of an inverter is given in Fig. 6–46. The threshold voltages of the NFET and PFET are +0.4 and –0.4 V, respectively. Determine the states of the two transistors (cut-off, linear, or saturation) at points A, B, C, and D, respectively. (Assume the output conductance of the transistor is very large.) Assume the two transistors have identical $\mu C_{ox}(W/L)$, $m = 1.333$.

	NFET operation mode	PFET operation mode
A		
B		
C		
D		

6.17 Consider the CMOS inverter shown in Fig. 6–47.

(a) Sketch the voltage transfer characteristics (VTC), i.e., a plot of V_0 vs. V_i for this inverter, if the threshold voltages of the N-channel and P-channel MOSFETs are V_{tn} and V_{tp}, respectively. Indicate the state (off, linear, or saturation) of each MOSFET as V_i is changed from 0 to V_{dd}. Indicate all points on the VTC where a MOSFET changes its conduction state.

(b) Calculate the voltage at all points indicated in part (a) if both MOSFETs $I_d - V_d$ are characterized by the square-law theory with the following parameters.
For the N-channel MOSFET: $\mu_{ns} C_{ox}(W/L) = 40$ mA $/V^{-2}$ and $V_{tn} = 1$ V.
For the P-channel MOSFET: $\mu_{ps} C_{ox}(W/L) = 35$ mA $/V^{-2}$ and $V_{tp} = 1$ V.
The supply voltage $V_{dd} = 5$ V.

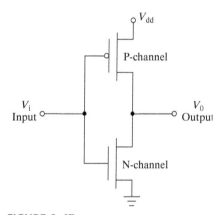

FIGURE 6–47

● Body Effect ●

6.18 P-channel MOSFET with heavily doped P-type poly-Si gate has a threshold voltage of −1.5 V with $V_{sb} = 0$ V. When a 5 V reverse bias is applied to the substrate, the threshold voltage changes to −2.3 V.

(a) What is the dopant concentration in the substrate if the oxide thickness is 100 nm?

(b) What is the threshold voltage if V_{sb} is −2.5 V?

● **Velocity-Saturation Effect** ●

6.19 The I_d–V_d characteristics of an NMOSFET are shown in Fig. 6–48.

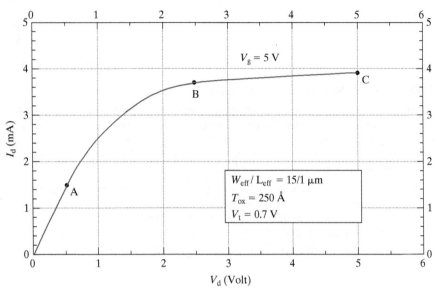

FIGURE 6–48

What are the velocities of the electrons near the drain and near the source at points A, B, and C? Use the following numbers in your calculations:

A: $I_{ds} = 1.5$ mA $\quad V_{ds} = 0.5$ V
B: $I_{ds} = 3.75$ mA $\quad V_{ds} = 2.5$ V
C: $I_{ds} = 4.0$ mA $\quad V_{ds} = 5.0$ V

(Hint: $I_d = W \times Q_{inv} \times v$.)

6.20 For an NMOS device with velocity saturation, indicate whether V_{dsat} and I_{dsat} increase, decrease, or remain unchanged when the following device parameters are reduced.

	T_{ox}	W	L	V_t	V_g
V_{dsat}					
I_{dsat}					

6.21 Verify Eq. (6.9.10) by equating Eqs. (6.9.3) and (6.9.9).

6.22 Verify Eq. (6.9.11) by substituting Eq. (6.9.10) into Eq. (6.9.3).

6.23 Consider a MOSFET with $\varepsilon_{sat} = 10^4$ V cm^{-1}. For $V_g - V_t = 2$ V, find V_{dsat} when:
  (a) $L = 0.1$ μm.
  (b) $L = 10$ μm.
  (c) For the device in part (a) with $I_{dsat} = 7$ mA, calculate the low field electron mobility if the gate capacitance is 10 fF.

6.24 An NMOSFET with a threshold voltage of 0.5 V and oxide thickness of 6 nm has a V_{dsat} of 0.75 V when biased at $V_{gs} = 2.5$ V. What is the channel length and saturation current per unit width of his device? (Hint: Use the universal mobility curve to find μ_s. From μ_s, you can determine

$$\varepsilon_{sat} = v_{sat}/2\mu_{ns} = \frac{8 \times 10^6 \text{cm/s}}{2\mu_{ns}}$$

6.25 The MOSFET drain current with velocity saturation is given as follows:

In linear region, I_{dlin}(velocity saturation) $= \dfrac{I_{dlin}(\text{no velocity saturation})}{1 + \dfrac{V_{ds}}{E_{sat}L}}$

In saturation region, I_{dsat}(velocity saturation) $= \dfrac{I_{dsat}(\text{no velocity saturation})}{1 + \dfrac{V_{gs} - V_t}{mE_{sat}L}}$

Consider a MOSFET with bulk charge factor $m = 1.2$, saturation velocity $v_{sat} = 8 \times 10^6$ cm/s and surface mobility $\mu_{ns} = 300$ cm^2/V $-$ s. Under what condition will velocity saturation cause the drain current to degrade by a factor of two? Assume $mV_{ds} > V_{gs} - V_t$

  (a) If $L = 100$ nm, $V_{gs} - V_{th} = $?
  (b) If $V_{gs} - V_t = 0.2$ V, $L = $?

● **Effective Channel Length** ●

6.26 The total resistance across the source and drain contacts of a MOSFET is $(R_s + R_d + R_{Channel})$, where R_s and R_d are source and drain series resistances, respectively, and $R_{Channel}$ is the channel resistance. Assume that V_{ds} is very small in this problem.

  (a) Write down an expression for $R_{Channel}$, which depends on V_{gs} (Hint: $R_{Channel} = V_{ds}/I_{ds}$).
  (b) Consider that $L_{effective} = L_{gate} - \Delta L$, where L_{gate} is the known gate length and ΔL accounts for source and drain diffusion, which extend beneath the gate. Define R_{sd} to be equal to $(R_s + R_d)$. Explain how you can find what R_{sd} and ΔL are. (Hint: Study the expression from part (a). Note that ΔL is the same for devices of all gate lengths. You may want to take measurements using a range of gate voltages and lengths.)
  (c) Prove that

$$I_{dsat} = \frac{I_{dsat0}}{1 + \dfrac{I_{dsat0}R_s}{(V_{gs} - V_t)}}$$

where I_{dsat0} is the saturation current in the absence of R_s.

  (d) Given $T_{ox} = 3$ nm, $W/L = 1/0.1$ μm, $V_{gs} = 1.5$ V and $V_t = 0.4$ V, what is I_{dsat} for $R_s = 0$, 100, and 1,000 Ω?

6.27 The drawn channel length of a transistor is in general different from the electrical channel length. We call the electrical channel length L_{eff}, while the drawn channel length is called L_{drawn}. Therefore the transistor I_d–V_d curves should be represented by

$$I_{\text{dsat}} = \frac{\mu_n C_{\text{ox}} W}{2 L_{\text{eff}}} (V_g - V_t)^2 \quad \text{for } V_d > V_{\text{dsat}}$$

$$I_{\text{dsat}} = \frac{\mu_n C_{\text{ox}} W}{L_{\text{eff}}} \left[(V_g - V_t) V_{\text{ds}} - \frac{V_{\text{ds}}^2}{2} \right] \quad \text{for } V_d < V_{\text{dsat}}$$

(a) How can you find the L_{eff}? (Hints: You may assume that several MOSFETs of different L_{drawn}, such as 1, 3, and 5 μm, are available. W and V_t are known.) Describe the procedure.

(b) Find the $\Delta L = L_{\text{drawn}} - L_{\text{eff}}$ and gate oxide thickness when you have three sets of I_{dsat} data measured at the same V_g as follows.

L (Drawn channel length)	1 (μm)	3 (μm)	5 (μm)
I_{dsat} (mA)	2.59	0.8	0.476

The channel width, W, is 10 μm, and the mobility, μ, is 300 cm²/V/s.

(c) If the I_{dsat} of the transistor is measured at V_{gs} = 2 V, what is the threshold voltage of the transistor with $L_{\text{drawn}} = 1$ μm?

● **Memory Devices** ●

6.28 (a) Qualitatively describe the differences among SRAM, DRAM, and flash memory in terms of closeness to the basic CMOS manufacturing technology, write speed, volatility, and cell size.

(b) What are the main applications of SRAM, DRAM, and flash memory? Why are each suitable for the applications. Hint: Consider your answers to (a).

6.29 (a) Match the six transistors in Fig. 6–34b to the transistors in Fig. 6–34a. (Hint: M_5 and M_6 usually have larger W than the transistors in the inverters.)

(b) Add the possible layout of the bit line and word line into Fig. 6–34b.

(c) Starting from the answer of (b), add another cell to the right and a third cell to the top of the original cell.

(d) Try to think of another way to arrange the six transistors (a new layout) that will pack them and the word line/bit lines into an even smaller cell area. (Hint: It is unlikely that you can pack them into a smaller area, although it should be fun spending 10 minutes trying. Furthermore, one cannot do this exercise fairly unless you know the detailed "design rules," which are the rules governing the size and spacing of all the features in a layout.)

● **REFERENCES** ●

1. Lilienfeld, J. E. "Method and Apparatus for Controlling Electronic Current." U.S. Patent 1,745,175 (1930).

2. Heil, O. "Improvements in or Relating to Electrical Amplifiers and Other Control Arrangements and Devices." British Patent 439,457 (1935).

3. Timp, G., et al. "The Ballistic Nano-transistor." *International Electron Devices Meeting Technical Digest* 1999, 55–58.

4. Chen, K., H. C. Wann, et al. "The Impact of Device Scaling and Power Supply Change on CMOS Gate Performance," *IEEE Electron Device Letters* 17 (5) (1996) 202–204.

5. Takagi, S., M. Iwase, and A. Toriumi. "On Universality of Inversion-Layer Mobility in N- and-P-channel MOSFETs." *International Electron Devices Meeting Technical Digest* (1988), 398–401.

6. Komohara, S., et al. MOSFET Carrier Mobility Model Based on the Density of States at the DC Centroid in the Quantized Inversion Layer. 5th International Conference on VLSI and CAD (1997), 398–401.

7. Chen, K., C. Hu, et al. "Optimizing Sub-Quarter Micron CMOS Circuit Speed Considering Interconnect Loading Effects." *IEEE Transactions on Electron Devices* 44 (9) (1997), 1556.

8. Assaderaghi, F., et al. "High-Field Transport of Inversion-Layer Electrons and Holes Including Velocity Overshoot." *IEEE Transactions on Electron Devices* 44 (4) (1997), 664–671.

9. Toh, K. Y., P. K. Ko, and R. G. Meyer. "An Engineering Model for Short-Channel MOS Devices." *IEEE Journal of Solid State Circuits* (1988), 23 (4), 950.

10. Hu, G. J., C. Chang, and Y. T. Chia. "Gate-Voltage Dependent Channel Length and Series Resistance of LDD MOSFETs." *IEEE Transactions on Electron Devices* 34 (1985), 2469.

11. Assad, F., et al. "Performance Limits of Silicon MOSFETs." *International Electron Devices Meeting Technical Digest* (1999) 547–550.

12. Hu, C. "A Compact Model for Rapidly Shrinking MOSFETs." *Electron Devices Meeting Technical Digest* (2001), 13.1.1–13.1.4.

13. Hu, C. "BSIM Model for Circuit Design Using Advanced Technologies." *VLSI Circuits Symposium Digest of Technical Papers* (2001), 5–10.

14. Hung, K. K., et al. "A Physics-Based MOSFET Noise Model for Circuit Simulations." *IEEE Transactions on Electron Devices Technical Digest* (1990), 1323–1333.

15. Fukuda, K., et al. "Random Telegraph Noise in Flash Memories—Model and Technology Scaling." *Electron Devices Meeting Technical Digest* (2007), 169–172.

16. Wu, S-Y., et al. "A 32 nm CMOS Low Power SoC Platform Technology for Foundry Applications with Functional High Density SRAM." *IEDM Technical Digest* (2007), 263–266.

17. Park, Y. K., et al. "Highly Manufacturable 90 nm DRAM Technology." *International Electron Devices Meeting Technical Digest* (2002), 819–822.

18. Brewer, J. E., and M. Gill, eds. *Nonvolatile Memory Technologies with Emphasis on Flash.* Hoboken, NJ: John Wiley & Sons, Inc., 2008.

19. Quader, K., et al. "Hot-Carrier Reliability Design Rules for Translating Device Degradation to CMOS Digital Circuit Degradation." *IEEE Transactions on Electron Devices* 41 (1994), 681–691.

● GENERAL REFERENCES ●

1. Taur, Y., and T. H. Ning. *Fundamentals of Modern VLSI Devices.* Cambridge, UK: Cambridge University Press, 1998.

2. Pierret, R. F. *Semiconductor Device Fundamentals.* Reading, MA: Addison-Wesley, 1996.

7

MOSFETs in ICs—Scaling, Leakage, and Other Topics

CHAPTER OBJECTIVES

How the MOSFET gate length might continue to be reduced is the subject of this chapter. One important topic is the off-state current or the leakage current of the MOSFETs. This topic complements the discourse on the on-state current conducted in the previous chapter. The major topics covered here are the subthreshold leakage and its impact on device size reduction, the trade-off between I_{on} and I_{off} and the effects on circuit design. Special emphasis is placed on the understanding of the opportunities for future MOSFET scaling including mobility enhancement, high-k dielectric and metal gate, SOI, multigate MOSFET, metal source/drain, etc. Device simulation and MOSFET compact model for circuit simulation are also introduced.

Metal–oxide–semiconductor (MOS) integrated circuits (ICs) have met the world's growing needs for electronic devices for computing, communication, entertainment, automotive, and other applications with continual improvements in cost, speed, and power consumption. These improvements in turn stimulated and enabled new applications and greatly improved the quality of life and productivity worldwide.

7.1 • TECHNOLOGY SCALING—FOR COST, SPEED, AND POWER CONSUMPTION •

In the forty-five years since 1965, the price of one bit of semiconductor memory has dropped 100 million times. The cost of a logic gate has undergone a similarly dramatic drop. This rapid price drop has stimulated new applications and semiconductor technology has improved the ways people carry out just about all human endeavors. The primary engine that powered the proliferation of electronics is "miniaturization." By making the transistors and the interconnects smaller, more circuits can be fabricated on each silicon wafer and therefore each circuit becomes cheaper. Miniaturization has also been instrumental to the improvements in speed and power consumption of ICs.

Gordon Moore made an empirical observation in 1965 that the number of devices on a chip doubles every 18 to 24 months or so. This **Moore's Law** is a succinct description of the rapid and persistent trend of miniaturization. Each time the minimum line width is reduced, we say that a new **technology generation** or **technology node** is introduced. Examples of technology generations are 0.18 µm, 0.13 µm, 90 nm, 65 nm, 45 nm … generations. The numbers refer to the minimum metal line width. Poly-Si gate length may be even smaller. At each new node, all the features in the circuit layout, such as the contact holes, are reduced in size to 70% of the previous node. This practice of periodic size reduction is called **scaling**. Historically, a new technology node is introduced every two to three years.

The main reward for introducing a new technology node is the reduction of circuit size by half. (70% of previous line width means ~50% reduction in area, i.e., $0.7 \times 0.7 = 0.49$.) Since nearly twice as many circuits can be fabricated on each wafer with each new technology node, the cost per circuit is reduced significantly. That drives down the cost of ICs.

● Initial Reactions to the Concept of the IC ●

Anecdote contributed by Dr. Jack Kilby, January 22, 1991

"Today the acceptance of the integrated circuit concept is universal. It was not always so. When the integrated circuit was first announced in 1959, several objections were raised. They were:

1) Performance of transistors might be degraded by the compromises necessary to include other components such as resistors and capacitors.

2) Circuits of this type were not producible. The overall yield would be too low.

3) Designs would be expensive and difficult to change.

Debate of the issues provided the entertainment at technical meetings for the next five or six years."

In 1959, Jack Kilby of Texas Instruments and Robert Noyce of Fairchild Semiconductor independently invented technologies of interconnecting multiple devices on a single semiconductor chip to form an electronic circuit. Following a 10 year legal battle, both companies' patents were upheld and Noyce and Kilby were recognized as the co-inventors of the IC. Dr. Kilby received a Nobel Prize in Physics in 2000 for inventing the integrated circuit. Dr. Noyce, who is credited with the layer-by-layer planar approach of fabricating ICs, had died in 1990.

Besides the line width, some other parameters are also reduced with scaling such as the MOSFET gate oxide thickness and the power supply voltage. The reductions are chosen such that the transistor current density (I_{on}/W) increases with each new node. Also, the smaller transistors and shorter interconnects lead to smaller capacitances. Together, these changes cause the circuit delays to drop (Eq. 6.7.1). Historically, IC speed has increased roughly 30% at each new technology node. Higher speed enables new applications such as wide-band data transmission via RF mobile phones.

Scaling does another good thing. Eq. (6.7.6) shows that reducing capacitance and especially the power supply voltage is effective in lowering the power consumption. Thanks to the reduction in C and V_{dd}, power consumption per chip has increased only modestly per node in spite of the rise in switching frequency, f and the doubling of transistor count per chip at each technology node. If there had been no scaling, doing the job of a single PC microprocessor chip (operating a billion transistors at 2 GHz) using 1970 technology would require the power output of an electrical power generation plant.

In summary, scaling improves cost, speed, and power consumption per function with every new technology generation. All of these attributes have been improved by 10 to 100 million times in four decades—an engineering achievement unmatched in human history! When it comes to ICs, small is beautiful.

7.1.1 Innovations Enable Scaling

Semiconductor researchers around the world have been meeting several times a year for the purpose of generating consensus on the transistor and circuit performance that will be required to fulfill the projected market needs in the future. Their annually updated document: **International Technology Roadmap for Semiconductors (ITRS)** only sets out the goals and points out the challenging problems but does not provide the solutions [1]. It tells the vendors of manufacturing tools and materials and the research community the expected roadblocks. The list of show stoppers is always long and formidable but innovative engineers working together and separately have always risen to the challenge and done the seemingly impossible.

Table 7–1 is a compilation of some history and some ITRS technology projection. High-performance (HP) stands for high-performance computer processor technology. LSTP stands for the technology for low standby-power products such as mobile phones. The physical gate length, L_g, is actually smaller than the technology node. Take the 90 nm node, for example; although lithography technology can only print 90 nm photoresist lines, engineers transfer the pattern into oxide lines and then isotropically etch (see Section 3.4) the oxide in a dry isotropic-etching tool to reduce the width (and the thickness) of the oxide lines. Using the narrowed oxide lines as the new etch mask, they produce the gate patterns by etching. Innumerable innovations by engineers at each node have enabled the scaling of the IC technology.

7.1.2 Strained Silicon and Other Innovations

I_{on} in Table 7–1 rises rapidly. This is only possible because of the **strained silicon** technology introduced around the 90 nm node [2]. The electron and hole mobility can be raised (or lowered) by carefully engineered mechanical strains. The strain changes the lattice constant of the silicon crystal and therefore the E–k relationship through the Schrodinger's wave equation. The E–k relationship, in turn, determines the effective mass and the mobility.

For example, the hole surface mobility of a PFET can be raised when the channel is compressively stressed. The compressive strain may be created in several ways. We illustrate one way in Fig. 7–1. After the gate is defined, trenches are etched into the silicon adjacent to the gate. The trenches are refilled by

TABLE 7–1 • Scaling from 90 nm to 22 nm and innovations that enable the scaling.

Year of Shipment	2003	2005	2007	2010	2013
Technology Node (nm)	90	65	45	32	22
L_g (nm) (HP/LSTP)	37/65	26/45	22/37	16/25	13/20
EOT_e (nm) (HP/LSTP)	1.9/2.8	1.8/2.5	1.2/1.9	0.9/1.6	0.9/1.4
V_{DD} (V) (HP/LSTP)	1.2/1.2	1.1/1.1	1.0/1.1	1.0/1.0	0.9/0.9
I_{on}, HP (μA/μm)	1100	1210	1500	1820	2200
I_{off}, HP (μA/μm)	0.15	0.34	0.61	0.84	0.37
I_{on}, LSTP (μA/μm)	440	465	540	540	540
I_{off}, LSTP (μA/μm)	1E-5	1E-5	3E-5	3E-5	2E-5
Innovations		⟶ Strained Silicon			
			⟶ High-k/metal-gate		
				⟶ Wet lithography	
					⟶ New Structure

HP: High-Performance technology. LSTP: Low Standby Power technology for portable applications. EOT_e: Equivalent electrical Oxide Thickness, i.e., equivalent T_{oxe}. I_{on}: NFET I_{on}.

epitaxial growth (see Section 3.7.3) of SiGe—typically a 20% Ge and 80% Si mixture. Because Ge atoms are larger than Si atoms and in epitaxial growth the number of atoms in the trench is equal to the original number of Si atoms, it is as if a large hand is forced into a small glove. A force is created that pushes on the channel (as shown in Fig. 7–1) region and raises the hole mobility. It is also attractive to incorporate a thin film of Ge material in the channel itself because Ge has higher carrier mobilities than Si [3].

In Table 7–1, EOT_e or the **electrical equivalent oxide thickness** is the total thickness of the gate dielectric, poly-gate depletion (if any), and the inversion layer expressed in equivalent SiO_2 thickness. It is improved (reduced) at the 45 nm node by a larger factor over the previous node. The enabling innovations are metal gate and high-k dielectric, which will be presented in Section 7.4.

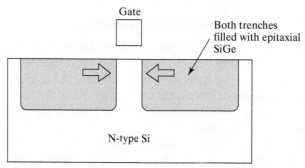

FIGURE 7–1 Example of strained-silicon MOSFET. Hole mobility can be raised with a compressive mechanical strain illustrated with the arrows pushing on the channel region.

At the 32 nm node, wet lithography (see Section 3.3.1) is used to print the fine patterns. At the 22 nm node, new transistor structures may be used to reverse the trend of increasing I_{off}, which is the source of a serious power consumption issue. Some new structures are presented in Section 7.8.

7.2 • SUBTHRESHOLD CURRENT—"OFF" IS NOT TOTALLY "OFF" •

Circuit speed improves with increasing I_{on}; therefore, it would be desirable to use a small V_t. Can we set V_t at an arbitrarily small value, say 10 mV? The answer is no.

At $V_{gs} < V_t$, an N-channel MOSFET is in the off state. However, a leakage current can still flow between the drain and the source. The MOSFET current observed at $V_{gs} < V_t$ is called the **subthreshold current**. This is the main contributor to the MOSFET **off-state current**, I_{off}. I_{off} is the I_d measured at $V_{gs} = 0$ and $V_{ds} = V_{dd}$. It is important to keep I_{off} very small in order to minimize the static power that a circuit consumes when it is in the standby mode. For example, if I_{off} is a modest 100 nA per transistor, a cell-phone chip containing one hundred million transistors would consume 10 A even in standby. The battery would be drained in minutes without receiving or transmitting any calls. A desktop PC processor would dissipate more power because it contains more transistors and face expensive problems of cooling the chip and the system.

Figure 7–2a shows a subthreshold current plot. It is plotted in a semi-log I_{ds} vs. V_{gs} graph. When V_{gs} is below V_t, I_{ds} is clearly a straight line, i.e., an exponential function of V_{gs}.

Figure 7–2b–d explains the subthreshold current. At V_{gs} below V_t, the inversion electron concentration (n_s) is small but nonetheless can allow a small leakage current to flow between the source and the drain. In Fig. 7–2b, a larger V_{gs} would pull the E_c at the surface closer to E_F, causing n_s and I_{ds} to rise. From the equivalent circuit in Fig. 7–2c, one can observe that

$$\frac{d\varphi_s}{dV_{gs}} = \frac{C_{oxe}}{C_{oxe} + C_{dep}} \equiv \frac{1}{\eta} \tag{7.2.1}$$

$$\eta = 1 + \frac{C_{dep}}{C_{oxe}} \tag{7.2.2}$$

Integrating Eq. (7.2.1) yields

$$\varphi_s = \text{constant} + V_g/\eta \tag{7.2.3}$$

I_{ds} is proportional to n_s, therefore

$$I_{ds} \propto n_s \propto e^{q\varphi_s/kT} \propto e^{q(\text{constant}+V_g/\eta)/kT} \propto e^{qV_g/\eta kT} \tag{7.2.4}$$

A practical and common definition of V_t is the V_{gs} at which $I_{ds} = 100 \text{ nA} \times W/L$ as shown in Fig. 6–12. (Some companies may use 200 nA instead of 100 nA.). Equation (7.2.4) may be rewritten as

$$I_{ds}(nA) = 100 \cdot \frac{W}{L} \cdot e^{q(V_{gs}-V_t)/\eta kT} \tag{7.2.5}$$

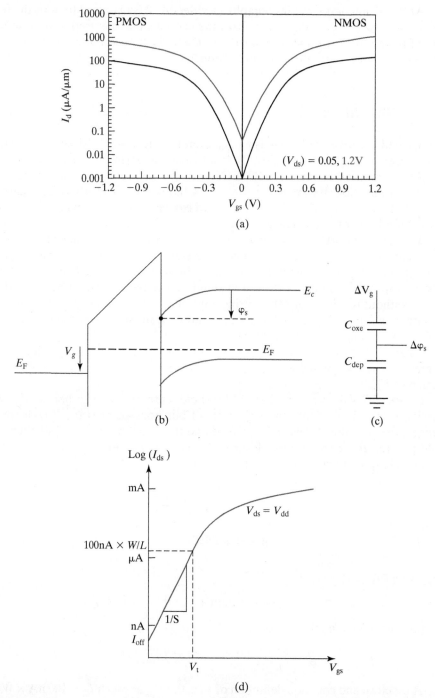

FIGURE 7–2 The current that flows at $V_{gs} < V_t$ is called the subthreshold current. $V_t \sim 0.2$ V. The lower/upper curves are for V_{ds} = 50 mV/1.2 V. After Ref. [2]. (b) When V_g is increased, E_c at the surface is pulled closer to E_F, causing n_s and I_{ds} to rise; (c) equivalent capacitance network; (d) subthreshold I-V with V_t and I_{off}. Swing, S, is the inverse of the slope in the subthreshold region.

Clearly, Eq. (7.2.5) agrees with the definition of V_t and Eq. (7.2.4). The simplicity of Eq. (7.2.5) is another reason for favoring the new V_t definition. At room temperature, the function $\exp(qV_{gs}/kT)$ changes by 10 for every 60 mV change in V_{gs}, therefore $\exp(qV_{gs}/\eta kT)$ changes by 10 for every $\eta \times 60$ mV. For example, if $\eta = 1.5$, Eq. (7.2.5) states that I_{ds} drops by ten times for every 90 mV of decrease in V_{gs} below V_t at room temperature. $\eta \times 60$ mV is called the **subthreshold swing** and represented by the symbol, S.

$$S(\text{mV/decade}) = \eta \cdot 60 \text{ mV} \cdot \frac{T}{300\text{K}} \quad (7.2.6)$$

$$I_{ds}(\text{nA}) = 100 \cdot \frac{W}{L} \cdot e^{q(V_{gs}-V_t)/\eta kT} = 100 \cdot \frac{W}{L} \cdot 10^{(V_{gs}-V_t)/S} \quad (7.2.7)$$

$$I_{off}(\text{nA}) = 100 \cdot \frac{W}{L} \cdot e^{-qV_t/\eta kT} = 100 \cdot \frac{W}{L} \cdot 10^{-V_t/S} \quad (7.2.8)$$

For given W and L, there are two ways to minimize I_{off} illustrated in Fig. 7–2 (d). The first is to choose a large V_t. This is not desirable because a large V_t reduces I_{on} and therefore degrades the circuit speed (see Eq. (6.7.1)). The preferable way is to reduce the subthreshold swing. S can be reduced by reducing η. That can be done by increasing C_{oxe} (see Eq. 7.2.2), i.e., using a thinner T_{ox}, and by decreasing C_{dep}, i.e., increasing W_{dep}.[1] An additional way to reduce S, and therefore to reduce I_{off}, is to operate the transistors at significantly lower than the room temperature. This last approach is valid in principle but rarely used because cooling adds considerable cost.

Besides the subthreshold leakage, there is another leakage current component that has becomes significant. That is the tunnel leakage through very thin gate oxide that will be presented in Section 7.4. The drain to the body junction leakage is the third leakage component.

● The Effect of Interface States ●

The subthreshold swing is degraded when interface states are present (see Section 5.7). Figure 7–3 shows that when φ_S changes, some of the interface traps move from above the Fermi level to below it or vice versa. As a result, these interface traps change from being empty to being occupied by electrons. This change of charge in response to change of voltage (φ_S) has the effect of a capacitor. The effect of the interface states is to add a parallel capacitor to C_{dep} in Fig. 7–2c. The subthreshold swing is poor unless the semiconductor-dielectric interface has low density of interface states such as carefully prepared Si-SiO$_2$ interface. The subthreshold swing is often degraded after a MOSFET is electrically stressed (see sidebar in Section 5.7) and new interface states are generated.

---

[1] According to Eq. 6.5.2 and Eq. 7.2.2, η should be equal to m. In reality, η is larger than m because C_{oxe} is smaller at low V_{gs} (subthreshold condition) than in inversion due to a larger T_{inv} as shown in Fig. 5–25. Nonetheless, η and m are closely related.

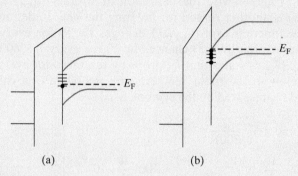

FIGURE 7–3 (a) Most of the interface states are empty because they are above E_F. (b) At another V_g, most of the interface states are filled with electrons. As a result, the interface charge density changes with V_g.

EXAMPLE 7–1 Subthreshold Leakage Current

An N-channel transistor has $V_t = 0.34$ V and $S = 85$ mV, $W = 10$ μm and $L = 50$ nm. (a) Estimate I_{off}. (b) Estimate I_{ds} at $V_g = 0.17$ V.

SOLUTION:

a. Use Eq. (7.2.6).

$$I_{off}(nA) = 100 \cdot \frac{W}{L} \cdot 10^{-V_t/S} = 100 \cdot \frac{10}{0.05} \cdot 10^{-0.34/0.0085} = 2 \text{ nA}$$

b. Use Eq. (7.2.7).

$$I_{ds} = 100 \cdot \frac{W}{L} \cdot 10^{(V_g - V_t)/S} = 100 \cdot \frac{10}{0.05} \cdot 10^{(0.17 - 0.34)/0.085} = 200 \text{ nA}$$

7.3 • V_t ROLL-OFF—SHORT-CHANNEL MOSFETs LEAK MORE •

The previous section pointed out that V_t must not be set too low; otherwise, I_{off} would be too large. The present section extends that analysis to show that the channel length (L) must not be too short. The reason is this: V_t drops with decreasing L as illustrated in Fig. 7–4. When V_t drops too much, I_{off} becomes too large and that channel length is not acceptable.

• Gate Length (L_g) vs. Electrical Channel Length (L) •

Gate length is the physical length of the gate and can be accurately measured with a scanning electron microscope (SEM). It is carefully controlled in the fabrication plant. The channel length, in comparison, cannot be determined very accurately and easily due to the lateral diffusion of the source and drain junctions. L tracks L_g but the difference between the two just cannot be quantified precisely in spite of efforts such as described in Section 6.11. As a result, L_g is widely used in lieu of L in data presentations as is done in Fig. 7–4. L is still a useful concept and is used in theoretical equations even though L cannot be measured precisely for small transistors.

7.3 • V_t Roll-Off—Short-Channel MOSFETs Leak More

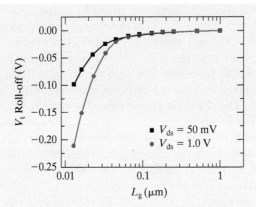

FIGURE 7–4 $|V_t|$ decreases at very small L_g. This phenomenon is called V_t roll-off. It determines the minimum acceptable L_g because I_{off} is too large when V_t becomes too low or too sensitive to L_g.

At a certain L_g, V_t becomes so low that I_{off} becomes unacceptable [see Eq. (7.2.8)]. Doping the bodies of the short-channel devices more heavily than the long-channel devices would raise their V_t. Still, at a certain L_g, V_t is so sensitive to the manufacturing caused variation in L that the worst case I_{off} becomes unacceptable. Device development engineers must design the device such that the V_t roll-off does not prevent the use of the targeted minimum L_g, e.g., those listed in the second row of Table 7–1.

Why does V_t decrease with decreasing L? Figure 7–5 illustrates a model for understating this effect. Figure 7–5a shows the energy band diagram along the semiconductor–insulator interface of a long channel device at $V_{gs} = 0$. Figure 7–5b shows the case at $V_{gs} = V_t$. In the case of (b), E_c in the channel is pulled lower than

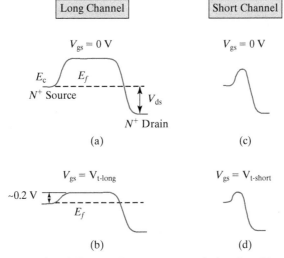

FIGURE 7–5 a–d: Energy band diagram from source to drain when $V_{gs} = 0$ V and $V_{gs} = V_t$. a–b long channel; c–d short channel.

in case (a) and therefore is closer to the E_c in the source. When the channel E_c is only ~0.2 eV higher than the E_c in the source (which is also ~E_{Fn}), n_s in the channel reaches ~10^{17} cm^3 and inversion threshold condition (I_{ds} = 100nA × W/L) is reached. We may say that a 0.2 eV potential barrier is low enough to allow the electrons in the N$^+$ source to flow into the channel to form the inversion layer. The following analogy may be helpful for understanding the concept of the energy barrier height. The source is a reservoir of water; the potential barrier is a dam; and V_{gs} controls the height of the dam. When V_{gs} is high enough, the dam is sufficiently low for the water to flow into the channel and the drain. That defines V_t.

Figure 7–5c shows the case of a short-channel device at $V_{gs} = 0$. If the channel is short enough, E_c will not be able to reach the same peak value as in Fig. 7–5a. As a result, a smaller V_{gs} is needed in Fig. 7–5d than in Fig. 7–5b to pull the barrier down to 0.2 eV. In other words, V_t is lower in the short channel device than the long channel device. This explains the V_t roll-off shown in Fig. 7–4.

We can understand V_t roll-off from another approach. Figure 7–6 shows a capacitor between the gate and the channel. It also shows a second capacitor, C_d, between the drain and the channel terminating at around the middle of the channel, where E_c peaks in Fig. 7–5d. As the channel length is reduced, the drain to source and the drain to "channel" distances are reduced; therefore, C_d increases. Do not be concerned with the exact definition or value of C_d. Instead, focus on the concept that C_d represents the capacitive coupling between the drain and the channel barrier point.

From this two-capacitor equivalent circuit, it is evident that the drain voltage has a similar effect on the channel potential as the gate voltage. V_{gs} and V_{ds}, together, determine the channel potential barrier height shown in Fig. 7–5. When V_{ds} is present, less V_{gs} is needed to pull the barrier down to 0.2 eV; therefore, V_t is lower by definition. This understanding gives us a simple equation for V_t roll-off,

$$V_t = V_{t\text{-long}} - V_{ds} \cdot \frac{C_d}{C_{oxe}} \qquad (7.3.1)$$

where $V_{t\text{-long}}$ is the threshold voltage of a long-channel transistor, for which $C_d = 0$. More accurately, V_{ds} should be supplemented with a constant that represents the combined effects of the 0.2 V built-in potentials between the N$^-$ inversion layer and both the N$^+$ drain and source at the threshold condition [4].

$$V_t = V_{t\text{-long}} - (V_{ds} + 0.4\text{ V}) \cdot \frac{C_d}{C_{oxe}} \qquad (7.3.2)$$

Using Fig. 7–6, one can intuitively see that as L decreases, C_d increases. Recall that the capacitance increases when the two electrodes are closer to each other. That intuition is correct for the two-dimensional geometry of Fig. 7–6, too. However, solution of the Poisson's equation (Section 4.1.3) indicates that C_d is an exponential function of L in this two-dimensional structure [5]. Therefore,

$$V_t = V_{t\text{-long}} - (V_{ds} + 0.4\text{ V}) \cdot e^{-L/l_d} \qquad (7.3.3)$$

where

$$l_d \propto \sqrt[3]{T_{oxe} W_{dep} X_j} \qquad (7.3.4)$$

X_j is the drain junction depth. Equation (7.3.3) provides a semi-quantitative model of the roll-off of V_t as a function of L and V_{ds}. It can serve as a guide for designing

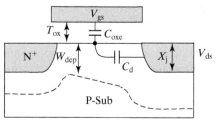

FIGURE 7–6 Schematic two-capacitor network in MOSFET. C_d models the electrostatic coupling between the channel and the drain. As the channel length is reduced, drain to "channel" distance is reduced; therefore, C_d increases.

small MOSFET and understanding new transistor structures. At a very large L, V_t is equal to $V_{t\text{-long}}$ as expected. The roll-off is an exponential function of L. The roll-off is also larger at larger V_{ds}, which can be as large as V_{dd}. The acceptable I_{off} determines the acceptable V_t through Eq. (7.2.8). This in turn determines the acceptable minimum L through Eq. (7.3.3). *The acceptable minimum L is several times of l_d.* The concept that the drain can lower the source–channel barrier and reduce V_t is called **drain-induced barrier lowering** or **DIBL**. l_d may be called the **DIBL characteristic length**. In order to support the reduction of L at each new technology node, l_d must be reduced in proportion to L. This means that we must reduce T_{ox}, W_{dep}, and/or X_j. In reality, all three are reduced at each node to achieve the desired reduction in l_d. Reducing T_{ox} increases the gate control or C_{oxe}. Reducing X_j decreases C_d by reducing the size of the drain electrode. Reducing W_{dep} also reduces C_d by introducing a ground plane (the neutral region of the substrate or the bottom of the depletion region) that tends to electrostatically shield the channel from the drain.

The basic message in Eq. (7.3.4) is that *the vertical dimensions in a MOSFET (T_{ox}, W_{dep}, X_j) must be reduced in order to support the reduction of the gate length.* As an example, Fig. 7–7 shows that the oxide thickness has been scaled roughly in proportion to the line width (gate length).

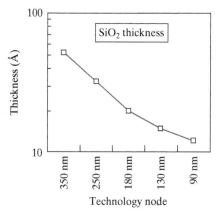

FIGURE 7–7 In the past, the gate oxide thickness has been scaled roughly in proportion to the line width.

7.4 • REDUCING GATE-INSULATOR ELECTRICAL THICKNESS AND TUNNELING LEAKAGE •

SiO_2 has been the preferred gate insulator since silicon MOSFET's beginning. The oxide thickness has been reduced over the years from 300 nm for the 10 µm technology to only 1.2 nm for the 65 nm technology. There are two reasons for the relentless drive to reduce the oxide thickness. First, a thinner oxide, i.e., a larger C_{ox} raises I_{on} and a large I_{on} raises the circuit speed [see Eq. (6.7.1)]. The second reason is to control V_t roll-off (and therefore the subthreshold leakage) in the presence of a shrinking L according to Eqs. (7.3.3) and (7.3.4). One must not underestimate the importance of the second reason. Figure 7–7 shows that the oxide thickness has been scaled roughly in proportion to the line width.

Thinner oxide is desirable. What, then, prevents engineers from using arbitrarily thin gate oxide films? Manufacturing thin oxide is not easy, but as Fig. 6–5 illustrates, it is possible to grow very thin and uniform gate oxide films with high yield. Oxide breakdown is another limiting factor. If the oxide is too thin, the electric field in the oxide can be so high as to cause destructive breakdown. (See the sidebar, "SiO_2 Breakdown Electric Field.") Yet another limiting factor is that long-term operation at high field, especially at elevated chip operating temperatures, breaks the weaker chemical bonds at the Si–SiO_2 interface thus creating oxide charge and V_t shift (see Section 5.7). V_t shifts cause circuit behaviors to change and raise reliability concerns.

For SiO_2 films thinner than 1.5 nm, tunneling leakage current becomes the most serious limiting factor. Figure 7–8a illustrates the phenomenon of gate leakage by tunneling (see Section 4.20). Electrons arrive at the gate oxide barrier at thermal velocity and emerge on the side of the gate with a probability given by Eq. (4.20.1). This is the cause of the gate leakage current. Figure 7–8b shows that the exponential rise of the SiO_2 leakage current with decreasing thickness agrees with the tunneling model prediction [6]. At 1.2 nm, SiO_2 leaks 10^3 A/cm². If an IC chip contains

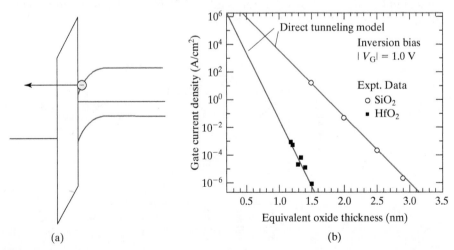

FIGURE 7–8 (a) Energy band diagram in inversion showing electron tunneling path through the gate oxide; (b) 1.2 nm SiO_2 conducts 10^3 A/cm² of leakage current. High-k dielectric such as HfO_2 allows several orders lower leakage current to pass. (After [6]. © 2003 IEEE.)

1 mm² total area of this thin dielectric, the chip oxide leakage current would be 10 A. This large leakage would drain the battery of a cell phone in minutes. The leakage current can be reduced by about 10 × with the addition of nitrogen into SiO_2.

Engineers have developed high-k dielectric technology to replace SiO_2. For example, HfO_2 has a relative dielectric constant (k) of ~24, six times larger than that of SiO_2. A 6 nm thick HfO_2 film is equivalent to 1 nm thick SiO_2 in the sense that both films produce the same C_{ox}. We say that this HfO_2 film has an **equivalent oxide thickness** or **EOT** of 1 nm. However, the HfO_2 film presents a much thicker (albeit lower) tunneling barrier to the electrons and holes. The consequence is that the leakage current through HfO_2 is several orders of magnitude smaller than that through SiO_2 as shown in Fig. 7–8b. Other attractive high-k dielectrics include ZrO_2 and Al_2O_3. The difficulties of adopting high-k dielectrics in IC manufacturing are chemical reactions between them and the silicon substrate, lower surface mobility than the Si–SiO_2 system, and more oxide charge. These problems are minimized by inserting a thin SiO_2 interfacial layer between the silicon substrate and the high-k dielectric.

Note that Eq. (7.3.4) contains the electrical oxide thickness, T_{oxe}, defined in Eq. (5.9.2). Besides T_{ox} or EOT, the poly-Si gate depletion layer thickness also needs to be minimized. Metal is a much better gate material in this respect. NFET and PFET gates may require two different metals (with metal work functions close to those of N^+ and P^+ poly-Si) in order to achieve the optimal V_ts [7].

In addition, T_{inv} is also part of T_{oxe} and needs to be minimized. The material parameters that determine T_{inv} is the electron or hole effective mass. A larger effective mass leads to a thinner T_{inv}. Unfortunately, a larger effective mass leads to a lower mobility, too (see Eq. (2.2.4)). Fortunately, the effective mass is a function of the spatial direction in a crystal. The effective mass in the direction normal to the oxide interface determines T_{inv}, while the effective mass in the direction of the current flow determines the surface mobility. It may be possible to build a transistor with a wafer orientation (see Fig. 1–2) that offers larger m_n and m_p normal to the oxide interface but smaller m_n and m_p in the direction of the current flow.

● **SiO_2 Breakdown Electric Field** ●

What is the breakdown field of SiO_2? There is no one simple answer because the breakdown field is a function of the test time. If a one second (1s) voltage pulse is applied to a 10 nm SiO_2 film, 15 V is needed to breakdown the film for a breakdown field of 15 MV/cm. The breakdown field is significantly lower if the same oxide is tested for one hour. The field is lower still if it is tested for a month. This phenomenon is called **time-dependent dielectric breakdown**. Most IC applications require a device lifetime of several years to over 10 years. Clearly, manufacturers cannot afford the time to actually measure the 10 year breakdown voltage for new oxide technologies. Instead, engineers predict the 10 year breakdown voltage based on hours- to month-long tests in combination with theoretical models of the physics of oxide breakdown. A wide range of breakdown field was predicted for SiO_2 by different models. In retrospect, the most optimistic of the predictions, 7 MV/cm for a 10 year operation, was basically right.

This breakdown model considers a sequence of events[8]. Carrier tunneling through the oxide at high field breaks up the weaker Si–O bonds in SiO_2, thus creating oxide defects. This process progresses more rapidly at those spots in the oxide sample where the densities of the weaker bonds happen to be statistically high. When the generated defects reach a critical density at any one spot, breakdown occurs. In a longer-term stress test, the breakdown field is lower because a lower rate of defect generation is sufficient to build up the critical defect density over the longer test time. A fortuitous fact is that the breakdown field increases in very thin oxide. The charge carriers gain less energy while traversing through a very thin oxide than a thick oxide film at a given electric field and are less able to create oxide defects.

7.5 • HOW TO REDUCE W_{dep} •

Equation (7.3.4) suggests that a small W_{dep} helps to control V_t roll-off and enable the use of a shorter L. W_{dep} can be reduced by increasing the substrate doping concentration, N_{sub}, because W_{dep} is proportional to $1/\sqrt{N_{sub}}$. However, Eq. (5.4.3), repeated here,

$$V_t = V_{fb} + \phi_{st} + \frac{\sqrt{qN_{sub}2\varepsilon_s\phi_{st}}}{C_{ox}} \quad (7.5.1)$$

dictates that, if V_t is not to increase, N_{sub} must not be increased unless C_{ox} is increased, i.e., T_{ox} is reduced. Equation (7.5.1) can be rewritten as Eq (7.5.2) by eliminating N_{sub} with Eq. (5.5.1). Clearly, W_{dep} can only be reduced in proportion to T_{ox}.

$$V_t = V_{fb} + \phi_{st}\left(1 + \frac{2\varepsilon_s T_{ox}}{\varepsilon_{ox} W_{dep}}\right) \quad (7.5.2)$$

This fact establishes T_{ox} as the main enabler of L reduction according to Eq. (7.3.4).

There is another way of reducing W_{dep}—adopt the steep retrograde doping profile illustrated in Fig. 6–12. In that case, W_{dep} is determined by the thickness of the lightly doped surface layer. It can be shown (see sidebar) that V_t of a MOSFET with ideal retrograde doping is

$$V_t = V_{fb} + \phi_{st}\left(1 + \frac{\varepsilon_s T_{ox}}{\varepsilon_{ox} T_{rg}}\right) \quad (7.5.3)$$

where T_{rg} is the thickness of the lightly doped thin layer. Again, T_{rg} in Eq. (7.5.3) can only be scaled in proportion to T_{ox} if V_t is to be kept constant. However, T_{rg}, the W_{dep} of an ideal retrograde device, can be about half the W_{dep} of a uniformly doped device [see Eq. (7.5.2)] and yield the same V_t. That is an advantage of the retrograde doping. Another advantage of retrograde doping is that ionized impurity scattering (see Section 2.2.2) in the inversion layer is reduced and the surface mobility can be higher. To produce a sharp retrograde profile with a very thin lightly doped layer, i.e., a very small W_{dep}, care must be taken to prevent dopant diffusion.

● **Derivation of Eq. (7.5.3)** ●

The energy diagram at the threshold condition is shown in Fig. 7–9.

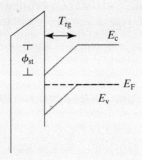

FIGURE 7–9 Energy diagram of a steep-retrograde doped MOSFET at the threshold condition.

The band bending, ϕ_{st}, is dropped uniformly over T_{rg}, the thickness of the lightly doped depletion layer, creating an electric field, $\mathscr{E}_s = \phi_{st}/T_{rg}$. Because of the continuity of the electric flux, the oxide field is $\mathscr{E}_{ox} = \mathscr{E}_s \cdot \varepsilon_s/\varepsilon_{ox}$. Therefore,

$$V_{ox} = T_{ox}\mathscr{E}_{ox} = \phi_{st}\frac{\varepsilon_s T_{ox}}{\varepsilon_{ox} T_{rg}} \tag{7.5.4}$$

From Eqs. (5.2.2), (7.5.4)

$$V_t = V_{fb} + \phi_{st}\left(1 + \frac{\varepsilon_s T_{ox}}{\varepsilon_{ox} T_{rg}}\right) \tag{7.5.5}$$

Here is an intriguing note about reducing W_{dep} further. A higher N_{sub} in Eq. (7.5.1) (and therefore a smaller W_{dep}) or a smaller T_{rg} in Eq. (7.5.3) can be used although it produces a large V_t than desired if this larger V_t is brought back down with a body (or well) to source bias voltage, V_{bs} (see Section 6.4). The required V_{bs} is a forward bias across the body–source junction. A forward bias is acceptable, i.e., the forward bias current is small, if V_{bs} is kept below 0.6 V.

● **Predicting the Ultimate Low Limit of Channel Length—A Retrospective** ●

When the channel length is too small, a MOSFET would have too large an I_{off} and it ceases to be a usable transistor for practical purposes. Assuming that lithography and etching technologies can produce as small features as one desires, what is the ultimate low limit of MOSFET channel length?

In the 1970s, the consensus in the semiconductor industry was that the ultimate lower limit of channel length is 500 nm. In the 1980s, the consensus was 250 nm. In the 1990s, it was 100 nm. Now it is much smaller. What made the experts underestimate the channel length scaling potential?

A review of the historical literature reveals that the researchers were mistaken about how thin the engineers can make the gate oxide in mass production. In the 1970s, it was thought that ~15 nm would be the limit. In the 1980s, it was 8 nm, and so on. Since the T_{ox} estimate was off, the estimates of the minimum acceptable W_{dep} and therefore the minimum L would be off according to Eq. (7.3.4).

7.6 ● SHALLOW JUNCTION AND METAL SOURCE/DRAIN MOSFET ●

Figure 7–10, first introduced as Fig. 6–24b, shows the cross-sectional view of a typical drain (and source) junction. Extra process steps are taken to produce the **shallow junction extension** between the deep N⁺ junction and the channel. This shallow junction is needed because the drain junction depth must be kept small according to Eq. (7.3.4). In order to keep this junction shallow, only very short annealing at the lowest necessary temperature is used to activate the dopants and anneal out the implantation damages in the crystal in 0.1S (flash annealing) or 1μS (laser annealing) (see Section 3.6). To further reduce dopant diffusion, the doping concentration in the **shallow junction extension** is kept much lower than the N⁺ doping density. Shallow junction and light doping combine to produce an undesirable parasitic resistance that reduces the precious I_{on}. That is a price to pay for suppressing V_t roll-off and the subthreshold leakage current. Farther away from the channel, as shown in Fig. 7–10, a deeper N⁺ junction is used to minimize total parasitic resistance. The width of the dielectric spacer in Fig. 7–10 should be as small as possible to minimize the resistance.

7.6.1 MOSFET with Metal Source/Drain

A **metal source/drain MOSFET** or **Schottky source/drain MOSFET** shown in Fig. 7–11a can have very shallow junctions (good for the short-channel effect) and low series-resistance because the silicide is ten times more conductive than N+ or

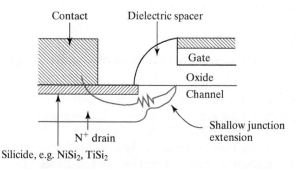

FIGURE 7–10 Cross-sectional view of a MOSFET drain junction. The shallow junction extension next to the channel helps to suppress the V_t roll-off.

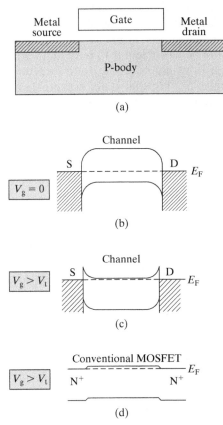

FIGURE 7–11 (a) Metal source/drain is the ultimate way to reduce the increasingly important parasitic resistance; (b) energy band diagrams in the off state; (c) in the on state there may be energy barriers impeding current flow. These barriers do not exist in the conventional MOSFET (d) and must be minimized.

P+ Si. The only problem is that the Schottky-S/D MOSFET would have a lower I_d than the regular MOSFET if ϕ_B is too large to allow easy flow of carriers (electrons for NFET) from the source into the channel.

Figure 7–11b shows the energy band diagram drawn from the source along the channel interface to the drain. V_{ds} is set to zero for simplicity. The energy diagram is similar to that of a conventional MOSFET at $V_g = 0$ in that a potential barrier stops the electrons in the source from entering the channel and the transistor is off. In the on state, Fig. 7–11c, channel E_c is pulled down by the gate voltage, but not at the source/drain edge, where the barrier height is fixed at ϕ_B (see Section 4.16). This barrier does not exist in a conventional MOSFET as shown in Fig. 7–11d, and they can degrade I_d of the metal S/D MOSFET.

To unleash the full potentials of Schottky S/D MOSFET, a very low-ϕ_B Schottky junction technology should be used (for NFETs). A thin N$^+$ region can be added between the metal and the channel. This minimizes the effect of the barriers on current flow as shown in Fig. 4–46. Attention must be paid to reduce the large reverse leakage current of a low-ϕ_{Bn} Schottky drain to body junction [9].

7.7 • TRADE-OFF BETWEEN I_{on} AND I_{off} AND DESIGN FOR MANUFACTURING •

Subthreshold I_{off} would not be a problem if V_t is set at a very high value. That is not acceptable because a high V_t would reduce I_{on} and therefore reduce circuit speed. Using a larger V_{dd} can raise I_{on}, but that is not acceptable either because it would raise the power consumption, which is already too large for comfort. Decreasing L can raise I_{on} but would also reduce V_t and raise I_{off}.

> QUESTION • Which, if any, of the following changes lead to both subthreshold leakage reduction and I_{on} enhancement? A larger V_t. A larger L. A smaller V_{dd}.

Figure 7–12 shows a plot of log I_{off} vs. I_{on} of a large number of transistors [2]. The trade-off between the two is clear. Higher I_{on} goes hand-in-hand with larger I_{off}. The spread in I_{on} (and I_{off}) is due to a combination of unintentional manufacturing variances in L_g and V_t and intentional difference in the gate length.

Techniques have been developed to address the strong trade-off between I_{on} and I_{off}, i.e., between speed and standby power consumption.

One technique gives circuit designers two or three (or even more) V_ts to choose from. A large circuit may be designed with only the high-V_t devices first. Circuit timing simulations are performed to identify those signal paths and circuits where speed must be tuned up. Intermediate-V_t devices are substituted into them. Finally, low-V_t devices are substituted into those few circuits that need even more help with speed. A similar strategy provides multiple V_{dd}. A higher V_{dd} is provided to a small number of circuits that need speed while a lower V_{dd} is used in the other circuits. The larger V_{dd} provides higher speed and/or allows a larger V_t to be used (to suppress leakage). Yet the dynamic power consumption (see Eq. (6.7.6)) can be kept low because most of the circuits operate at the lower V_{dd}.

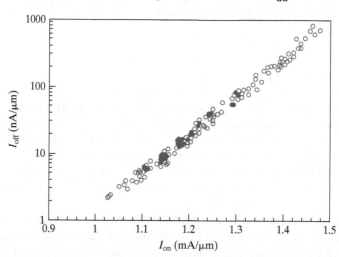

FIGURE 7–12 Log I_{off} vs. linear I_{on}. The spread in I_{on} (and I_{off}) is due to the presence of several slightly different drawn L_gs and unintentional manufacturing variations in L_g and V_t. (After [2]. © 2003 IEEE.)

In a large circuit such as a microprocessor, only some circuit blocks need to operate at high speed at a given time and other circuit blocks operate at lower speed or are idle. V_t can be set relatively low to produce large I_{on} so that circuits that need to operate at high speed can do so. A well bias voltage, V_{sb} in Eq. (6.4.6), is applied to the other circuit blocks to raise the V_t and suppress the subthreshold leakage. This technique requires intelligent control circuits to apply V_{sb} where and when needed.

This well bias technique also provides a way to compensate for the chip-to-chip and block-to-block variations in V_t that results from nonuniformity among devices due to inevitable variations in manufacturing equipment and process. Many techniques at the border between manufacturing and circuit design can help to ease the problem of manufacturing variations. These techniques are collectively known as **design for manufacturing** or **DFM**. A major cause of the device variations is the imperfect control of L_g in the lithography process. Some of the variation is more or less **random variation** in nature. The other part is more or less predictable, called **systematic variation**. One example of the systematic variations is the distortion in photolithography due to the interference of neighboring patterns of light and darkness. Elaborate mathematical optical proximity correction or OPC (see Section 3.3) reshapes each pattern in the photomask to compensate for the effect of the neighboring patterns. Another example is that the carrier mobility and therefore the current of a MOSFET is changed by the mechanical **stress effect** (see Section 7.1.1) created by nearby structures, e.g., shallow trench isolation or other MOSFETs. Sophisticated simulation tools can analyze the mechanical strain and predict the I_{on} based on the neighboring structures and feed the I_{on} information to circuit simulators to obtain more accurate simulation results. An example of random variation is the **gate edge roughness** or waviness caused by the graininess of the photoresist and the poly-crystalline Si. Yet another example of random variation is the **random dopant fluctuation** phenomenon. The statistical variation of the number of dopant atoms and their location in small size MOSFET creates significant variations in the threshold voltage. It requires complex design methodologies to include the intra-chip and inter-chip random variations in circuit design.

7.8 ● ULTRA-THIN-BODY SOI AND MULTIGATE MOSFETs ●

There are alternative MOSFET structures that are less susceptible to V_t roll-off and allow gate length scaling beyond the limit of conventional MOSFET. Figure 7–6 gives a simple description of the competition between the gate and the drain over the control of the channel barrier height shown in Fig. 7–5. We want to maximize the gate-to-channel capacitance and minimize the drain-to-channel capacitance. To do the former, we reduce T_{ox} as much as possible. To accomplish the latter, we reduce W_{dep} and X_j as much as possible. It is increasingly difficult to make these dimensions smaller. The real situation is even worse. In the subthreshold region, T_{ox} may be a small part T_{oxe} in Eq. (7.3.4) because the inversion-layer thickness, T_{inv} (see Sec. 5.9), is large. Imagine that T_{ox} could be made infinitesimally small. This would give the gate a perfect control over the potential barrier height—but only right at the Si surface. The drain could still have more control than the gate along other leakage current paths that are some distance below the Si surface as shown in Fig. 7–13. At this submerged location, the gate is far away and the gate control is weak. The drain voltage can pull the

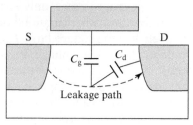

FIGURE 7–13 The drain could still have more control than the gate along another leakage current path that is some distance below the Si surface.

potential barrier down and allow leakage current to flow along this submerged path. There are two transistor structures that can eliminate the leakage paths that are far away from the gate [10]. One is called the **ultra-thin-body MOSFET** or **UTB MOSFET**. The other is **multigate MOSFET**. They are presented next.

7.8.1 Ultra-Thin-Body MOSFET and SOI

There are two ways to eliminate these submerged leakage paths. One is to use an ultra-thin-body structure as shown in Fig. 7–14 [11]. This MOSFET is built in a thin Si film on an insulator (SiO_2). Since the Si film is very thin, perhaps less than 10 nm, no leakage path is very far from the gate. (The worst-case leakage path is along the bottom of the Si film.) Therefore, the gate can effectively suppress the leakage. Figure 7–15 shows that the subthreshold leakage is reduced as the Si film is made thinner. It can be shown that the thin Si thickness should take the places of W_{dep} and X_j in Eq. (7.3.4) such that L_g can be scaled roughly in proportion to T_{Si}, the Si thickness. T_{Si} should be thinner than about one half of the gate length in order to reap the benefit of the UTB MOSFET concept to sustain scaling. UTB MOSFETs, as the multigate MOSFETs of the next section, offer additional device benefits. Because small l_d (Eq. (7.3.4)) can be obtained without heavy channel doping, carrier mobility is improved. The body effect that is detrimental to circuit speed (see Section 6.4) is eliminated because the body is **fully depleted** and floating and has no fixed voltage. One challenge posed by UTB MOSFETs is the large source/drain resistance due to their thinness. The solution is to thicken the source and drain with epitaxial deposition. These **raised source/drains** are visible in Figs. 7–14 and 7–15.

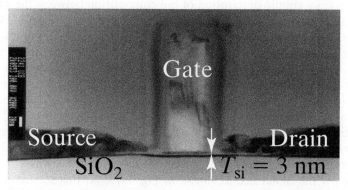

FIGURE 7–14 The SEM cross section of UTB device. (After [11]. © 2000 IEEE.)

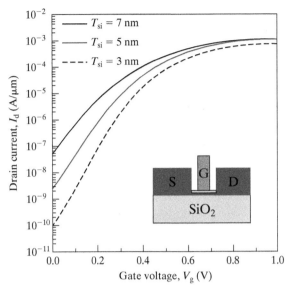

FIGURE 7–15 The subthreshold leakage is reduced as the Si film (transistor body) is made thinner. L_g = 15 nm. (After [11]. © 2000 IEEE.)

● **SOI-Silicon on Insulator** ●

Figure 7–16 shows the steps of making an SOI or **silicon-on-Insulator** wafer [12]. (The conventional wafer is sometimes called **bulk silicon** wafer for clarity.) Step 1 is to implant hydrogen into a silicon wafer that has a thin SiO_2 film at the surface. The hydrogen concentration peaks at a distance D below the surface. Step 2 is to place the first wafer, upside down, over a second plain wafer. The two wafers adhere to each other by the atomic bonding force. A low temperature annealing causes the two wafers to fuse together. Step 3 is to apply another annealing step that causes the implanted hydrogen to coalesce and form a large number of tiny hydrogen bubbles at depth D. This creates sufficient mechanical stress to break the wafer at that plane. The final step, Step 4, is to polish the surface. Now the SOI wafer is ready for use.

The Si film is of high quality and suitable for IC manufacturing. Even without using an ultra-thin body, SOI provides a speed advantage because the source/drain to body junction capacitance is practically eliminated as the source and drain diffusion regions extend vertically to the buried oxide. The cost of an SOI wafer is higher than an ordinary Si wafer and increases the cost of IC chips. For these reasons, only some microprocessors, which command high prices and compete on speed, have employed this technology so far. Figure 7–17 shows the cross-sectional SEMs of an SOI product. SOI also finds other compelling applications because it offers extra flexibility for making novel structures such as the ultra-thin-body MOSFET and some multigate MOSFET structures that can be scaled to smaller gate length beyond the capability of bulk MOSFETs.

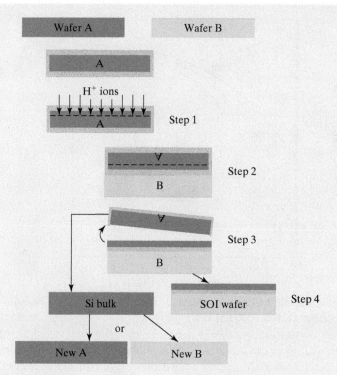

FIGURE 7–16 Steps of making an SOI wafer. (After [12].)

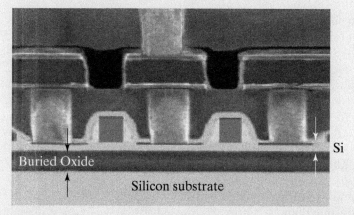

FIGURE 7–17 The cross-sectional electron micrograph of an SOI integrated circuit. The lower level structures are transistors and contacts. The upper two levels are the vias and the interconnects, which employ multiple layers of materials to achieve better reliability and etch stops.

7.8.2 FinFET - Multigate MOSFET

The second way of eliminating deep submerged leakage paths is to provide gate control from more than one side of the channel as shown in Fig. 7–18. The Si film is

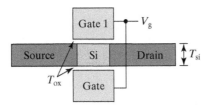

FIGURE 7–18 A schematic sketch of a double-gate MOSFET with gates connected.

very thin so that no leakage path is far from one of the gates. (The worst-case path is along the center of the Si film.) Therefore, the gate(s) can suppress leakage current more effectively than the conventional MOSFET. Because there are more than one gate, the structure may be called **multigate MOSFET**. The structure shown in Fig. 7–18 is a **double-gate MOSFET**. Shrinking T_{Si} automatically reduces W_{dep} and X_j in Eq. (7.3.4) and V_t roll-off can be suppressed to allow L_g to shrink to as small as a few nm. Because the top and bottom gates are at the same voltage and the Si film is fully depleted, the Si surface potential moves up and down with V_g mV for mV in the subthreshold region. The voltage divider effect illustrated in Fig. 7–1c does not exist and η in Eq. (7.2.4) is the desired unity and I_{off} is very low. There is no need for heavy doping in the channel to reduce W_{dep}. This leads to low vertical field and less impurity scattering; as a result the mobility is higher (see Section 6.3). Finally, there are two channels (top and bottom) to conduct the transistor current. For these reasons, a multigate MOSFET can have shorter L_g, lower I_{off}, and larger I_{on} than a single-gate MOSFET. But, there is one problem—how to fabricate the multigate MOSFET structure.

There is a multigate structure that is attractive for its simplicity of fabrication and it is illustrated in Fig. 7–19. Consider the center structure in Fig. 7–19. The process starts with an SOI wafer or a bulk Si wafer. A thin fin of Si is created by lithography and etching. Gate oxide is grown over the exposed surfaces of the fin. Poly-Si gate material is deposited over the fin and the gate is patterned by lithography and etching. Finally, source/drain implantation is

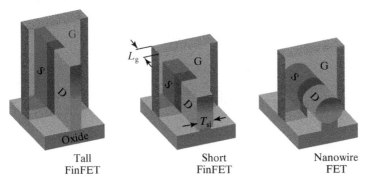

Tall FinFET

Short FinFET

Nanowire FET

FIGURE 7–19 Variations of FinFET. Tall FinFET has the advantage of providing a large W and therefore large I_{on} while occupying a small footprint. Short FinFET has the advantage of less challenging lithography and etching. Nanowire FET gives the gate even more control over the transistor body by surrounding it. FinFETs can also be fabricated on bulk Si substrates.

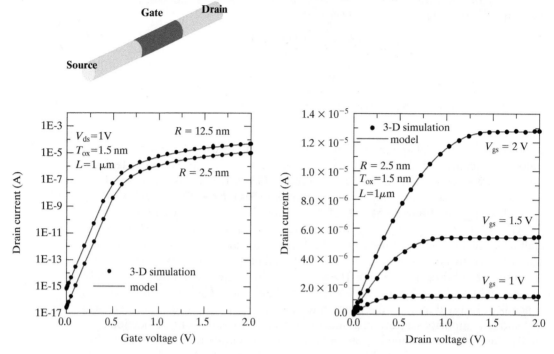

FIGURE 7–20 Simulated I–V curves of a nanowire MOSFET. R is the nanowire radius. (After [16].)

performed. The final structure in Fig. 7–19 is basically the multigate structure in Fig. 7–18 turned on its side. This structure is called the **FinFET** because its Si body resembles the back fin of a fish [13]. The channel consists of the two vertical surfaces and the top surface of the fin. The channel width, W, is the sum of twice the fin height and the width of the fin.

Several variations of FinFET are shown in Fig. 7–19 [14,15]. A tall FinFET has the advantage of providing a large W and therefore large I_{on} while occupying a small footprint. A short FinFET has the advantage of less challenging etching. In this case, the top surface of the fin contributes significantly to the suppression of V_t roll-off and to leakage control. This structure is also known as a **triple-gate MOSFET**. The third variation gives the gate even more control over the Si wire by surrounding it. It may be called a nanowire FET and its behaviors shown in Fig. 7–20 can be modeled with the same methods and concepts used to model the basic MOSFETs. FinFETs with L_g as small as 3 nm have been experimentally demonstrated. It will allow transistor scaling beyond the scaling limit of the conventional planar transistor.

7.9 • OUTPUT CONDUCTANCE •

Output conductance limits the transistor voltage gain. It has been introduced in Section 6.13. However, its cause and theory are intimately related to those of V_t roll-off. Therefore, the present chapter is a fitting place to explain it.

What device design parameters determine the output conductance? Let us start with Eq. (6.13.1),

$$g_{ds} \equiv \frac{dI_{dsat}}{dV_{ds}} = \frac{dI_{dsat}}{dV_t} \cdot \frac{dV_t}{dV_{ds}} \quad (7.9.1)$$

Since I_{ds} is a function of $V_{gs} - V_t$ [see Eq. (6.9.11)], it is obvious that

$$\frac{dI_{dsat}}{dV_t} = \frac{-dI_{dsat}}{dV_{gs}} = -g_{msat} \quad (7.9.2)$$

The last step is the definition of g_{msat} given in Eq. (6.6.8). Now, Eq. (7.9.1) can be evaluated with the help of Eq. (7.3.3).

$$g_{ds} = g_{msat} \times e^{-L/l_d} \quad (7.9.3)$$

$$\text{Instrinsic voltage gain} = \frac{g_{msat}}{g_{ds}} = e^{-L/l_d} \quad (7.9.4)$$

Intrinsic voltage gain was introduced in Eq. (6.13.5). Equation (7.3.3) states that increasing V_{ds} would reduce V_t. That is why I_{ds} continues to increase without saturation. *The output conductance is caused by the drain/channel capacitive coupling, the same mechanism that is responsible for V_t roll-off.* This is why g_{ds} is larger in a MOSFET with shorter L. To reduce g_{ds} or to increase the intrinsic voltage gain, we can use a large L and/or reduce l_d. Circuit designers routinely use much larger L than the minimum value allowed for a given technology node when the circuits require large voltage gains. Reducing l_d is the job of device designers and Eq. (7.3.4) is their guide. Every design change that improves the suppression of V_t roll-off also suppresses g_{ds} and improves the voltage gain.

V_t dependence on V_{ds} is the main cause of output conductance in very short MOSFETs. For larger L and V_{ds} close to V_{dsat}, another mechanism may be the dominant contributor to g_{ds}—**channel length modulation**. A voltage, $V_{ds}-V_{dsat}$, is dissipated over a finite (non-zero) distance next to the drain. This distance increases with increasing V_{ds}. As a result, the effective channel length decreases with increasing V_{ds}. I_{ds}, which is inversely proportional to L, thus increases without true saturation. It can be shown that g_{ds}, due to the channel length modulation, is approximately

$$g_{ds} = \frac{l_d \cdot I_{dsat}}{L(V_{ds} - V_{dsat})} \quad (7.9.5)$$

where l_d is given in Eq. (7.3.4). This component of g_{ds} can also be suppressed with larger L and smaller T_{ox}, X_j, and W_{dep}.

7.10 • DEVICE AND PROCESS SIMULATION •

There are commercially available computer simulation suites [17] that solve all the equations presented in this book with few or no approximations (e.g., Fermi–Dirac statistics is used rather than Boltzmann approximation). Most of these equations are solved simultaneously, e.g., Fermi–Dirac probability, incomplete ionization of dopants, drift and diffusion currents, current continuity equation, and Poisson equation. Device simulation is an important tool that provides the engineers with quick feedback about device behaviors. This narrows down the number of variables that need to be checked with expensive and time-consuming experiments. Examples of simulation results are shown in Figs. 7–15 and 7–20. Each of the figures takes from minutes to several hours of simulation time to generate.

Related to device simulation is process simulation. The input that a user provides to the process simulation program are the lithography mask pattern, implantation dose and energy, temperatures and times for oxide growth and annealing steps, etc. The process simulator then generates a two- or three-dimensional structure with all the deposited or grown and etched thin films and doped regions. This output may be fed into a device simulator together with the applied voltages and the operating temperature as the input to the device simulator.

7.11 ● MOSFET COMPACT MODEL FOR CIRCUIT SIMULATION ●

Circuit designers can simulate the operation of circuits containing up to hundreds of thousands or even more MOSFETs accurately, efficiently, and robustly. Accuracy must be delivered for DC as well as RF operations, analog as well digital circuits, memory as well as processor ICs. In circuit simulations, MOSFETs are modeled with analytical equations much like the ones introduced in this and the previous two chapters. More details are included in the model equations than this textbook can introduce. These models are called **compact models** to highlight their computational efficiency in contrast with the device simulators described in Section 7.10.

It could be said that the compact model (and the layout design rules) is the link between two halves of the semiconductor industry—technology/manufacturing on the one side and design/product on the other. A compact model must capture all the subtle behaviors of the MOSFET over wide ranges of voltage, L, W, and temperature and present them to the circuit designers in the form of equations. Some circuit-design methodologies, such as analog circuit design, use circuit simulations directly. Other design methodologies use **cell libraries**. A cell library is a collection of hundreds of small building blocks of circuits that have been carefully designed and characterized beforehand using circuit simulations.

At one time, nearly every company developed its own compact models. In 1997, an industry standard setting group selected **BSIM** [18] as the first industry standard model. If the I_{ds} equation of BSIM is printed out on paper, it will fill several pages.

Figure 7–21 shows selected comparisons of a compact model and measured device data to illustrate the accuracy of the compact model [19]. It is also important for the compact model to accurately model the transistor behaviors for any L and W that a circuit designer may specify. Figure 7–22 illustrates this capability. Finally, a good compact model should provide fast simulation times by using simple model equations. In addition to the IV of N-channel and P-channel transistors, the model also includes capacitance models, gate dielectric leakage current model, and source

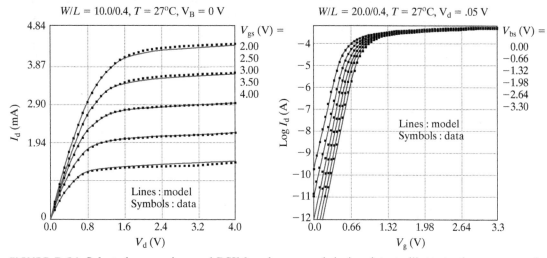

FIGURE 7–21 Selected comparisons of BSIM and measured device data to illustrate the accuracy of a compact model. (After [18].)

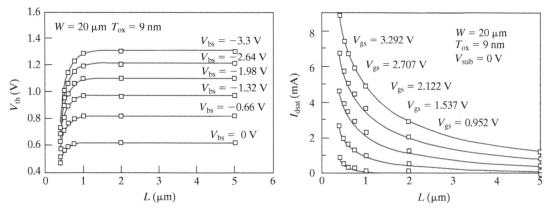

FIGURE 7–22 A compact model needs to accurately model the transistor behaviors for any L and W that circuit designers may specify. (After [19]. © 1997 IEEE.)

and drain junction diode model. Noise and high-frequency models are usually provided, too.

7.12 • CHAPTER SUMMARY •

To reduce cost and improve speed in order to open up new applications, transistors and interconnects are downsized periodically. Very small MOSFETs are prone to have excessive leakage current called I_{off}. The basic component of I_{off} is the **subthreshold current**

$$I_{\text{off}}(\text{nA}) = 100 \cdot \frac{W}{L} \cdot e^{-qV_t/\eta kT} = 100 \cdot \frac{W}{L} \cdot 10^{-V_t/S} \tag{7.2.8}$$

S is the **subthreshold swing**. To keep I_{off} below a given level, there is a minimum acceptable V_t. Unfortunately, a larger V_t is deleterious to I_{on} and speed. Therefore, it is important to reduce S by reducing the ratio $T_{\text{oxe}}/W_{\text{dep}}$. Furthermore, V_t decreases with L, a fact known as V_t **roll-off**, caused by DIBL.

$$V_t = V_{t\text{-long}} - (V_{ds} + 0.4\text{V}) \cdot e^{-L/l_d} \tag{7.3.3}$$

where

$$l_d \propto \sqrt[3]{T_{\text{oxe}} W_{\text{dep}} X_j} \tag{7.3.4}$$

Since V_t is a sensitive function of L, even the small (a few nm) manufacturing variations in L can cause problematic variations in V_t, I_{off}, and I_{on}. To allow L reduction, Eq. (7.3.3) states that l_d must be reduced, i.e., T_{oxe}, W_{dep}, and/or X_j must be reduced.

T_{ox} reduction is limited mostly by **gate tunneling leakage**, which can be suppressed by replacing SiO_2 with a **high-k dielectric** such as HfO_2. Metal gate can reduce T_{oxe} by eliminating the poly-Si gate depletion effect.

W_{dep} can be reduced with retrograde body doping. X_j can be reduced with mS flash annealing or the metal source–drain MOSFET structure. X_j and W_{dep} can also be reduced with the ultra-thin-body SOI device structure or the multigate MOSFET structure. More importantly, these new structures eliminate the more vulnerable leakage paths, which are the farthest from the gate.

Equation (7.3.3) also provides a theory for output conductance of the short channel transistors.

$$g_{ds} = g_{\text{msat}} \times e^{-L/l_d} \tag{7.9.3}$$

● PROBLEMS ●

● **Subthreshold Leakage Current** ●

7.1 Assume that the gate oxide between an n+ poly-Si gate and the p-substrate is 11 Å thick and $N_a = 1\text{E}18$ cm^{-3}.

(a) What is the V_t of this device?

(b) What is the subthreshold swing, S?

(c) What is the maximum leakage current if $W = 1$ μm, $L = 18$ nm? (Assume $I_{ds} = 100\ W/L$ (nA) at $V_g = V_t$.)

● **Field Oxide Leakage** ●

7.2 Assume the field oxide between an n+ poly-Si wire and the p-substrate is 0.3 μm thick and that $N_a = 5\text{E}17$ cm^{-3}.

(a) What is the V_t of this field oxide device?

(b) What is the subthreshold swing, S?

(c) What is the maximum field leakage current if $W = 10$ μm, $L = 0.3$ μm, and $V_{dd} = 2.0$ V?

● **V_t Roll-off** ●

7.3 Qualitatively sketch $\log(I_{ds})$ vs. V_g (assume $V_{ds} = V_{dd}$) for the following:

(a) $L = 0.2$ μm, $N_a = 1\text{E}15$ cm^{-3}.

(b) $L = 0.2$ μm, $N_a = 1\text{E}17$ cm^{-3}.

(c) $L = 1$ μm, $N_a = 1E15$ cm^{-3}.

(d) $L = 1$ μm, $N_a = 1E17$ cm^{-3}.

Please pay attention to the positions of the curves relative to each other and label all curves.

● **Trade-off between I_{off} and I_{on}** ●

7.4 Does each of the following changes increase or decrease I_{off} and I_{on}? A larger V_t. A larger L. A shallower junction. A smaller V_{dd}. A smaller T_{ox}. Which of these changes contribute to leakage reduction without reducing the precious I_{on}?

7.5 There is a lot of concern that we will soon be unable to extend Moore's Law. In your own words, explain this concern and the difficulties of achieving high I_{on} and low I_{off}.

  (a) Answer this question in one paragraph of less than 50 words.

  (b) Support your description in (a) with three hand-drawn sketches of your choice.

  (c) Why is it not possible to maximize I_{on} and minimize I_{off} by simply picking the right values of T_{ox}, X_j, and W_{dep}? Please explain in your own words.

  (d) Provide three equations that help to quantify the issues discussed in (c).

7.6 (a) Rewrite Eq. (7.3.4) in a form that does not contain W_{dep} but contains V_t. Do so by using Eqs. (5.5.1) and (5.4.3) assuming that V_t is given.

  (b) Based on the answer to (a), state what actions can be taken to reduce the minimum acceptable channel length.

7.7 (a) What is the advantage of having a small W_{dep}?

  (b) For given L and V_t, what is the impact of reducing W_{dep} on I_{dsat} and gate? (Hint: consider the "m" in Chapter 6)

Discussion: Overall, smaller W_{dep} is desirable because it is more important to be able to suppress V_t roll-off so that L can be scaled.

● **MOSFET with Ideal Retrograde Doping Profile** ●

7.8 Assume an N-channel MOSFET with an N$^+$ poly gate and a substrate with an idealized retrograde substrate doping profile as shown in Fig. 7–23.

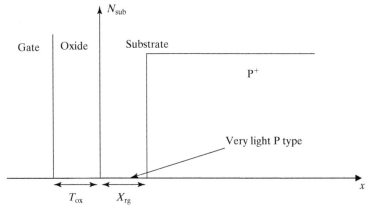

FIGURE 7–23

(a) Draw the energy band diagram of the MOSFET along the x direction from the gate through the oxide and the substrate, when the gate is biased at threshold voltage. (Hint: Since the P region is very lightly doped you may assume that the field in this region is constant or $d\varepsilon/dx = 0$). Assume that the Fermi level in the P$^+$ region coincides with E_v and the Fermi level in the N$^+$ gate coincides with E_c. Remember to label E_c, E_v, and E_F.

(b) Find an expression for V_t of this ideal retrograde device in terms of V_{ox}. Assume V_{ox} is known. (Hint: Use the diagram from (a) and remember that V_t is the difference between the Fermi levels in the gate and in the substrate. At threshold, E_c of Si coincides with the Fermi level at the Si–SiO$_2$ interface).

(c) Now write an expression for V_t in terms of X_{rg}, T_{ox}, ε_{ox}, ε_{si} and any other common parameters you see fit, but not in terms of V_{ox}. Hint: Remember N_{sub} in the lightly doped region is almost 0, so if your answer is in terms of N_{sub}, you might want to rethink your strategy. Maybe $\varepsilon_{ox}\varepsilon_{ox} = \varepsilon_{si}\varepsilon_{si}$ could be a starting point.

(d) Show that the depletion layer width, W_{dep} in an ideal retrograde MOSFET can be about half the X_{dep} of a uniformly doped device and still yield the same V_t.

(e) What is the advantage of having a small W_{dep}?

(f) For given L and V_t, what is the impact of reducing W_{dep} on I_{dsat} and inverter delay?

● REFERENCES ●

1. International Technology Roadmap for Semiconductors (http://public.itrs.net/)
2. Ghani, T., et al. "A 90 Nm High Volume Manufacturing Logic Technology Featuring Novel 45 nm Gate Length Strained Silicon CMOS Transistors," *IEDM Technical Digest*. 2003, 978–980.
3. Yeo, Y-C., et al. "Enhanced Performance in Sub-100nm CMOSFETs Using Strained Epitaxial Si-Ge." *IEDM Technical Digest*. 2000, 753–756.
4. Liu, Z. H., et al. "Threshold Voltage Model for Deep-Submicrometer MOSFETs." *IEEE Trans. on Electron Devices*. 40, 1 (January 1993), 86–95.
5. Wann, C. H., et al. "A Comparative Study of Advanced MOSFET Concepts." *IEEE Transactions on Electron Devices*. 43, 10 (October 1996), 1742–1753.
6. Yeo, Yee-Chia, et al. "MOSFET Gate Leakage Modeling and Selection Guide for Alternative Gate Dielectrics Based on Leakage Considerations." *IEEE Transactions on Electron Devices*. 50, 4 (April 2003), 1027–1035.
7. Lu, Q., et al. "Dual-Metal Gate Technology for Deep-Submicron CMOS Transistor," Symp. on VLSI Technology Digest of Technical Papers, 2000, 72–73.
8. Chen, I. C., et al. "Electrical Breakdown in Thin Gate and Tunneling Oxides." *IEEE Trans. on Electron Devices*. ED-32 (February 1985), 413–422.
9. Kedzierski, J., et al. "Complementary Silicide Source/Drain Thin-Body MOSFETs for the 20 nm Gate Length Regime." *IEDM Technical Digest*, 2000, 57–60.
10. Hu, C. "Scaling CMOS Devices Through Alternative Structures," *Science in China (Series F)*. February 2001, 44 (1) 1–7.
11. Choi, Y-K., et al. "Ultrathin-body SOI MOSFET for Deep-sub-tenth Micron Era," *IEEE Electron Device Letters*. 21, 5 (May 2000), 254–255.
12. Celler, George, and Michael Wolf. "Smart Cut™ A Guide to the Technology, the Process, the Products," *SOITEC*. July 2003.

13. Huang, X., et al. "Sub 50-nm FinFET: PMOS." *IEDM Technical Digest*, (1999), 67–70.
14. Yang, F-L, et al. "25 nm CMOS Omega FETs." *IEDM Technical Digest*. (1999), 255–258.
15. Yang, F-L, et al. "5 nm-Gate Nanowire FinFET." VLSI Technology, 2004. Digest of Technical Papers, 196–197.
16. Lin, C-H., et al. "Corner Effect Model for Compact Modeling of Multi-Gate MOSFETs." 2005 SRC TECHCON.
17. Taurus Process, Synoposys TCAD Manual, Synoposys Inc., Mountain View, CA.
18. http://www-device.eecs.berkeley.edu/~bsim3/bsim4.html
19. Cheng, Y., et al. "A Physical and Scalable I-V Model in BSIM3v3 for Analog/Digital Circuit Simulation." *IEEE Trans. on Electron Devices*. 44, 2, (February 1997), 277–287.

● **GENERAL REFERENCES** ●

1. Taur, Y., and T. H. Ning. *Fundamentals of Modern VLSI Devices*. Cambridge, UK: Cambridge University Press, 1998.
2. Wolf, S. *VLSI Devices*. Sunset Beach, CA: Lattice Press, 1999.

8

Bipolar Transistor

CHAPTER OBJECTIVES

This chapter introduces the bipolar junction transistor (BJT) operation and then presents the theory of the bipolar transistor I-V characteristics, current gain, and output conductance. High-level injection and heavy doping induced band narrowing are introduced. SiGe transistor, transit time, and cutoff frequency are explained. Several bipolar transistor models are introduced, i.e., Ebers–Moll model, small-signal model, and charge control model. Each model has its own areas of applications.

The **bipolar junction transistor** or **BJT** was invented in 1948 at Bell Telephone Laboratories, New Jersey, USA. It was the first mass produced transistor, ahead of the MOS field-effect transistor (MOSFET) by a decade. After the introduction of metal-oxide-semiconductor (MOS) ICs around 1968, the high-density and low-power advantages of the MOS technology steadily eroded the BJT's early dominance. BJTs are still preferred in some high-frequency and analog applications because of their high speed, low noise, and high output power advantages such as in some cell phone amplifier circuits. When they are used, a small number of BJTs are integrated into a high-density complementary MOS (CMOS) chip. Integration of BJT and CMOS is known as the **BiCMOS technology**.

The term **bipolar** refers to the fact that both electrons and holes are involved in the operation of a BJT. In fact, minority carrier diffusion plays the leading role just as in the PN junction diode. The word **junction** refers to the fact that PN junctions are critical to the operation of the BJT. BJTs are also simply known as **bipolar transistors**.

8.1 ● INTRODUCTION TO THE BJT ●

A BJT is made of a heavily doped **emitter** (see Fig. 8–1a), a P-type **base**, and an N-type **collector**. This device is an **NPN** BJT. (A **PNP** BJT would have a P^+ emitter, N-type base, and P-type collector.) NPN transistors exhibit higher transconductance and

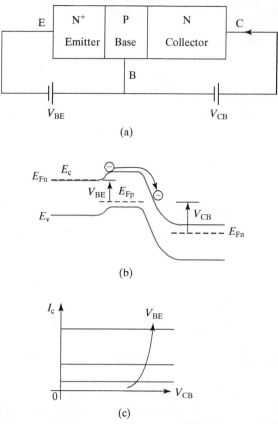

FIGURE 8–1 (a) Schematic NPN BJT and normal voltage polarities; (b) electron injection from emitter into base produces and determines I_C; and (c) I_C is basically determined by V_{BE} and is insensitive to V_{CB}.

speed than PNP transistors because the electron mobility is larger than the hole mobility. BJTs are almost exclusively of the NPN type since high performance is BJTs' competitive edge over MOSFETs.

Figure 8–1b shows that when the base–emitter junction is forward biased, electrons are injected into the more lightly doped base. They diffuse across the base to the reverse-biased base–collector junction (edge of the depletion layer) and get swept into the collector. This produces a **collector current**, I_C. I_C is independent of V_{CB} as long as V_{CB} is a reverse bias (or a small forward bias, as explained in Section 8.6). Rather, I_C is determined by the rate of electron injection from the emitter into the base, i.e., determined by V_{BE}. You may recall from the PN diode theory that the rate of injection is proportional to $e^{qV_{BE}/kT}$. These facts are obvious in Fig. 8–1c.

Figure 8–2a shows that the emitter is often connected to ground. (The emitter and collector are the equivalents of source and drain of a MOSFET. The base is the equivalent of the gate.) Therefore, the I_C curve is usually plotted against V_{CE} as shown in Fig. 8–2b. For V_{CE} higher than about 0.3 V, Fig. 8–2b is identical to Fig. 8–1c but with a shift to the right because $V_{CE} = V_{CB} + V_{BE}$. Below $V_{CE} \approx 0.3$ V,

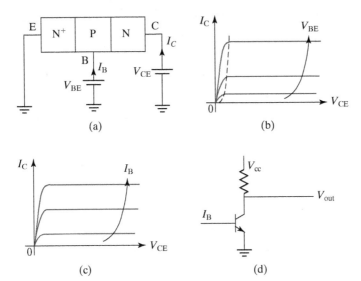

FIGURE 8–2 (a) Common-emitter convention; (b) I_C vs. V_{CE}; (c) I_B may be used as the parameter instead of V_{BE}; and (d) circuit symbol of an NPN BJT and an inverter circuit.

the base–collector junction is strongly forward biased and I_C decreases as explained in Section 8.6. Because of the parasitic IR drops, it is difficult to accurately ascertain the true base–emitter junction voltage. For this reason, the easily measurable base current, I_B, is commonly used as the variable parameter in lieu of V_{BE} (as shown in Fig. 8–2c). We will see later that I_C is proportional to I_B.

8.2 • COLLECTOR CURRENT •

The collector current is the output current of a BJT. Applying the electron diffusion equation [Eq. (4.7.7)] to the base region,

$$\frac{d^2 n'}{dx^2} = \frac{n'}{L_B^2} \tag{8.2.1}$$

$$L_B \equiv \sqrt{\tau_B D_B} \tag{8.2.2}$$

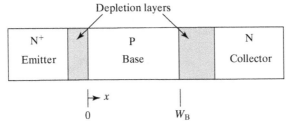

FIGURE 8–3 $x = 0$ is the edge of the BE junction depletion layer. W_B is the width of the base neutral region.

τ_B and D_B are the recombination lifetime and the minority carrier (electron) diffusion constant in the base, respectively. The boundary conditions are [Eq. (4.6.3)]

$$n'(0) = n_{B0}(e^{qV_{BE}/kT} - 1) \tag{8.2.3}$$

$$n'(W_B) = n_{B0}(e^{qV_{BC}/kT} - 1) \approx -n_{B0} \approx 0 \tag{8.2.4}$$

where $n_{B0} = n_i^2/N_B$, and N_B is the base doping concentration. V_{BE} is normally a forward bias (positive value) and V_{BC} is a reverse bias (negative value). The solution of Eq. (8.2.1) is

$$n'(x) = n_{B0}(e^{qV_{BE}/kT} - 1)\frac{\sinh\left(\frac{W_B - x}{L_B}\right)}{\sinh(W_B/L_B)} \tag{8.2.5}$$

Equation (8.2.5) is plotted in Fig. 8–4.

Modern BJTs have base widths of about 0.1 μm. This is much smaller than the typical diffusion length of tens of microns (see Example 4–4 in Section 4.8). In the case of $W_B \ll L_B$, Eq. (8.2.5) reduces to a straight line as shown in Fig. 8–4.

$$\boxed{\begin{aligned} n'(x) &= n'(0)(1 - x/W_B) \\ &= \frac{n_{iB}^2}{N_B}(e^{qV_{BE}/kT} - 1)\left(1 - \frac{x}{W_B}\right) \end{aligned}} \tag{8.2.6}$$

n_{iB} is the intrinsic carrier concentration of the base material. The subscript, B, is added to n_i because the base may be made of a different semiconductor (such as SiGe alloy, which has a smaller band gap and therefore a larger n_i than the emitter and collector material).

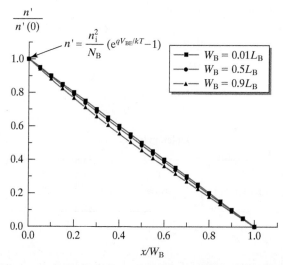

FIGURE 8–4 When $W_B \ll L_B$, the excess minority carrier concentration in the base is approximately a linear function of x.

As explained in the PN diode analysis, the minority-carrier current is dominated by the diffusion current. The sign of I_C is defined in Fig. 8–2a and is positive.

$$I_C = \left| A_E q D_B \frac{dn}{dx} \right| = A_E q D_B \frac{n'(0)}{W_B}$$

$$= A_E q \frac{D_B n_{iB}^2}{W_B N_B}(e^{qV_{BE}/kT} - 1) \tag{8.2.7}$$

A_E is the area of the BJT, specifically the emitter area. Notice the similarity between Eq. (8.2.7) and the PN diode IV relation [Eq. (4.9.4)]. Both are proportional to $(e^{qV/kT} - 1)$ and to Dn_i^2/N. In fact, the only difference is that dn'/dx has produced the $1/W_B$ term in Eq. (8.2.7) due to the linear n' profile. Equation (8.2.7) can be condensed to

$$I_C = I_S(e^{qV_{BE}/kT} - 1) \tag{8.2.8}$$

where I_S is the saturation current. Equation (8.2.7) can be rewritten as

$$I_C = A_E \frac{q n_i^2}{G_B}(e^{qV_{BE}/kT} - 1) \tag{8.2.9}$$

In the special case of Eq. (8.2.7)

$$G_B = \frac{n_i^2}{n_{iB}^2}\frac{N_B}{D_B}W_B = \frac{n_i^2}{n_{iB}^2}\frac{p}{D_B}W_B \tag{8.2.10}$$

where p is the majority carrier concentration in the base. It can be shown that Eq. (8.2.9) is valid even for nonuniform base and high-level injection condition if G_b is generalized to [1]

$$G_B \equiv \int_0^{W_B} \frac{n_i^2}{n_{iB}^2}\frac{p}{D_B} dx \tag{8.2.11}$$

G_B has the unusual dimension of s/cm^4 and is known as the **base Gummel number**. In the special case of $n_{iB} = n_i$, D_B is a constant, and $p(x) = N_B(x)$ (low-level injection),

$$G_B = \frac{1}{D_B}\int_0^{W_B} N_B(x)dx = \frac{1}{D_B} \times \text{base dopant atoms per unit area} \tag{8.2.12}$$

Equation (8.2.12) illustrates that the base Gummel number is basically proportional to the base dopant density per area. The higher the base dopant density is, the lower the I_C will be for a given V_{BE} as given in Eq. (8.2.9).

The concept of a Gummel number simplifies the I_C model because it [Eq. (8.2.11)] contains all the subtleties of transistor design that affect I_C: changing base material through $n_{iB}(x)$, nonconstant D_B, nonuniform base dopant concentration through $p(x) = N_B(x)$, and even the **high-level injection** condition (see Sec. 8.2.1), where $p > N_B$. Although many factors affect G_B, G_B can be easily determined from the **Gummel plot** shown in Fig. 8–5. The (inverse) slope of the

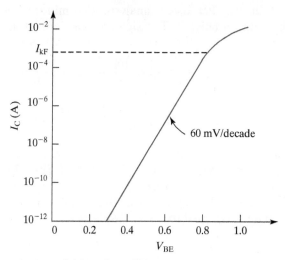

FIGURE 8–5 I_C is an exponential function of V_{BE}.

straight line in Fig. 8–5 can be described as 60 mV per decade. The extrapolated intercept of the straight line and $V_{BE} = 0$ yields I_S [Eq. (8.2.8)]. G_B is equal to $A_E q n_i^2$ divided by the intercept.

8.2.1 High-Level Injection Effect

The decrease in the slope of the curve in Fig. 8–5 at high I_C is called the **high-level injection effect**. At large V_{BE}, n' in Eq. (8.2.3) can become larger than the base doping concentration N_B

$$n' = p' \gg N_B \tag{8.2.13}$$

The first part of Eq. (8.2.13) is simply Eq. (2.6.2) or charge neutrality. The condition of Eq. (8.2.13) is called **high-level injection**. A consequence of Eq. (8.2.13) is that in the base

$$n \approx p \tag{8.2.14}$$

From Eqs. (8.2.14) and (4.9.6)

$$n \approx p \approx n_i e^{qV_{BE}/2kT} \tag{8.2.15}$$

Equations (8.2.15) and (8.2.11) yield

$$G_B \propto n_i e^{qV_{BE}/2kT} \tag{8.2.16}$$

Equations (8.2.16) and (8.2.9) yield

$$I_c \propto n_i e^{qV_{BE}/2kT} \tag{8.2.17}$$

Therefore, at high V_{BE} or high I_c, $I_c \propto e^{qV_{BE}/2kT}$ and the (inverse) slope in Fig. 8–5 becomes 120 mV/decade. I_{kF}, the **knee current**, is the current at which the slope changes. It is a useful parameter in the BJT model for circuit simulation. The IR drop in the parasitic resistance significantly increases V_{BE} at very high I_C and further flattens the curve.

8.3 • BASE CURRENT •

Whenever the base–emitter junction is forward biased, some holes are injected from the P-type base into the N$^+$ emitter. These holes are provided by the base current, I_B.[1] I_B is an undesirable but inevitable side effect of producing I_C by forward biasing the BE junction. The analysis of I_B, the base to emitter injection current, is a perfect parallel of the I_C analysis. Figure 8–6b illustrates the mirror equivalence. At an ideal ohmic contact such as the contact of the emitter, the equilibrium condition holds and $p' = 0$ similar to Eq. (8.2.4). Analogous to Eq. (8.2.9), the base current can be expressed as

$$I_B = A_E \frac{qn_i^2}{G_E}(e^{qV_{BE}/kT} - 1) \qquad (8.3.1)$$

$$G_E = \int_0^{W_E} \frac{n_i^2}{n_{iE}^2} \frac{n}{D_E} dx \qquad (8.3.2)$$

G_E is the **emitter Gummel number**. As an exercise, please verify that in the special case of a uniform emitter, where n_{iE}, N_E (emitter doping concentration) and D_E are not functions of x,

$$I_B = A_E q \frac{D_E n_{iE}^2}{W_E N_E}(e^{qV_{BE}/kT} - 1) \qquad (8.3.3)$$

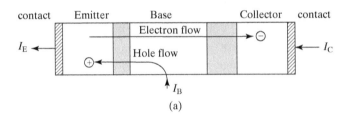

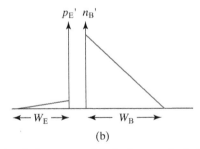

FIGURE 8–6 (a) Schematic of electron and hole flow paths in BJT; (b) hole injection into emitter closely parallels electron injection into base.[2]

---

[1] In older transistors with VERY long bases, I_B also supplies holes at a significant rate for recombination in the base. Recombination is negligible in the narrow base of a typical modern BJT.

[2] A good metal–semiconductor ohmic contact (at the end of the emitter) is an excellent source and sink of carriers. Therefore, the excess carrier concentration is assumed to be zero.

8.4 • CURRENT GAIN •

Perhaps the most important DC parameter of a BJT is its **common-emitter current gain,** β_F.

$$\boxed{\beta_F \equiv \frac{I_C}{I_B}} \tag{8.4.1}$$

Another current ratio, the **common-base current gain,** is defined by

$$I_C = \alpha_F I_E \tag{8.4.2}$$

$$\alpha_F \equiv \frac{I_C}{I_E} = \frac{I_C}{I_B + I_C} = \frac{I_C/I_B}{1 + I_C/I_B} = \frac{\beta_F}{1 + \beta_F} \tag{8.4.3}$$

α_F is typically very close to unity, such as 0.99, because β_F is large. From Eq. (8.4.3), it can be shown that

$$\beta_F = \frac{\alpha_F}{1 - \alpha_F} \tag{8.4.4}$$

I_B is a load on the input signal source, an undesirable side effect of forward biasing the BE junction. I_B should be minimized (i.e., β_F should be maximized). Dividing Eq. (8.2.9) by Eq (8.3.1),

$$\boxed{\beta_F = \frac{G_E}{G_B}} = \frac{D_B W_E N_E n_{iB}^2}{D_E W_B N_B n_{iE}^2} \tag{8.4.5}$$

A typical good β_F is 100. D and W in Eq. (8.4.5) cannot be changed very much. The most obvious way to achieve a high β_F, according to Eq. (8.4.5), is to use a large N_E and a small N_B. A small N_B, however, would introduce too large a base resistance, which degrades the BJT's ability to operate at high current and high frequencies. Typically, N_B is around 10^{18} cm^{-3}.

An emitter is said to be efficient if the emitter current is mostly the useful electron current injected into the base with little useless hole current (the base current). The **emitter efficiency** is defined as

$$\gamma_E = \frac{I_E - I_B}{I_E} = \frac{I_C}{I_C + I_B} = \frac{1}{1 + G_B/G_E} \tag{8.4.6}$$

EXAMPLE 8–1 Current Gain

A BJT has $I_C = 1$ mA and $I_B = 10$ µA. What are I_E, β_F, and α_F?

$$I_E = I_C + I_B = 1\,\text{mA} + 10\,\mu\text{A} = 1.01\,\text{mA}$$

$$\beta_F = \frac{I_C}{I_B} = \frac{1\,\text{mA}}{10\,\mu\text{A}} = 100$$

$$\alpha_F = \frac{I_C}{I_E} = \frac{1\,\text{mA}}{1.01\,\text{mA}} = 0.9901$$

SOLUTION:

Using this example, we can confirm Eqs. (8.4.3) and (8.4.4).

$$\frac{\beta_F}{1+\beta_F} = \frac{100}{101} = 0.9901 = \alpha_F$$

$$\frac{\alpha_F}{1-\alpha_F} = \frac{0.9901}{0.0099} = 100 = \beta_F$$

8.4.1 Emitter Band Gap Narrowing

To raise β_F, N_E is typically made larger than 10^{20} cm^{-3}. Unfortunately, when N_E is very large, n_{iE}^2 becomes larger than n_i^2. This is called the **heavy doping effect**. Recall Eq. (1.8.12)

$$n_i^2 = N_c N_v e^{-E_g/kT} \tag{8.4.7}$$

Heavy doping can modify the Si crystal sufficiently to reduce E_g and cause n_i^2 to increase significantly.[3] Therefore, the heavy doping effect is also known as **band gap narrowing**.

$$n_{iE}^2 = n_i^2 e^{\Delta E_{gE}/kT} \tag{8.4.8}$$

ΔE_{gE} is the narrowing of the emitter band gap relative to lightly doped Si and is negligible for $N_E < 10^{18}$ cm^{-3}, 50 meV at 10^{19} cm^{-3}, 95 meV cm^{-3} at 10^{20} cm^{-3}, and 140 meV at 10^{21} cm^{-3} [2].

8.4.2 Narrow Band-Gap Base and Heterojunction BJT

To further elevate β_F, we can raise n_{iB} by using a base material that has a smaller band gap than the emitter material. Si$_{1-\eta}$Ge$_\eta$ is an excellent base material candidate for an Si emitter. With $\eta = 0.2$, E_{gB} is reduced by 0.1 eV. In an SiGe BJT, the base is made of high-quality P-type epitaxial SiGe. In practice, η is graded such that $\eta = 0$ at the emitter end of the base and 0.2 at the drain end to create a built-in field that improves the speed of the BJT (see Section 8.7.2).

Because the emitter and base junction is made of two different semiconductors, the device is known as a **heterojunction bipolar transistor** or **HBT**. HBTs made of InP emitter ($E_g = 1.35$ eV) and InGaAs base ($E_g = 0.68$ eV) and GaAlAs emitter with GaAs base are other examples of well-studied HBTs. The ternary semiconductors are used to achieve lattice constant matching at the heterojunction (see Section 4.13.1).

---

[3] Heavy doping also affects n_i by altering N_c and N_v in a complex manner. It is customary to lump all these effects into an effective narrowing of the band gap.

EXAMPLE 8–2 Emitter Band-Gap Narrowing and SiGe Base

Assuming $D_B = 3D_E$, $W_E = 3W_B$, $N_B = 10^{18}$ cm^{-3}, and $n_{iB}^2 = n_i^2$. What is β_F for (a) $N_E = 10^{19}$ cm^{-3}, (b) $N_E = 10^{20}$ cm^{-3}, and (c) $N_E = 10^{20}$ cm^{-3} and the base is substituted with SiGe with a band narrowing of $\Delta E_{gB} = 60$ meV?

SOLUTION:

a. At $N_E = 10^{19}$ cm^{-3}, $\Delta E_{gE} \approx 50$ meV

$$n_{iE}^2 = n_i^2 e^{\Delta E_{gE}/kT} = n_i^2 e^{50/26 \text{ meV}} = n_i^2 e^{1.92} = 6.8 n_i^2$$

From Eq. (8.4.5), $\beta_F = \dfrac{D_B W_E}{D_E W_B} \times \dfrac{N_E n_i^2}{N_B n_{iE}^2} = \dfrac{9 \cdot 10^{19} \cdot n_i^2}{10^{18} \cdot 6.8 n_i^2} = 13$

b. At $N_E = 10^{20}$ cm^{-3}, $\Delta E_{gE} \approx 95$ meV

$$n_{iE}^2 = n_i^2 e^{\Delta E_{gE}/kT} = n_i^2 e^{95/26 \text{ meV}} = n_i^2 e^{3.65} = 38 n_i^2$$

$$\beta_F = \dfrac{D_B W_E}{D_E W_B} \times \dfrac{N_E n_i^2}{N_B n_{iE}^2} = \dfrac{9 \cdot 10^{20} \cdot n_i^2}{10^{18} \cdot 38 n_i^2} = 24$$

Increasing N_E from 10^{19} cm^{-3} to 10^{20} cm^{-3} does not increase β_F by anywhere near 10× because of band-gap narrowing. β_F can be raised of course by reducing N_B at the expense of a higher base resistance, which is detrimental to device speed (see Eq. 8.9.6).

c.
$$n_{iB}^2 = n_i^2 e^{\Delta E_{gB}/kT} = n_i^2 e^{60/26 \text{ meV}} = 10 n_i^2$$

$$\therefore \beta_F = \dfrac{D_B W_E}{D_E W_B} = 9 \times \dfrac{N_E n_{iB}^2}{N_B n_{iE}^2} = \dfrac{9 \cdot 10^{20} \cdot 10 n_{iB}^2}{10^{18} \cdot 39 n_i^2} = 237$$

8.4.3 Poly-Silicon Emitter

Whether the base material is SiGe or plain Si, a high-performance BJT would have a relatively thick (>100 nm) layer of As doped N$^+$ poly-Si film in the emitter (as shown in Fig. 8–7). Arsenic is thermally driven into the "base" by ~20 nm and converts that single-crystalline layer into a part of the N$^+$ emitter. This way, β_F is larger due to the large W_E, mostly made of the N$^+$ poly-Si. This is the **poly-Silicon emitter** technology. The simpler alternative, a deeper implanted or diffused N$^+$ emitter without the poly-Si film, is known to produce a higher density of crystal defects in the thin base (causing excessive emitters to collector leakage current or even shorts in a small number of the BJTs).

8.4.4 Gummel Plot and β_F Fall-Off at High and Low I_C

High-speed circuits operate at high I_C, and low-power circuits may operate at low I_C. Current gain, β, drops at both high I_C and at low I_C. Let us examine the causes.

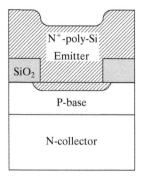

FIGURE 8–7 Schematic illustration of a poly-Si emitter, a common feature of high-performance BJTs.

We have seen in Fig. 8–5 (Gummel plot) that I_C flattens at high V_{BE} due to the high-level injection effect in the base. That I_C curve is replotted in Fig. 8–8. I_B, arising from hole injection into the emitter, does not flatten due to this effect (Fig. 8–8) because the emitter is very heavily doped, and it is practically impossible to inject a higher density of holes than N_E.

Over a wide mid-range of I_C in Fig. 8–8, I_C and I_B are parallel, indicating that the ratio of I_C/I_B, i.e., β_F, is a constant. This fact is obvious in Fig. 8–9. Above 1 mA, the slope of I_c in Fig. 8–8 drops due to high-level injection. Consequently, the I_c/I_B ratio or β_F decreases rapidly as shown in Fig. 8–9. This fall-off of current gain unfortunately degrades the performance of BJTs at high current where the BJT's speed is the highest (see Section 8.9). I_B in Fig. 8–8 is the base–emitter junction forward-bias current. As shown in Fig. 4–22, forward-bias current slope decreases at low V_{BE} or very low current due to the space-charge region (SCR) current (see Section 4.9.1). A similar slope change is sketched in Fig. 8–8. As a result, the I_c/I_B ratio or β_F decreases at very low I_C. The weak V_{BC} dependence of β_F in Fig. 8–9 is explained in the next section.

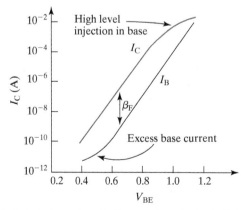

FIGURE 8–8 Gummel plot of I_C and I_B indicates that β_F ($= I_C/I_B$) decreases at high and low I_C.

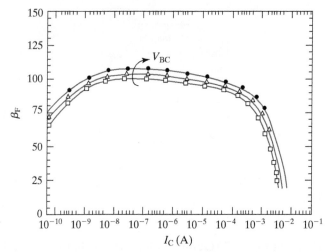

FIGURE 8–9 Fall-off of current gain at high- and low-current regions. $A_E = 0.6 \times 4.8\ \mu m^2$. From top to bottom: $V_{BC} = 2, 1$ and 0 V. Symbols are data. Lines are from a BJT model for circuit simulation. (From [3].)

8.5 ● BASE-WIDTH MODULATION BY COLLECTOR VOLTAGE ●

Instead of the flat I_C–V_{CE} characteristics shown in Fig. 8–2c, Fig. 8–10a (actual I_C–V_{CE} data) clearly indicates the presence of finite slopes. As in MOSFETs, a large **output conductance**, $\partial I_C/\partial V_{CE}$, of BJTs is deleterious to the voltage gain of circuits. The cause of the output conductance is **base-width modulation**, explained in Fig. 8–11. The thick vertical line indicates the location of the base-collector junction. With increasing V_{ce}, the base-collector depletion region widens and the neutral base width decreases. This leads to an increase in I_C as shown in Fig. 8–11.

If the I_C – V_{CE} curves are extrapolated as shown in Fig. 8–10b, they intercept the $I_C = 0$ axis at approximately the same point. Figure 8–10b defines the **Early voltage**, V_A. V_A is a parameter that describes the flatness of the I_C curves. Specifically, the output resistance can be expressed as V_A/I_C:

$$r_0 \equiv \left(\frac{\partial I_C}{\partial V_{CE}}\right)^{-1} = \frac{V_A}{I_C} \tag{8.5.1}$$

A large V_A (i.e., a large r_0) is desirable for high voltage gains. A typical V_A is 50 V. V_A is sensitive to the transistor design. Qualitatively, we can expect V_A and r_0 to increase (i.e., expect the base-width modulation to be a smaller fraction of the base width) if we:

(a) increase the base width

(b) increase the base doping concentration, N_B, or

(c) decrease the collector doping concentration, N_C.

Clearly, (a) would reduce the sensitivity to any given ΔW_B (see Fig. 8–11). (b) would reduce the depletion region thickness on the base side because the depletion region penetrates less into the more heavily doped side of a PN junction

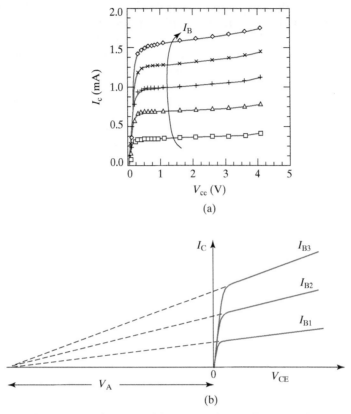

FIGURE 8–10 BJT output conductance: (a) measured BJT characteristics. I_B = 4, 8, 12, 16, and 20 µA. (From [3].); (b) schematic drawing illustrates the definition of Early voltage, V_A.

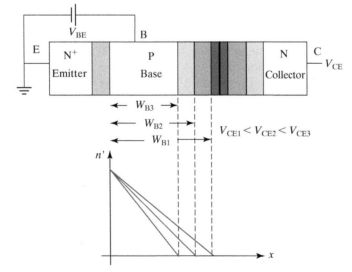

FIGURE 8–11 *As V_C increases, the BC depletion layer width increases and W_B decreases causing dn'/dx and I_C to increase.* In reality, the depletion layer in the collector is usually much wider than that in the base.

[see Eq. (4.2.5)]. For the same reason, (c) would tend to move the depletion region into the collector and thus reduce the depletion region thickness on the base side, too. Both (a) and (b) would depress β_F [see Eq. (8.4.5)]. (c) is the most acceptable course of action. It also reduces the base–collector junction capacitance, which is a good thing. Therefore, the collector doping is typically ten times lighter than base doping. In Fig. 8–10, the larger slopes at $V_{CE} > 3V$ are caused by impact ionization (Section 4.5.3). The rise of I_c due to base-width modulation is known as the **Early effect**, after its discoverer.

● **Early on Early Voltage** ●

Anecdote contributed by Dr. James Early, November 10, 1990

"In January, 1952, on my way to a Murray Hill Bell Labs internal meeting, I started to think about how to model the collector current as a function of the collector voltage. Bored during the meeting, I put down the expression for collector current $I_C = \beta_F I_B$. Differentiating with respect to V_C while I_B was held constant gave:

$$\frac{\partial I_C}{\partial V_C} = I_B \frac{\partial \beta_F}{\partial V_C}$$

How can β_F change with V_C? Of course! The collector depletion layer thickens as collector voltage is raised. The base gets thinner and current gain rises. Obvious! And necessarily true.

Why wasn't this found sooner? Of those who had thought about it at all before, none was educated in engineering analysis of electron devices, used to setting up new models, and bored at a meeting."

8.6 ● EBERS–MOLL MODEL ●

So far, we have avoided examining the part of the I–V curves in Fig. 8–12 that is close to $V_{CE} = 0$. This portion of the I–V curves is known as the **saturation region** because the base is saturated with minority carriers injected from both the emitter and the collector. (Unfortunately the MOSFET saturation region is named in exactly the opposite manner.) The rest of the BJT operation region is known as the **active region** or the linear region because that is where BJT operates in active circuits such as the linear amplifiers.

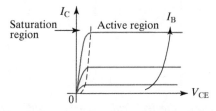

FIGURE 8–12 In the saturation region, I_C drops because the collector–base junction is significantly forward biased.

8.6 • Ebers–Moll Model

The Ebers–Moll model is a way to visualize as well as to mathematically describe both the active and the saturation regions of BJT operation. It is also the basis of BJT SPICE models for circuit simulation. The starting point is the idea that I_C is driven by two forces, V_{BE} and V_{BC}, as shown in Fig. 8–13. Let us first assume that a V_{BE} is present but $V_{BC} = 0$. Using Eq. (8.2.8),

$$I_C = I_S(e^{qV_{BE}/kT} - 1) \tag{8.6.1}$$

$$I_B = \frac{I_S}{\beta_F}(e^{qV_{BE}/kT} - 1) \tag{8.6.2}$$

Now assume that the roles of the collector and emitter are reversed, i.e., a (possibly forward bias) V_{BC} is present and $V_{BE} = 0$. Electrons would be injected from the collector into base and flow to the emitter. The collector now functions as the emitter and the emitter functions as the collector[4]

$$I_E = I_S(e^{qV_{BC}/kT} - 1) \tag{8.6.3}$$

$$I_B = \frac{I_S}{\beta_R}(e^{qV_{BC}/kT} - 1) \tag{8.6.4}$$

$$I_C = -I_E - I_B = -I_S\left(1 + \frac{1}{\beta_R}\right)(e^{qV_{BC}/kT} - 1) \tag{8.6.5}$$

β_R is the **reverse current gain**. (This is why β_F has F as the subscript. β_F is the **forward current gain**.) While β_F is usually quite large, β_R is small because the doping concentration of the collector, which acts as the "emitter" in the reverse mode, is not high. When both V_{BE} and V_{BC} are present, Eqs. (8.6.1) and (8.6.5) are superimposed as are Eqs. (8.6.2) and (8.6.4).

$$I_C = I_S(e^{qV_{BE}/kT} - 1) - I_S\left(1 + \frac{1}{\beta_R}\right)(e^{qV_{BC}/kT} - 1) \tag{8.6.6}$$

$$I_B = \frac{I_S}{\beta_F}(e^{qV_{BE}/kT} - 1) + \frac{I_S}{\beta_R}(e^{qV_{BC}/kT} - 1) \tag{8.6.7}$$

Equations (8.6.6) and (8.6.7) compromise the Ebers–Moll model as commonly used in SPICE models. These two equations can generate I_C vs. V_{CE} plots with excellent agreement with measured data as shown in Fig. 8–14.

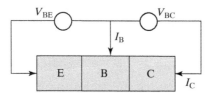

FIGURE 8–13 I_C is driven by two voltage sources, V_{BE} and V_{BC}.

---

[4] When the emitter and collector roles are interchanged, the upper and lower limits of integration in Eq. (8.2.11) are interchanged with no effect on G_B or I_S.

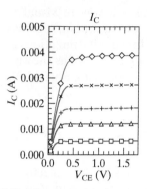

FIGURE 8–14 Ebers–Moll model (line) agrees with the measured data (symbols) in both the saturation and linear regions. I_B = 4.3, 11, 17, 28, and 43 µA. High-speed SiGe-base BJT. $A_E = 0.25 \times 5.75$ µm^2. (From [3].)

What causes I_C to decrease at low V_{CE}? In this region, both the BE and BC junctions are forward biased. (For example: V_{BE} = 0.8 V, V_{BC} = 0.6 V, thus V_{CE} = 0.2 V.) A forward-biased V_{BC} causes the n' at x = W_B to rise in Fig. 8–4. This depresses dn'/dx and therefore I_C.

8.7 ● TRANSIT TIME AND CHARGE STORAGE ●

Static IV characteristics are only one part of the BJT story. Another part is its dynamic behavior or its speed. When the BE junction is forward biased, excess holes are stored in the emitter, the base, and even the depletion layers. We call the sum of all the excess hole charges everywhere Q_F. Q_F is the stored **excess carrier charge**. If Q_F = 1 pC (pico coulomb), there is +1 pC of excess hole charge and –1 pC of excess electron charge stored in the BJT.[5] The ratio of Q_F to I_C is called the **forward transit time**, τ_F.

$$\boxed{\tau_F \equiv \frac{Q_F}{I_C}} \qquad (8.7.1)$$

Equation (8.7.1) states the simple but important fact that I_C and Q_F are related by a constant ratio, τ_F. Some people find it more intuitive to think of τ_F as the **storage time**. In general, Q_F and therefore τ_F are very difficult to predict accurately for a complex device structure. However, τ_F can be measured experimentally (see Sec. 8.9) and once τ_F is determined for a given BJT, Eq. (8.7.1) becomes a powerful conceptual and mathematical tool giving Q_F as a function of I_C, and vice versa. τ_F sets a high-frequency limit of BJT operation.

8.7.1 Base Charge Storage and Base Transit Time

To get a sense of how device design affects the transit time, let us analyze the excess hole charge in the base, Q_{FB}, from which we will obtain the base transit time, τ_{FB}.

Q_{FB} is qA_E times the area under the line in Fig. 8–15.

---

[5] This results from Eq. (2.6.2), n' = p'.

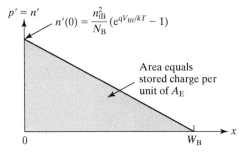

FIGURE 8–15 Excess hole and electron concentrations in the base. They are equal due to charge neutrality [Eq. (2.6.2)].

$$Q_{FB} = qA_E n'(0)W_B/2 \tag{8.7.2}$$

Dividing Q_{FB} by I_C and using Eq. (8.2.7),

$$\boxed{\frac{Q_{FB}}{I_C} \equiv \tau_{FB} = \frac{W_B^2}{2D_B}} \tag{8.7.3}$$

To reduce τ_{FB} (i.e., to make a faster BJT), it is important to reduce W_B.

EXAMPLE 8–3 Base Transit Time

What is τ_{FB} if $W_B = 70$ nm and $D_B = 10$ cm²/s?

SOLUTION:

$$\tau_{FB} = \frac{W_B^2}{2D_B} = \frac{(7 \times 10^{-6} \text{cm})^2}{2 \times 10 \text{ cm}^2/\text{s}} = 2.5 \times 10^{-12} \text{s} = 2.5 \text{ ps}$$

2.5 ps is a very short time. Since light speed is 3×10^8 m/s, light travels less than 1 mm in 2.5 ps.

8.7.2 Drift Transistor–Built-In Base Field

The base transit time can be further reduced by building into the base a drift field that aids the flow of electrons from the emitter to the collector. There are two ways of accomplishing this. The classical method is to use graded base doping, i.e., a large N_B near the EB junction, which gradually decreases toward the CB junction as shown in Fig. 8–16a.

Such a doping gradient is automatically achieved if the base is produced by dopant diffusion. The changing N_B creates a dE_v/dx and a dE_c/dx. This means that there is a drift field [Eq. (2.4.2)]. Any electrons injected into the base would drift toward the collector with a base transit time shorter than the **diffusion transit time**, $W_B^2/2D_B$.

Figure 8–16b shows a more effective technique. In a **SiGe BJT**, P-type epitaxial $Si_{1-\eta}Ge_\eta$ is grown over the Si collector with a constant N_B and η linearly varying from about 0.2 at the collector end to 0 at the emitter end [4]. A large

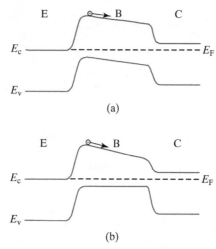

FIGURE 8–16 Two ways of building dE_C/dx into the base. (a) E_{gB} fixed, N_B decreasing from emitter end to collector end; (b) N_B fixed, E_{gB} decreasing from emitter end to collector end.

dE_c/dx can be produced by the grading of E_{gB}. These high-speed BJTs are used in high-frequency communication circuits. Drift transistors can have a base transit time several times less than $W_B^2/2D_B$, as short as 1 ps.

8.7.3 Emitter-to-Collector Transit Time and Kirk Effect[6]

The total forward transit time, τ_F, is also known as the **emitter-to-collector transit time**. τ_{FB} is only one portion of τ_F. The base transit time typically contributes about half of τ_F. To reduce the transit (or storage) time in the emitter and collector, the emitter and the depletion layers must be kept thin. τ_F can be measured, and an example of τ_F is shown in Fig. 8–17. τ_F starts to increase at a current density where the electron density corresponding to the dopant density in the collector ($n = N_C$) is insufficient to support the collector current even if the dopant-induced electrons move at the saturation velocity (see Section 6.8). This intriguing condition of too few dopant atoms and too much current leads to a reversal of the sign of the charge density in the "depletion region."

$$I_C = A_E q n v_{sat} \tag{8.7.4}$$

$$\rho = qN_C - qn$$

$$= qN_C - \frac{I_C}{A_E v_{sat}} \tag{8.7.5}$$

$$\frac{d\mathscr{E}}{dx} = \rho/\varepsilon_s \tag{4.1.5}$$

---

[6] This section may be omitted in an accelerated course.

8.7 • Transit Time and Charge Storage

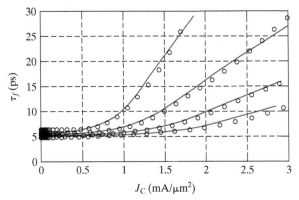

FIGURE 8–17 Transit time vs. I_C/A_E. From top to bottom: $V_{CE} = 0.5, 0.8, 1.5$, and 3 V. The rise at high I_C is due to base widening (Kirk effect). (From [3].)

When I_C is small, $\rho = qN_C$ as expected from the PN junction analysis (see Section 4.3), and the electric field in the depletion layer is shown in Fig. 8–18a. The shaded area is the potential across the junction, $V_{CB} + \phi_{bi}$. The N^+ collector is always present to reduce the series resistance (see Fig. 8–22). No depletion layer is

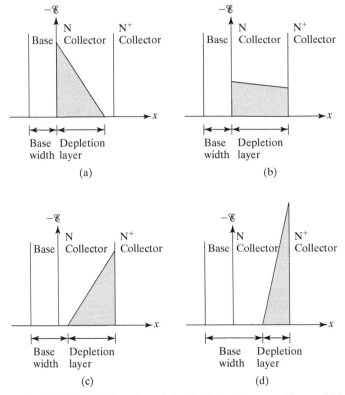

FIGURE 8–18 Electric field $\mathscr{E}(x)$, location of the depletion layer, and base width at (a) low I_C such as 0.1 mA/μm² in Fig. 8–17; (b) larger I_C; (c) even larger I_C (such as 1 mA/μm²) and base widening is evident; and (d) very large I_C with severe base widening.

shown in the base for simplicity because the base is much more heavily doped than the collector. As I_C increases, ρ decreases [Eq. (8.7.5)] and $d\mathscr{E}/dx$ decreases as shown in Fig. 8–18b. The electric field drops to zero in the very heavily doped N⁺ collector as expected. Note that the shaded area under the $\mathscr{E}(x)$ line is basically equal to the shaded area in the Fig. 8–18a because V_{CB} is kept constant. In Fig. 8–18c, I_C is even larger such that ρ in Eq. (8.7.5) and therefore $d\mathscr{E}/dx$ has changed sign. The size of the shaded areas again remains unchanged. In this case, the high-field region has moved to the right-hand side of the N collector. As a result, the base is effectively widened. In Fig. 8–18d, I_C is yet larger and the base become yet wider. Because of the **base widening**, τ_F increases as a consequence [see Eq. (8.7.3)]. This is called the **Kirk effect**. Base widening can be reduced by increasing N_C and V_{CE}. *The Kirk effect limits the peak BJT operating speed* (see Fig. 8–21).

8.8 • SMALL-SIGNAL MODEL •

Figure 8–19 is an equivalent circuit for the behavior of a BJT in response to a small input signal, e.g., a 10 mV sinusoidal signal, superimposed on the DC bias. BJTs are often operated in this manner in **analog circuits**.

If V_{BE} is not close to zero, the "1" in Eq. (8.2.8) is negligible; in that case

$$I_C = I_s e^{qV_{BE}/kT} \tag{8.8.1}$$

When a signal v_{BE} is applied to the BE junction, a collector current $g_m v_{BE}$ is produced. g_m, the **transconductance**, is

$$g_m \equiv \frac{dI_C}{dV_{BE}} = \frac{d}{dV_{BE}}(I_s e^{qV_{BE}/kT})$$

$$= \frac{q}{kT} I_s e^{qV_{BE}/kT} = I_C / \frac{kT}{q} \tag{8.8.2}$$

$$\boxed{g_m = I_C / \frac{kT}{q}} \tag{8.8.3}$$

At room temperature, for example, $g_m = I_C/26$ mV. The transconductance is determined by the collector bias current, I_C.

The input node, the base, appears to the input drive circuit as a parallel RC circuit as shown in Fig. 8–19.

$$\frac{1}{r_\pi} = \frac{dI_B}{dV_{BE}} = \frac{1}{\beta_F}\frac{dI_C}{dV_{BE}} = \frac{g_m}{\beta_F} \tag{8.8.4}$$

$$r_\pi = \beta_F / g_m \tag{8.8.5}$$

Q_F in Eq. (8.7.1) is the excess carrier charge stored in the BJT. If $Q_F = 1$ pC, there is +1 pC of excess holes and –1 pC of excess electrons in the BJT. All the excess hole charge, Q_F, is supplied by the base current, I_B. Therefore, the base presents this capacitance to the input drive circuit:

$$C_\pi = \frac{dQ_F}{dV_{BE}} = \frac{d}{dV_{BE}} \tau_F I_C = \tau_F g_m \tag{8.8.6}$$

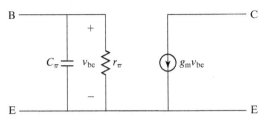

FIGURE 8–19 A basic small-signal model of the BJT.

The capacitance in Eq. (8.8.6) may be called the **charge-storage capacitance**, better known as the **diffusion capacitance**. In addition, there is one charge component that is not proportional to I_C and therefore cannot be included in Q_F [see Eq. (8.7.1)]. That is the junction depletion-layer charge. Therefore, a complete model of C_π should include the BE junction depletion-layer capacitance, C_{dBE}

$$C_\pi = \tau_F g_m + C_{dBE} \tag{8.8.7}$$

EXAMPLE 8–4 Small-Signal Model Parameters

The BJT represented in Figs. 8–9 and 8–17 is biased at $I_C = 1$ mA and $V_{CE} = 3$ V. $T = 300$ K and $A_E = 5.6$ μm². Find (a) g_m, (b) r_π, and (c) C_π.

SOLUTION:

a. $g_m \equiv I_C / \dfrac{kT}{q} = \dfrac{1 \text{ mA}}{26 \text{ mV}} = 39 \dfrac{\text{mA}}{\text{V}} = 39$ mS (milli siemens)

b. From Fig. 8–9, $\beta_F = 90$ at $I_C = 1$ mA and $V_{CB} = 2$ V. ($V_{CB} = V_{CE} + V_{EB} = 3\text{ V} + V_{EB} \approx 3\text{ V} - 1\text{ V} = 2\text{ V}$.)

$$r_\pi = \beta_F / g_m = \dfrac{90}{39 \text{ mS}} = 2.3 \text{ k}\Omega$$

c. From Fig. 8–17, at $J_C = I_C/A_E = 1$ mA/5.6 μm² = 0.18 mA/μm² and $V_{CE} = 3$ V, we find $\tau_F = 5$ ps.

$C_\pi = \tau_F g_m = 5 \times 10^{-12} \times 0.039 \approx 2.0 \times 10^{-13}$ F = 200 fF (femtofarad).

Once the parameters in Fig. 8–19 have been determined, one can use the small-signal model to analyze circuits with arbitrary signal-source impedance network (comprising resistors, capacitors, and inductors) and load impedance network as illustrated in Fig. 8–20a. The next section on cutoff frequency presents an example of the use of the small signal model.

While Fig. 8–20a is convenient for hand analysis, SPICE circuit simulation can easily use the more accurate small-signal model shown in Fig. 8–20b.

Some of the new parameters in Fig. 8–20b have familiar origins. For example, r_0 is the intrinsic output resistance, V_A/I_C (Section 8.5). C_μ also arises from base width modulation; when V_{BC} varies, the base width varies; therefore, the base stored charge (area of the triangle in Fig. 8–11) varies, thus giving rise to $C_\mu = dQ_{FB}/dV_{CB}$. C_{dBC} is the CB junction depletion-layer capacitance. Model

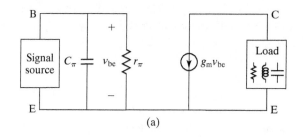

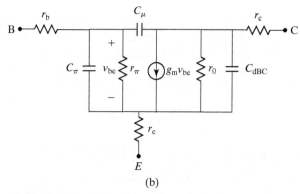

FIGURE 8–20 (a) The small-signal model can be used to analyze a BJT circuit by hand; (b) a small-signal model for circuit simulation by computer.

parameters are difficult to predict from theory with the accuracy required for commercial circuit design. Therefore, the parameters are routinely determined through comprehensive measurement of the BJT AC and DC characteristics.

8.9 • CUTOFF FREQUENCY •

Consider a special case of Fig. 8–20a. The load is a short circuit. The signal source is a current source, i_b, at a frequency f. At what frequency does the AC current gain $\beta\,(\equiv i_c/i_b)$ fall to unity?

$$v_{be} = \frac{i_b}{\text{input admittance}} = \frac{i_b}{1/r_\pi + j\omega C_\pi} \tag{8.9.1}$$

$$i_c = g_m v_{be} \tag{8.9.2}$$

Using Eqs. (8.9.1), (8.9.2), (8.8.7), and (8.8.3)

$$\beta(\omega) \equiv \left|\frac{i_c}{i_b}\right| = \frac{g_m}{|1/r_\pi + j\omega C_\pi|} = \frac{1}{|1/g_m r_\pi + j\omega \tau_F + j\omega C_{dBE}/g_m|}$$

$$= \frac{1}{|1/\beta_F + j\omega \tau_F + j\omega C_{dBE} kT/qI_C|} \tag{8.9.3}$$

At $\omega = 0$, i.e., DC, Eq. (8.9.3) reduces to β_F as expected. As ω increases, β drops. By carefully analyzing the $\beta(\omega)$ data, one can determine τ_F. If $\beta_F \gg 1$ so that $1/\beta_F$ is negligible, Eq. (8.9.3) shows that $\beta(\omega) \propto 1/\omega$ and $\beta = 1$ at

$$f_T = \frac{1}{2\pi(\tau_F + C_{dBE}kT/qI_C)} \tag{8.9.4}$$

Using a more complete small-signal model similar to Fig. 8–20b, it can be shown that

$$f_T = \frac{1}{2\pi[\tau_F + (C_{dBE} + C_{dBC})kT/(qI_C) + C_{dBC}(r_e + r_c)]} \tag{8.9.5}$$

f_T is the cutoff frequency and is commonly used to compare the speed of transistors. Equations (8.9.4) and (8.9.5) predict that f_T rises with increasing I_C due to increasing g_m, in agreement with the measured f_T shown in Fig. 8–21. At very high I_C, τ_F increases due to base widening (Kirk Effect, Fig. 8–17), and therefore, f_T falls. *BJTs are often biased near the I_C where f_T peaks in order to obtain the best high-frequency performance.*

f_T is the frequency of unity current gain. The frequency of unity power gain, called the **maximum oscillation frequency**, can be shown to be [5]

$$f_{max} = \left(\frac{f_T}{8\pi r_b C_{dBC}}\right)^{1/2} \tag{8.9.6}$$

It is therefore important to reduce the base resistance, r_b.

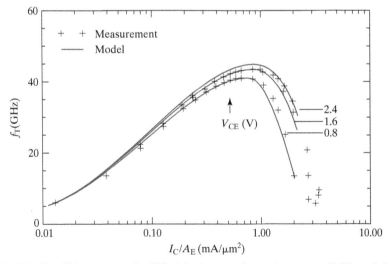

FIGURE 8–21 Cutoff frequency of a SiGe bipolar transistor. A compact BJT model matches the measured f_T well. (From [6]. © 1997 IEEE.)

● BJT Structure for Minimum Parasitics and High Speed ●

While MOSFET scaling is motivated by the need for high packing density and large I_{dsat}, BJT scaling is often motivated by the need for high f_T and f_{max}. This involves the reduction of τ_F (thin base, etc.) and the reduction of parasitics (C_{dBE}, C_{dBC}, r_b, r_e, and r_c). Figure 8–22 is a schematic of a high-speed BJT.

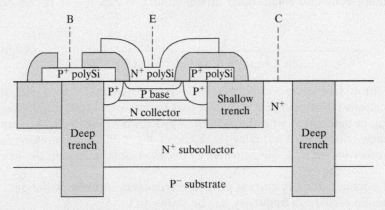

FIGURE 8–22 Schematic of a BJT with poly-Si emitter, self-aligned base, and deep-trench isolation. The darker areas represent electrical insulator regions.

An N$^+$ poly-Si emitter and a thin base are clearly seen in Fig. 8–22. The base is contacted through two small P$^+$ regions created by boron diffusion from a P$^+$ poly-Si film. The film also provides a low-resistance electrical connection to the base without introducing a large P$^+$ junction area and junction capacitance. To minimize the base series resistance, the emitter opening in Fig. 8–22 is made very narrow. The lightly doped epitaxial (see Section 3.7.3) N-type collector is contacted through a heavily doped subcollector in order to minimize the collector series resistance. The substrate is lightly doped to minimize the collector capacitance. Both the **shallow trench** and the **deep trench** are filled with dielectrics (SiO$_2$) and serve the function of electrical **isolation**. The deep trench forms a rectangular moat that completely surrounds the BJT. It isolates the collector of this transistor from the collectors of neighboring transistors. The structure in Fig. 8–22 incorporates many improvements that have been developed over the past decades and have greatly reduced the device size from older BJT designs. Still, a BJT is a larger transistor than a MOSFET.

8.10 ● CHARGE CONTROL MODEL[7] ●

The small-signal model is ideal for analyzing circuit response to small sinusoidal signals. What if the input signal is large? For example, what $I_C(t)$ is produced by a step-function I_B switching from zero to 20 μA or by any $I_B(t)$? The response can be

---

[7] This section may be omitted. Charge control model is used for analysis of digital switching operations.

conveniently analyzed with the **charge control model,** a simple extension of the charge storage concept (Eq. (8.7.1)).

$$I_C = Q_F/\tau_F \tag{8.10.1}$$

Assume that Eq. (8.10.1) holds true even if Q_F varies with time

$$\boxed{I_C(t) = Q_F(t)/\tau_F} \tag{8.10.2}$$

$I_C(t)$ becomes known if we can solve for $Q_F(t)$. (τ_F has to be characterized beforehand for the BJT being used.) Equation (8.10.2) suggests the concept that I_C is controlled by Q_F, hence the name of the charge control model. From Eq. (8.10.1), at DC condition,

$$I_B = I_C/\beta_F = Q_F/\tau_F\beta_F \tag{8.10.3}$$

Equation (8.10.3) has a straightforward physical meaning: *In order to sustain a constant excess hole charge in the transistor, holes must be supplied to the transistor through I_B to replenish the holes that are lost to recombination. Therefore, DC I_B is proportional to Q_F.* When holes are supplied by I_B at the rate of $Q_F/\tau_F\beta_F$, the rate of hole supply is exactly equal to the rate of hole loss to recombination and Q_F remains at a constant value. What if I_B is larger than $Q_F/\tau_F\beta_F$? In that case, holes flow into the BJT at a higher rate than the rate of hole loss–and the stored hole charge (Q_F) increases with time.

$$\boxed{\frac{dQ_F}{dt} = I_B(t) - \frac{Q_F}{\tau_F\beta_F}} \tag{8.10.4}$$

Equations (8.10.4) and (8.10.2), together constitute the basic charge control model. For any given $I_B(t)$, Eq. (8.10.4) can be solved for $Q_F(t)$ analytically or by numerical integration. Once $Q_F(t)$ is found, $I_C(t)$ becomes known from Eq. (8.10.2). We may interpret Eq. (8.10.4) with the analogy of filling a very leaky bucket from a faucet shown in Fig. 8–23. Q_F is the amount of water in the bucket, and $Q_F/\tau_F\beta_F$ is the rate

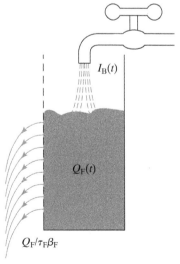

FIGURE 8–23 Water analogy of the charge control model. Excess hole charge (Q_F) rises (or falls) at the rate of supply (I_B) minus loss ($\propto Q_F$).

of water leakage. I_B is the rate of water flowing from the faucet into the bucket. If the faucet is turned fully open, the water level rises in the bucket; if it is turned down, the water level falls.

EXAMPLE 8-5 Finding $I_C(t)$ for a Step $I_B(t)$

QUESTION: τ_F and β_F of a BJT are given. $I_B(t)$ is a step function rising from zero to I_{B0} at $t = 0$ as shown in Fig. 8-24. Find $I_C(t)$.

SOLUTION:
At $t \geq 0$, $I_B(t) = I_{B0}$ and the solution of Eq. (8.10.4)

$$\frac{dQ_F}{dt} = I_B(t) - \frac{Q_F}{\tau_F \beta_F} \tag{8.10.4}$$

is $Q_F(t) = \tau_F \beta_F I_{B0}(1 - e^{-t/\tau_F \beta_F})$ (8.10.5)

Please verify that Eq. (8.10.5) is the correct solution by substituting it into Eq. (8.10.4). Also verify that the initial condition $Q_F(0) = 0$ is satisfied by Eq. (8.10.5). $I_C(t)$ follows Eq. (8.10.2).

$$I_C(t) = Q_F(t)/\tau_F = \beta_F I_{B0}(1 - e^{-t/\tau_F \beta_F}) \tag{8.10.6}$$

$I_C(t)$ is plotted in Fig. 8-24. At $t \to \infty$, $I_C = \beta_F I_{B0}$ as expected. $I_C(t)$ can be determined for any given $I_B(t)$ by numerically solving Eq. (8.10.4).

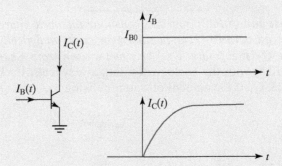

FIGURE 8-24 From the given step-function $I_B(t)$, charge control analysis can predict $I_C(t)$.

What we have studied in this section is a basic version of the charge control model. For a more exact analysis, one would introduce the junction depletion-layer capacitances into Eq. (8.10.4). Diverting part of I_B to charge the junction capacitances would produce an additional delay in $I_C(t)$.

8.11 • MODEL FOR LARGE-SIGNAL CIRCUIT SIMULATION •

The BJT model used in circuit simulators such as SPICE can accurately represent the DC and dynamic currents of the transistor in response to $V_{BE}(t)$ and $V_{CE}(t)$. A typical **circuit simulation model** or **compact model** is made of the Ebers–Moll model (with V_{BE} and V_{BC} as the two driving forces for I_C and I_B) plus additional enhancements for high-level injection, voltage-dependent capacitances that accurately represent the charge storage in the transistor, and parasitic resistances as shown in Fig. 8–25. This BJT model is known as the **Gummel–Poon model**.

The two diodes represent the two I_B terms due to V_{BE} and V_{BC} similar to Eq. (8.6.7). The capacitor labeled Q_F is voltage dependent such that the charge stored in it is equal to the Q_F described in Section 8.7. Q_R is the counterpart of Q_F produced by a forward bias at the BC junction. Inclusion of Q_R makes the dynamic response of the model accurate even when V_{BC} is sometimes forward biased. C_{BE} and C_{BC} are the junction depletion-layer capacitances. C_{CS} is the collector-to-substrate capacitance (see Fig. 8–22).

$$I_C = I_S'(e^{qV_{BE}/kT} - e^{qV_{BC}/kT})\left(1 + \frac{V_{CB}}{V_A}\right) - \frac{I_S}{\beta_R}(e^{qV_{BC}/kT} - 1) \quad (8.11.1)$$

The similarity between Eqs. (8.11.1) and (8.6.6) is obvious. The $1 + V_{CB}/V_A$ factor is added to represent the Early effect—I_C increasing with increasing V_{CB}. I_S' differs from I_S in that I_S' decreases at high V_{BE} due to the high-level injection effect in accordance with Eq. (8.2.11) and as shown in Fig. 8–5.

$$I_B = \frac{I_S}{\beta_F}(e^{qV_{BE}/kT} - 1) + \frac{I_S}{\beta_R}(e^{qV_{BC}/kT} - 1) + I_{SE}(e^{qV_{BE}/n_E kT} - 1) \quad (8.11.2)$$

Equation (8.11.2) is identical to Eq. (8.6.7) except for the additional third term, which represents the excess base junction current shown in Fig. 8–8. I_{SE} and n_E parameters are determined from the measured BJT data as are all of the several dozens of model parameters. The continuous curves in Figs. 8–9, 8–10a, and 8–17 are all examples of compact models. The excellent agreement between the models and the discrete data points in the same figures are necessary conditions for the

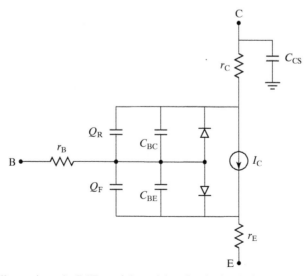

FIGURE 8–25 Illustration of a BJT model used for circuit simulation.

circuit simulation results to be accurate. The other necessary condition is that the capacitance in Fig. 8–24 be modeled accurately.

8.12 ● CHAPTER SUMMARY ●

The base–emitter junction is usually forward biased while the base–collector junction is reverse biased (as shown in Fig. 8–1b). V_{BE} determines the rate of electron injection from the emitter into the base, and thus uniquely determines the collector current, I_C, regardless of the reverse bias, V_{CB}

$$I_C = A_E \frac{qn_i^2}{G_B}(e^{qV_{BE}/kT} - 1) \qquad (8.2.9)$$

$$G_B \equiv \int_0^{W_B} \frac{n_i^2}{n_{iB}^2} \frac{p}{D_B} dx \qquad (8.2.11)$$

G_B is the base Gummel number, which represents all the subtleties of BJT design that affect I_C: base material, nonuniform base doping, nonuniform material composition, and the high-level injection effect.

An undesirable but unavoidable side effect of the application of V_{BE} is a hole current flowing from the base, mostly into the emitter. This base (input) current, I_B, is related to I_C by the common-emitter current gain, β_F.

$$\beta_F = \frac{I_C}{I_B} \approx \frac{G_E}{G_B} \qquad (8.4.1), (8.4.5)$$

where G_E is the emitter Gummel number. The common-base current gain is

$$\alpha_F \equiv \frac{I_C}{I_E} = \frac{\beta_F}{1 + \beta_F} \qquad (8.4.3)$$

The Gummel plot, Fig. 8–8, indicates that β_F falls off in the high I_C region due to high-level injection in the base and also in the low I_C region due to excess base current.

Base-width modulation by V_{CB} results in a significant slope of the I_C–V_{CE} curve in the active region. This is the Early effect. The slope, called the output conductance, limits the voltage gain that can be produced with a BJT. The Early effect can be suppressed with a lightly doped collector. A heavily doped subcollector (see Fig. 8–22) is routinely used to reduce the collector resistance.

Due to the forward bias, V_{BE}, a BJT stores a certain amount of excess hole charge, which is equal but of opposite sign to the excess electron charge. Its magnitude is called the excess carrier charge, Q_F. Q_F is linearly proportional to I_C.

$$Q_F \equiv I_C \tau_F \qquad (8.7.1)$$

τ_F is the forward transit time. If there were no excess carriers stored outside the base

$$\tau_F = \tau_{FB} = \frac{W_B^2}{2D_B} \qquad (8.7.3)$$

τ_{FB} is the base transit time. In general, $\tau_F > \tau_{FB}$ because excess carrier storage in the emitter and in the depletion layer are also significant. All these regions should be made small in order to minimize τ_F. Besides minimizing the base width, W_B, τ_{FB} may be reduced by building a drift field into the base with graded base doping (or better, with graded Ge content in a SiGe base). τ_{FB} is significantly increased at large I_C due to base widening, also known as the **Kirk effect**.

For computer simulation of circuits, the Gummel–Poon model, shown in Fig. 8–25, is widely used. Both the DC and the dynamic (charge storage) currents are well modeled. The Early effect and high-level injection effect are included. Simpler models consisting of R, C, and current source are used for hand analysis of circuits. The small-signal models (Figs. 8–19 and 8–20b) employ parameters such as transconductance

$$g_m \equiv \frac{dI_C}{dV_{BE}} = I_C \Big/ \frac{kT}{q} \qquad (8.8.2)$$

input capacitance

$$C_\pi = \frac{dQ_F}{dV_{BE}} = \tau_F g_m \qquad (8.8.6)$$

and input resistance.

$$r_\pi = \frac{dV_{BE}}{dI_B} = \beta_F / g_m \qquad (8.8.5)$$

The BJT's unity-gain cutoff frequency (at which β falls to unity) is f_T. In order to raise device speed, device density, or current gain, a modern high-performance BJT usually employs (see Fig. 8–22) poly-Si emitter, self-aligned poly-Si base contacts, graded Si-Ge base, shallow oxide trench, and deep trench isolation. High-performance BJTs excel over MOSFETs in circuits requiring the highest device g_m and speed.

● **PROBLEMS** ●

● Energy Band Diagram of BJT ●

8.1 A Silicon PNP BJT with $N_{aE} = 5 \times 10^{18}$ cm^{-3}, $N_{dB} = 10^{17}$ cm^{-3}, $N_{aC} = 10^{15}$ cm^{-3}, and $W_B = 3$ µm is at equilibrium at room temperature.
   (a) Sketch the energy band diagram for the device, properly positioning the Fermi level in the three regions.
   (b) Sketch (i) the electrostatic potential, setting $V = 0$ in the emitter region, (ii) the electric field, and (iii) the charge density as a function of position inside the BJT.
   (c) Calculate the net built-in potential between the collector and the emitter.
   (d) Determine the quasi-neutral region width of the base.
   Bias voltages of $V_{EB} = 0.6$ V and $V_{CB} = -2$ V are now applied to the BJT.
   (e) Sketch the energy band diagram for the device, properly positioning the Fermi level in the three regions.
   (f) On the sketches completed in part (b), sketch the electrostatic potential, electric field, and charge density as a function of position inside the biased BJT.

● **IV Characteristics and Current Gain** ●

8.2 Derive Eq. (8.4.4) from the definitions of β_F (Eq. 8.4.1) and α_F (Eq. 8.4.2).

8.3 Consider a conventional NPN BJT with uniform doping. The base–emitter junction is forward biased, and the base–collector junction is reverse biased.

  (a) Qualitatively sketch the energy band diagram.

  (b) Sketch the minority carrier concentrations in the base, emitter, and collector regions.

  (c) List all the causes contributing to the base and collector currents. You may neglect thermal recombination–generation currents in the depletion regions.

8.4 Neglect all the depletion region widths. The emitter, base, and collector of an NPN transistor have doping concentrations 10^{19}, 10^{17}, and 10^{15} cm^{-3} respectively. $W_E = 0.8$ μm, $W_B = 0.5$ μm, and $W_C = 2.2$ μm as shown in Fig. 8–26. Assume $\exp(qV_{BE}/kT) = 10^{10}$ and the base–collector junction is reverse biased. Assume that the device dimensions are much smaller than the carrier diffusion lengths throughout.

  (a) Find and plot the electron current density, $J_n(x)$, and hole current density, $J_p(x)$, in each region (J_p in the base is rather meaningless since it is three-dimensional in reality).

  (b) What are γ_E and β_F (assume $L_B = 10$ μm)?

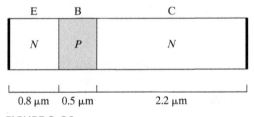

FIGURE 8–26

8.5 For the following questions, answer in one or two sentences.

  (a) Why should the emitter be doped more heavily than the base?

  (b) "The base width is small" is often stated in device analysis. What is it being compared with?

  (c) If the base width, W_B, were made smaller, explain how it would affect the base width modulation.

  (d) Why does β_F increase with increasing I_C at small values of collector current?

  (e) Explain why β_F falls off at large values of collector current.

  (f) For a PNP device, indicate the voltage polarity (+ or –) for the following:

Region of operation	V_{EB}	V_{CB}
Active		
Saturation		

● **Schottky Emitter and Collector** ●

8.6 The emitter of a high-β_F BJT should be heavily doped.

  (a) Is it desirable to replace the emitter in BJT with a metal?

  (b) Considering a metal on N–Si junction. The minority-carrier injection ratio is the number per second of minority carriers injected into the semiconductor divided by the majority carrier injected per second from the semiconductor into the metal when the device is forward biased. The ratio is I_{diff} / I_{te}, where I_{diff} and I_{te} are respectively the hole diffusion current flowing into the semiconductor and the thermionic emission current of electrons flowing into the metal. Estimate the minority carrier injection ratio in an Si Schottky diode where K = 140 A/cm^{-2}, Φ_B = 0.72 eV, $N_d = 10^{16}$ cm^{-3}, $\tau_p = 10^{-6}$ s and T = 300 K. I_{diff} in the given diode is the same as the hole diffusion current into the N side of a P$^+$-N step junction diode with the same N_d & τ_p.

(c) If the collector in BJT is replaced with a metal, would it still function as a BJT? (Hint: Compare the energy diagrams of the two cases.)

● **Gummel Plot** ●

8.7 Consider an NPN transistor with W_E = 0.5 μm, W_B = 0.2 μm, W_C = 2 μm, D_B = 10 cm^2/s.

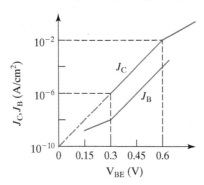

FIGURE 8–27

(a) Find the peak β_F from Fig. 8–27.
(b) Estimate the base doping concentration N_B.
(c) Find the V_{BE} at which the peak minority carrier concentration in the base is about to $N_B = 10^{17}$ cm^{-3}.
(d) Find the base transit time.

● **Ebers–Moll Model** ●

8.8 Consider the excess minority-carrier distribution of a PNP BJT as shown in Fig. 8–28. (The depletion regions at junctions are not shown.) Assume all generation–recombination current components are negligible and each region is uniformly doped. Constant D_n = 30 cm^2/s and D_p = 10 cm^2/s are given. This device has a cross-section area of 10^{-5} cm^2 and $N_E = 10^{18}$ cm^{-3}.

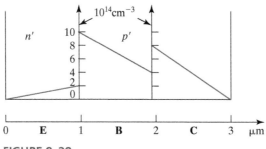

FIGURE 8–28

(a) Find N_C, i.e., the dopant concentration in the collector.

(b) In what region of the IV characteristics is this BJT operating? Explain your answer. (Hints: Are the BE and BE junctions forward or reverse biased?)

(c) Calculate the total stored excess carrier charge in the base (in Coulombs).

(d) Find the emitter current I_E.

(e) Calculate β_F, i.e., the common-emitter current gain when the BJT is operated in the nonsaturation region (i.e., $V_{EB} >$; 0.7 V and $V_{CE} >$; 0.3 V. Neglect base-width modulation).

8.9 An NPN BJT is biased so that its operating point lies at the boundary between active mode and saturation mode.

(a) Considering the Ebers–Moll of an NPN transistor, draw the simplified equivalent circuit for the transistor at the given operating point.

(b) Employing the simplified equivalent circuit of part (a), or working directly with Ebers–Moll equations, obtain an expression for V_{EC} at the specified operating point. Your answer should be in terms of I_B and the Ebers–Moll parameters.

● **Drift-Base Transistors** ●

8.10 An NPN BJT with a $Si_{0.8}Ge_{0.2}$ base has an E_{gB}, which is 0.1 eV smaller than an Si-base NPN BJT.

(a) At a given V_{BE}, how do I_B and I_C change when a SiGe base is used in place of an Si base? If there is a change, indicate whether the currents are larger or smaller.

(b) To reduce the base transit time and increase β, the percentage of Ge in an $Si_{1-y}Ge_y$ base is commonly graded in order to create a drift field for electrons across the base. Assume that E_g is linearly graded and that $y = 0$ at the emitter–base junction and $y = 0.2$ at the base–collector junction. What is $\beta(SiGe)/\beta(Si)$? (Hint: $n_{iB}^2 = n_{i,Si}^2 \exp[(\Delta E_{g,Si_{0.8}Ge_{0.2}}/kT)(x/W_B)]$, where W_B is the base width.)

8.11 An NPN transistor is fabricated such that the collector has a uniform doping of 5×10^{15} cm^{-3}. The emitter and base doping profiles are given by $N_{dE} = 10^{20} e(-x/0.106)$ cm^{-3}. And $N_{aB} = 4 \times 10^{18} (-x/0.19)$ cm^{-3}, where x is in micrometers.

(a) Find the intercept of N_{dE} and N_{aB} and the intercept of N_{aB} and N_c. What is the difference between the two intercepts? What is the base width ignoring the depletion widths, known as the **metallurgical base width**?

(b) Find base and emitter Gummel numbers. Ignore the depletion widths for simplicity.

(c) Find the emitter injection efficiency γ_E.

(d) Now considering only the N_{aB} doping in the base (ignore the other doping), sketch the energy band diagram of the base and calculate the built-in electric field, defined as $\mathscr{E}_{bi} = (1/q)(dE_c/dx)$, where E_c is the conduction band level.

● **Kirk Effect** ●

8.12 Derive an expression for the "base width" in Fig. 8–18c or Fig. 8–18d as a function of I_C, V_{CB}, and N-collector width, W_C. Assume all common BJT parameters are known.

● **Charge Control Model** ●

8.13 Solve the problem for the step-function I_B in Example 8–5 in Section 8.10 on your own without copying the provided solution.

8.14 A step change in base current occurs as shown in Fig. 8–29. Assuming forward active operation, estimate the collector current $i_C(t)$ for all time by application of the charge control model and reasonable approximations. Depletion region capacitance can be neglected. The following parameters are given: $\alpha_F = 0.9901$, $\tau_F = 10$ ps, $i_{B1} = 100$ µA, and $i_{B2} = 10$ µA.

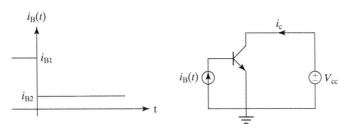

FIGURE 8–29

● **Cutoff Frequency** ●

8.15 After studying Section 8.9, derive expressions for $\beta(\omega)$ and f_T.

● **REFERENCES** ●

1. Taur, Y., and T. Ning. *Fundamentals of VLSI Devices*. Cambridge, UK: Cambridge University Press, 1998, Ch. 6.

2. del Alamo, J., S. Swirhum, and R. M. Swanson. "Simultaneous Measurement of Hole Lifetime, Mobility, and Bandgap Narrowing in Heavily Doped N-type Silicon." *International Electron Devices Meeting Technical Digest*. (1985), 290–293.

3. Paasschens, J., W. Kloosterman, and D.B.M. Klaassen. "Mextram 504." Presentation at Compact Model Council, September 29, 1999. http://www.eigroup.org/cmc/minutes/wk092999.pdf

4. Harame, D. L., et al. "Si/SiGe Epitaxial-Base Transistors." *IEEE Transactions on Electron Devices*, 42, 3 (1995), 455–482.

5. Roulston, D. J. *Bipolar Semiconductor Devices*. New York: McGraw Hill, 1990.

6. Tran, H. Q., et al. "Simultaneous Extraction of Thermal and Emitter Series-Resistances in Bipolar Transistors." Proceedings of the IEEE Bipolar/BiCMOS Circuits and Technology Meeting, Minneapolis, MN, 1997.

● **GENERAL REFERENCES** ●

1. Roulston, D. J. *Bipolar Semiconductor Devices*. New York: McGraw-Hill, 1990.

2. Taur, Y., and T. Ning. *Fundamentals of VLSI Devices*. Cambridge, UK: Cambridge University Press, 1998.

Derivation of the Density of States

In this appendix, we will derive the density of states (Eq. 1.6.2). Consider a cube of semiconductor crystal with length L on each side. The electron waves in the crystal are standing waves. In the x direction, the wavelength is

$$\lambda_x = \frac{L}{n_x} \tag{I.1}$$

n_x is equal to 1, 2, 3 ... The wavelength is related to the electron momentum in the x direction, p_x, through the de Broglie relationship.

$$p_x = \pm \frac{h}{\lambda_x} \tag{I.2}$$

h is the Planck's constant and + and − represent the momentum in the x and the $-x$ directions. Since λ_x only takes on a set of discrete values, so does p_x.

$$p_x = \pm \frac{n_x h}{L}, \tag{I.3}$$

The increment between the allowable p_xs is h/L. Similarly p_y and p_z can only take on discrete values with increments of h/L. Figure I–1 shows a three-dimensional space with axes p_x, p_y, and p_z. Allowed energy states occupy points separated from one another by h/L in p_x, p_y, and p_z. There are two allowed states (the factor of 2 accounting for the two spin directions) for every cube of h^3/L^3 volume in the momentum space. Each state therefore occupies a volume of $h^3/2L^3$.

Figure I–2 shows the same momentum space as Fig. I–1 but in a very much zoomed-out scale. The allowable states are now semicontinuous. Nonetheless, each allowed state still occupies a volume of $h^3/2L^3$. A sphere in this space represents a constant total momentum, p, and therefore a constant kinetic energy, E.

$$E = \frac{p^2}{2m_n} \tag{I.4}$$

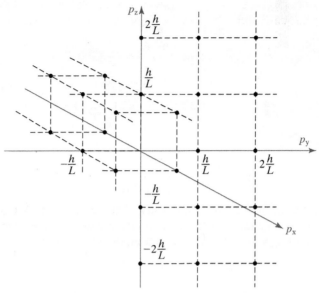

FIGURE I-1 The allowable states in the momentum space from a large uniform 3-D grid. Only a few grid points are shown for simplicity.

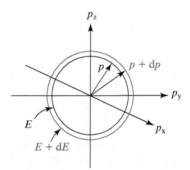

FIGURE I-2 Each sphere in the momentum space represents a constant-energy surface.

where m_n is the electron effective mass. Using Eq. (I.4) twice, we find that

$$\frac{dE}{dp} = \frac{p}{m_n} = \frac{\sqrt{2m_n E}}{m_n} = \sqrt{\frac{2E}{m_n}} \tag{I.5}$$

According to Eq. (I.5), two spheres that differ in energy dE have two radii that differ by

$$dp = \sqrt{\frac{m_n}{2E}}\, dE \tag{I.6}$$

The volume of the shell that is between these two spheres is the surface area times dp.

$$\text{Volume} = 4\pi p^2\, dp = 4\pi\, 2 m_n E\, dp$$

$$= 8\pi m_n \sqrt{\frac{m_n E}{2}}\, dE \tag{I.7}$$

where we have first used Eq. (I.4) and then Eq. (I.6). The number of states contained in this shell between E and $E + dE$ is the volume of the shell divided by $h^3/2L^3$.

$$8 \pi m_n \sqrt{\frac{m_n E}{2}} \times \frac{2L^3}{h^3} dE \qquad (I.8)$$

Therefore the number of states per unit volume (the volume of the sample is L^3) per unit energy (the definition of the *density* of states according to Eq. 1.6.1) is

$$D_c(E) = \frac{8\pi m_n \sqrt{2 m_n E}}{h^3} \qquad (I.9)$$

E in Eq. (I.9) represents the kinetic energy, which is $E - E_c$ in Eq. (1.6.2a).

Derivation of the Fermi–Dirac Distribution Function

In this appendix, we will derive the Fermi–Dirac distribution function, Eq. (1.7.1). First, we derive the number of ways of arranging n identical balls in g compartments, each of which may contain either one ball or none. $g \geq n$. The problem is illustrated in Fig. II–1.

There are g ways (positions) to place the 1st ball into any of the g compartments.
There are $g-1$ ways to place the 2nd ball since only $g-1$ compartments are empty.
There are $g-2$ ways to place the 3rd ball.
$\vdots$
There are $g-(n-1)$ ways to place the final and nth ball.
The total number of ways to arrange the balls is

$$W = \frac{g(g-1)(g-2)\ldots[g-(n-1)]}{n!}$$

$$= \frac{g!}{(g-n)!\,n!} \quad \text{(II.1)}$$

Where the factorial function $x!$ is

$$x! \equiv x(x-1)(x-2)\ldots 2\cdot 1 \quad \text{(II.2)}$$

$n!$ appears in the denominator of Eq. (II.1) because the above algorithm generates $n!$ ways to create each unique final placements of the n balls. All the $n!$ ways produce one and the same set of n occupied compartments. Which ball occupies which of these n compartments is immaterial because the balls are identical to one another.

Next, we consider a system of electron states such as those in the conduction band in Fig. 1–5(a). To facilitate the analysis, the states are lumped into k (k being very large) discrete energy values, $E_1, E_2, E_3 \ldots E_k$ as shown in Fig II–2. There are g_1 electron states at E_1 and g_2 states at E_2, etc. Each of the electron states may be occupied by one electron or empty as the compartments in Fig. II–1.

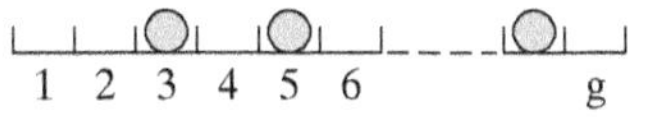

FIGURE II–1 Arranging n identical balls in g compartments.

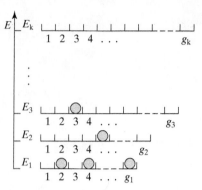

FIGURE II–2 An electron energy system. There are g_i states at E_i.

Assume that there are n_1 identical electrons in the E_1 state and n_2 identical electrons in the E_2 states, etc. The numbers of ways of distributing the electrons among the states at $E_1 \ldots E_k$ are

$$W_1 = \frac{g_1!}{(g_1 - n_1)!\, n_1!} \tag{II.3}$$

$$\vdots$$

$$W_k = \frac{g_k!}{(g_k - n_k)!\, n_k!} \tag{II.4}$$

The total number of ways of distributing the electrons among the states is simply the product of the above series.

$$W = W_1 \times W_2 \times \ldots \times W_k \tag{II.5}$$

The ratio of n_i/g_i is the probability of a state at E_i being occupied. Our goal is to find this ratio as a function of E. There are two constraints on how the electrons may be allocated among $n_1, n_2 \ldots n_k$. The total number of electrons in the system is conserved and the total energy of the electrons is conserved.

$$\Sigma_k\, n_i = \text{constant} \tag{II.6}$$

$$\Sigma_k\, n_i\, E_i = \text{constant} \tag{II.7}$$

Within the constraints of Eq. (II.6) and Eq. (II.7), there are a large number of ways to distribute the electrons among $n_1, n_2 \ldots n_k$. The equilibrium distribution is the most probable distribution, i.e., the distribution that maximizes W in Eq. (II.5). In other words, it is the most disordered distribution and it maximizes the entropy. Maximizing W is equivalent to maximizing $\ln W$ or equating $d\,(\ln W)/d\, n_i$ to zero.

$$\ln W = \Sigma_k \ln W_i = \Sigma_k\, [\ln g_i! - \ln (g_i - n_i)! - \ln n!] \tag{II.8}$$

Equation (II.8) can be simplified with Sterling's approximation for factorials of large numbers.

$$\ln x! \approx x \ln x - x \tag{II.9}$$

$$\ln W = \Sigma_k\, [g_i \ln g_i - (g_i - n_i) \ln (g_i - n_i) - n_i \ln n_i] \tag{II.10}$$

Appendix II • Derivation of the Fermi–Dirac Distribution Function

$$d \ln W = \Sigma_k \left[\frac{g_i - n_i}{g_i - n_i} + \ln(g_i - n_i) - \frac{n_i}{n_i} - \ln n_i \right] d n_i$$

$$= 0 \tag{II.11}$$

$$\Sigma_k [\ln(g_i - n_i) - \ln n_i] d n_i = 0 \tag{II.12}$$

Equations (II.6) and (II.7) can be rewritten as

$$\Sigma_k d n_i = 0 \tag{II.13}$$

$$\Sigma_k E_i d n_i = 0 \tag{II.14}$$

The last three equations represent all the conditions that have to be met and they can be literally summed up into one condition using the Legrange undetermined multipliers method.

$$\Sigma_k \left[\ln \frac{g_i - n_i}{n_i} + \alpha - \beta E_i \right] d n_i = 0 \tag{II.15}$$

where α and β are constants to be determined. The solution for Eq. (II.15) is simply

$$\ln \frac{g_i - n_i}{n_i} = -\alpha + \beta E_i \tag{II.16}$$

$$\frac{g_i - n_i}{n_i} = e^{-\alpha + \beta E_i} \tag{II.17}$$

$$\frac{g_i}{n_i} = 1 + e^{-\alpha + \beta E_i} \tag{II.18}$$

The ratio n_i/g_i is the Fermi–Dirac distribution function, $f(E_i)$

$$f(E_i) \equiv \frac{n_i}{g_i} = \frac{1}{1 + e^{(E_i - \alpha/\beta)\beta}} \tag{II.19}$$

It can be shown [see Prob. 1.10(d)] that Eq. (II.19) leads to the conclusion that the average energy of a dilute electron gas is $3/2\ \beta$. Since it is known from thermodynamics that the average energy is $3kT/2$, one can conclude that $\beta = 1/kT$. Furthermore, we introduce E_F to represent α/β, which is simply a constant. Finally, E_i simply represents the energy of the various energy states and can be replaced with the continuous variable, E.

$$f(E) = \frac{1}{1 + e^{(E - E_F)/kT}} \tag{II.20}$$

Equation (II.20) is Eq. (1.7.1). Two important conclusions may be drawn for the Fermi energy, E_F. First it is a constant (at equilibrium) because α/β is a constant in the derivation. Second, E_F varies with the number of electrons in the system (which may be determined by the donor density, for example) because α is introduced through Eq. (II.13), which represents the constancy of the number of electrons in the system.

III

Self-Consistencies of Minority Carrier Assumptions

In Chapter 4 the PN junction's current is analyzed by neglecting the minority drift current. By neglecting that current component, the minority current continuity equation becomes easily solvable. This appendix illustrates the self-consistency of that assumption with a numerical example. Along the way we will find the majority drift and diffusion currents as well as the minority drift and diffusion currents.

For simplicity, consider a long-base, silicon abrupt P$^+$N junction diode with uniform cross section and a constant 10^{16} cm^{-3} donor concentration on the N side of the junction. The carrier lifetime is $\tau_p = 10^{-8}$s in the N-type region.

We will carry out the analysis of the currents on the N side with the following procedure.

(a) Calculate the density of the excess minority carriers as a function of x (distance from the junction) when the applied voltage is $23 \times kT/q$, which is 0.589 V at room temperature. Ignore the minority drift current.

(b) Find minority (diffusion) currents, the total current, and the majority current as functions of x.

(c) Find the majority-carrier diffusion current as a function of x.

(d) Use the results of parts (b) and (c) to find the majority-carrier drift current, $J_{n\,\mathrm{drift}}$.

(e) Find electric field $\mathcal{E}(x)$ from $J_{n\,\mathrm{drift}}$.

(f) Find the minority drift current $J_{n\,\mathrm{drift}}$ and show that $J_{n\,\mathrm{drift}} \ll J_{n\,\mathrm{drift}}$.

(g) Justify the assumption of $n' = p'$.

(a) For $N_d = 10^{16}$ cm^{-3}, $D_n = 30$ cm^2 sec^{-1}, $L_n = 5.5$ μm, $D_p = 10$ cm^2 sec^{-1}, $L_p = 3.2$ μm

$$p'(x) = \frac{n_i^2}{N_d}(e^{qV_a/kT} - 1)\,e^{-x/L_p} = 10^{14}\,e^{-x/L_p}\ \mathrm{cm}^{-3}. \quad \text{(III.1)}$$

(b) Minority (diffusion) current:

$$J_p(x) = -qD_p\frac{dp'(x)}{dx} = q\frac{n_i^2}{N_d L_p}D_p(e^{qV_a/kT} - 1)\,e^{-x/L_p} = 0.5 e^{-x/L_p}\ \mathrm{A/cm}^2. \quad \text{(III.2)}$$

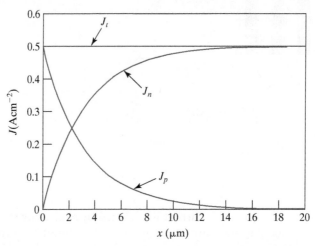

FIGURE III–1 Minority current density, majority current density, and the total current density.

The total current, J_t, is equal to the minority current at $x = 0$, at which location the majority (electron) current is zero because these is negligible electron injection into the P+ region. J_t is not a function of x because of the continuity of the total current.

Majority current:

$$J_n(x) = J_t - J_p(x) = J_p(0) - J_p(x) = 0.5\,(1 - e^{-x/L_p})\ \text{A/cm}^2 \tag{III.3}$$

(c) $J_{n\text{diff}}(x) = qD_n \dfrac{dn'(x)}{dx} = -J_p(x) \dfrac{D_n}{D_p} = 3J_p(x) = -1.5e^{-x/L_p}$ \hfill (III.4)

Here we have used the charge neutrality condition, $n' = p'$. See (g) for justification.

(d) $J_{n\text{drift}}(x) = J_n(x) - J_{n\text{diff}}(x) = 0.5 + 1 \times e^{-x/L_p}\ \text{A/cm}^2.$ \hfill (III.5)

Please note that this majority drift current is comparable to the majority diffusion current in (c). Clearly neither is negligible. This makes it difficult to solve the continuity equation of the majority current.

(e) The electric field is

$$\mathscr{E}(x) = \dfrac{J_{n\text{drift}}}{nq\mu_n} = 0.25 + 0.5 \times e^{-x/L_p}\ \text{V/cm}. \tag{III.6}$$

(f) The minority drift current density is

$$J_{p\text{drift}}(x) = pq\mu_p\,\mathscr{E}(x) \approx p'q\mu_p\,\mathscr{E}(x) = 1.6 \times 10^{-3}\,e^{-x/L_p}(1 + 2e^{-x/L_p})\ \text{A/cm}^2 \tag{III.7}$$

Comparing this minority drift current with the minority diffusion current in part (b), it is obvious that

$$J_{p\text{drift}} \ll J_{p\text{diff}} \Rightarrow J_p \approx J_{p\text{diff}}. \tag{III.8}$$

The minority drift current and the minority diffusion current are compared graphically in Fig. III–2.

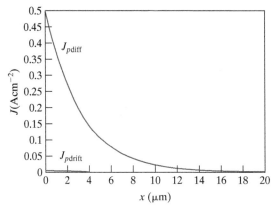

FIGURE III–2 The minority drift and diffusion currents. The drift current is negligibly small compared to the diffusion current.

(g) From Eq. (III.5) and the Possion equation, Eq. (4.1.5), it can be shown that the charge density, ρ, is $1.7 \times 10^{-9} \, e^{-x/L_p}$ C/cm^{-3}. This negligibly small compared with the excess hole charge density qp' in (a), which is $1.6 \times 10^{-5} \, e^{-x/L_p}$ C/cm^{-3}. This shows that $p' = n'$ is a good assumption because $\rho = qp' - qn'$.

Answers to Selected Problems

● **CHAPTER 1** ●

1.1 (a) 8 atoms

1.3 (a) ½

(b) $E_F = \dfrac{E_c + E_v}{2}$

1.8 (a) $n = \dfrac{\sqrt{\pi}}{2} A(kT)^{3/2} \, e^{-(E_c - E_F)/kT}$;

$p = \dfrac{\sqrt{\pi}}{2} B(kT)^{3/2} \, e^{-(E_F - E_v)/kT}$

(b) 0.009 eV below the mid-bandgap of the semiconductor

1.11 (a) 10^{15} cm^{-3}

(c) $n = 4.01 \times 10^{14}$ cm^{-3},
$p = 2.49 \times 10^{5}$ cm^{-3} or
$n = 2.20 \times 10^{14}$ cm^{-3},
$p = 4.55 \times 10^{5}$ cm^{-3}

1.13 (a) P-type

(b) At $T = 300$ K, $p = 4 \times 10^{16}$ cm^{-3},
$n = 2{,}500$ cm^{-3}

At $T = 600$ K, $p = 4.12 \times 10^{16}$ cm^{-3},
$n = 3.27 \times 10^{13}$ cm^{-3}

(d) 0.34 eV above the valence band

1.15 (a) Arsenic: 777 K;
Boron: 635 K

(b) Arsenic: $p = 2.1 \times 10^{4}$ cm^{-3};
Boron: $n = 2.1 \times 10^{5}$ cm^{-3}

(c) Arsenic: $= E_F - E_v = 0.88$ eV;
Boron: $= E_F - E_v = 0.24$ eV

(d) $n = 9 \times 10^{15}$ cm^{-3};
$p = 1.11 \times 10^{4}$ cm^{-3};
$E_F - E_v = 0.90$ eV

● **CHAPTER 2** ●

2.1 (a) 2.85×10^{-13} s

(b) 0.14 nm

2.3 (b) 502 cm^2/V s

(c) 84 A/cm^2

2.5 (a) Sample 1: $p = 10^{2}$ cm^{-3}
Sample 2: $n = 10^{5}$ cm^{-3}
Sample 3: $p = 10^{2}$ cm^{-3}

(b) Sample 1: 12/Ω-cm
Sample 2: 12/Ω-cm
Sample 3: 11.88/Ω-cm

(c) Sample 1: $E_c - E_F = 0.15$ eV
Sample 2: $E_F - E_v = 0.24$ eV
Sample 3: $E_c - E_F = 0.15$ eV

2.7 (a) $n = 4.5 \times 10^{17}$ cm^3; $p = 222$ cm^3;
hole is the minority carrier

(b) $E_F = E_c - 0.107$ eV

(c) 467 Ω

2.9 (a) $\mathscr{E} = -\dfrac{dV}{dx} = \dfrac{1}{q}\dfrac{dE_v}{dx} = \dfrac{1}{qL}\Delta = \dfrac{\Delta}{qL}$

(b) $n(x) = n_0 e^{-\Delta x / LkT}$

CHAPTER 3

3.6 (a) 0.256 μm
(b) Linear: 156% error
Quadratic: 25% error

3.8 (b) $\Delta x_j = 1.42 \times 10^{-9}$ μm

3.9 (a) 1.97×10^{-4} cm
(b) $\Delta x_j = 3.75 \times 10^{-35}$ cm ≈ 0

CHAPTER 4

4.3 (a) N-type region will be fully depleted first.
(b) P-type region will be fully depleted first
(c) For (a): 8.63×10^{-9} F/cm^2
For (b): 7.85×10^{-9} F/cm^2

4.5 (a) $x_n = 0.032$ μm; $x_p = 0.019$ μm;
$W_{dep} = 0.551$ μm
(b) 148 V/cm
(c) 0.78 V
(d) The breakdown voltage of the P-I-N structure is ~10 V higher than the P-N structure

4.8 (a) 34.2 μm;
(b) (i) $p_N' = n_N' = 0$ cm^{-3},
(ii) $p_N' = n_N' = 4.8 \times 10^{10}$ cm^{-3}

4.12 (a) $\dfrac{dV}{dT} = \dfrac{k}{q} \ln\left(\dfrac{I}{I_0} + 1\right)$
(b) 0.00167 V/K

4.20 (a) 0.282 pF
(b) −0.813 V

4.23 (b) Schottky diode: 58.5 μA
PN diode : 0.32 fA
(c) Schottky diode: 0.36 V
PN diode : 1.03 V

4.24 (a) 8.85×10^{18} cm^{-3}

4.25 (b) 0.368 V, maximum

CHAPTER 5

5.5 (a) 8.62 μm
(b) 2.9×10^{16} cm^{-3}
(c) −1 V
(d) 0.345 V

5.8 (a) $C_{ox} = 3.45 \times 10^{-7}$ F/cm^2
$\phi_B = 0.4$ V
$V_t = 0.18$ V
(b) 3.45×10^{-7} C/cm^2
(c) $Q_{dep} = -1.1 \times 10^{-7}$ C/cm^2
$Q_{inv} = -6.3 \times 10^{-7}$ C/cm^2

5.11 (a) −0.80 V
(b) 2.65×10^{-9} F/cm^2
(e) -5.3×10^{-9} C/cm^2
(f) 2.17×10^{-9} F/cm^2
(g) 7.22 V

5.12 Negative; -3.52×10^{-8} C/cm^2

5.20 (b) 205 nm
(c) $\sim 10^{15}$ cm^{-3}

5.21 (f) $V_t = -0.055$ V;
$Q_{inv} = 2.20 \times 10^{-6}$ C/cm^2
(g) 1.90×10^{-6} C/cm^2

CHAPTER 6

6.2 (a) 0.12 μm
(b) −1.82 V

6.3 (a) 0.52 V
(b) −1.64 V
(c) −0.52 V
(d) $V_b = 0$ V; $V_s = 0$ V;
$V_d = 2.5$ V; $V_g = 2.5$ V
(e) $V_b = 2.5$ V; $V_s = 2.5$ V;
$V_d = 0$ V; $V_g = 0$ V
(f) (c)

6.5 (a) 1 V
(b) 361 cm^2/V s

6.7 (a) 2.5 V
(b) 0
(c) 2.25×10^{-3} A

6.8 (a) −0.064 V
(b) 1.21 mA
(c) 1.17 ms
(d) 1.13 ms

Answers to Selected Problems **339**

6.18 (a) 8.9×10^{14} cm^{-3}
(b) 2.0 V

6.19 A: source side: 1.7×10^6 cm/s
drain side: 1.9×10^6 cm/s
B: source side: 4.2×10^6 cm/s
drain side: 8×10^6 cm/s
C: source side: 4.5×10^6 cm/s
drain side: 8×10^6 cm/s

6.23 (a) 0.54 V
(b) 1.83 V
(c) 480 cm^2/V/s

6.24 $L \sim 0.19$ μm; $I_{dsat} = 575$ μA/μm width.

6.27 (b) $\Delta L \cong 0.1$ μm
(c) 0.5 V

● **CHAPTER 7** ●

7.1 (a) 0.109 V
(b) 65.6 mV/dec
(c) 121 nA

7.4 (i) Larger V_t: Decrease I_{off} and decrease I_{on}
(ii) Larger L: Decrease I_{off} and decrease I_{on}
(iii) Shallower junction: Decrease I_{off} and somewhat decrease I_{on}
(iv) Smaller V_{dd}: Decrease I_{off} and decrease I_{on}
(v) Smaller T_{ox}: Decrease I_{off} and increase I_{on}
(vi) A smaller T_{ox} contributes to leakage reduction and increases the precious I_{on}.

7.6 (a) $l_d \propto \sqrt[3]{\dfrac{\varepsilon_{ox}(V_t - V_{fb} - 2\phi_B)}{qN_a}} X_j$

7.8 (c) $V_t = (E_g/q)\dfrac{\varepsilon_{si}T_{ox}}{\varepsilon_{ox}X_{rg}}$

● **CHAPTER 8** ●

8.1 (c) 0.221 V
(d) 2.88 μm

8.4 (b) $\gamma_E = 0.999325$; $\beta_F = 519$

8.5 (a) 10^{17} cm^{-3}
(c) 1.12×10^{-13} C
(d) 0.192 mA
(e) $\beta = 1.7$

8.11 (a) Intercept of N_{dE} and N_{aB}: 0.77 μm;
Intercept of N_{aB} and N_c: 1.27 μm;
Base width: 0.5 μm
(b) Base dopant per unit area and approximate Gummel number: 1.23×10^{12} cm^{-2} and 4×10^{10} s-cm^{-4};
Emitter dopant per unit area and approximate Gummel number: 1.06×10^{15} cm^{-2} and 2×10^{14} s-cm^{-4}
(c) $\gamma = 0.99993$
(d) $\mathscr{E}_{bi} = 3.16 \times 10^5$ V/m

Index

A

Abrupt junction, 90
Absorption coefficient, light, 119
Acceptors, 5, 10
   band model, 10–11
Accumulation charge, 161
Accumulation charge-layer thicknesses, 176–179
Accumulation layer, 161
Active region, MOSFET, 211
Amorphous solids, 76
Amphoteric dopants, 7
Analog circuit issues
   BJT cutoff frequency, 312–313
   BJT small-signal model, 310–312
   diode small-signal model, 116–117
   MOSFET flicker noise, 235–237
   MOSFET high-frequency performance, 230–232
   MOSFET output conductance, 229–230, 282–283
   MOSFET thermal noise, 233–235
Anisotropic etching, 69–70
Antenna effect, 70
Asher, 64
Avalanche breakdown, PN junction, 102–104
Avalanche photodiode, 138

B

Back gate, 210
Back-end process, 80–82
Ballistic transport, 43–44
Band gap, 8, 299
Bardeen, John, 2

Base current, BJT, 297
Base Gummel number, 295
Base widening, BJT, 309–310
Base-width modulation by collector voltage, BJT, 302–304
BiCMOS technology, 291
Binary semiconductors, LEDs, 125
Bipolar junction transistor (BJT), 291–323
   base current, 297
   base-width modulation by collector voltage, 302–304
   charge control model, 315
   charge storage, 306–310
      base charge storage, 306–307
   circuit simulation model, 316–318
   collector current, 292
   current gain, 298–302
   cutoff frequency, 313
   Early effect, 302–304
   Early voltage, 303
   Ebers–Moll model, 304–306
   Gummel number, 295, 297
   heterojunction, 299
   maximum oscillation frequency, 313
   NPN BJT, 291–292
   SiGe, 309
   small-signal model, 310–312
   structure, 314
   transit time, 306–310
      base transit time, 306–307
      drift transistor, 307–308
      emitter-to-collector transit time, 308–310
Bit line, 238–240
Blu-ray, 132

Body effect, MOSFET, 207–209
Boltzmann approximation, 17
Boltzmann tail, 105
Bond model
  electrons, 4–8
  GaAs, 7
  holes, 4–8
Brattain, Walter H., 2
Breakdown, PN junction, 100–104
Breakdown electric field of SiO_2, 271–274
Built-in potential, PN junction, 91–93
Bulk silicon, 279
Bulk-charge effect, 210
Burn-in, *See* Reliability

C

Capacitance–voltage (C–V)
  characteristics, 98–100
  MOS capacitor, 168–172
    high-frequency, 171
    low-frequency, 171
    quasi-static 169–170
Carrier concentrations, 28–29
Carrier injection
  forward bias, 105–107
  lasers, 131
  LED, 124
  quasi-equilibrium boundary condition, 105–107
Carrier mobility, 39
Carrier recombination, 50–51
  radiative, 51, 125
    lasers, 128–130
    LED, 124–125
  through traps, 51
Carrier scattering, mechanisms of, 40–44
  ballistic transport, 43–44
  ionized impurity scattering, 40
  phonon scattering, 40
  velocity overshoot, 220, 228
  velocity saturation, 43–44, 219–225
CCD, *See* Charge-coupled device
Cell libraries, 284
Channel length modulation, 230, 283
Charge control model, BJT, 315
Charge neutrality, 26, 50, 335
Charge storage, BJT, 306–307
Charge storage, PN junction, 115–116

Charge-coupled device (CCD), 179–182
  two-dimensional array, 182
Charge-storage capacitance, 117, 311
Charge-trap NVM, 244
Chemical vapor deposition (CVD), 75, 77–78
  high-temperature oxide (HTO), 78
Chemical-mechanical polishing (CMP), 81
Circuit simulation model
  BJT, 316–318
  MOSFET, 284–285
CMOS imagers, 178–184
CMP, *See* Chemical mechanical polishing
Code storage, NVM, 242
Collector current, BJT, 292
  high-level injection effect, 296
Collisions, thermal motion, 36
Color imagers, 184
Compact fluorescent lamp, 128
Compact models, MOSFET, 284–285
Complementary MOS (CMOS) technology, 198–200
  imager, 178–184
  inverter, 214–216
    speed, 216–218
    Voltage Transfer Curve (VTC), 214–216
  power consumption, CMOS, 218
  pull-down device, 200
  pull-up device, 200
Compound semiconductors, 6–8
  LEDs, 125
Concentrations, carrier, 19–28
  dopant effect, 25–28
  intrinsic, 23
Conduction band, 8
Conduction electrons, 5
Conductivity, 44–46
Conductivity effective mass, 21
Conductors, 11–12
Contact resistance, 226
  specific, 144–145
Continuity equation, current, 107–108
Coulombic scattering, 235
Cross talk, 232
Crystal structure, silicon, 1–4
  crystal planes, 4
  unit cell, 1–3
Crystalline solids, 76

Current continuity equation, PN junction, 107–109
Current gain, BJT, 298–302
 common-base current gain, 298
 common-emitter current gain, 298
 emitter band gap narrowing, 299
 emitter efficiency, 298
 Gummel plot, 301
 narrow band-gap base, 299–300
 poly-silicon emitter, 300
Cutoff frequency, 230
 BJT, 313
 MOSFET, 230–232
CVD, *See* Chemical vapor deposition
Cyclotron resonance, 14

D

Damascene process, 81–82
Data storage, 242
DBR, *See* Distributed Bragg reflector
De Broglie relationship, 325
Deep-depletion, 179–182
Density of states, 15–16, 325–327
 effective mass, 21
Depletion-layer capacitance, 99
Depletion-layer model, PN junction, 94–97
Depletion-mode transistor, 205
Depletion region, PN junction, 90–91, 94, 114–115
Design for manufacturing (DFM), 276–277
Device fabrication technology, 59–87, *See also* Etching; Lithography; Silicon device fabrication process
 interconnect, 80–82
 ion implantation, 70–72
 oxidation, 61–64
 pattern transfer, 68–70
DFM, *See* Design for manufacturing
Diamond crystal structure, 2
DIBL, *See* Drain-induced barrier lowering
Diffusion
 carrier, 122
  capacitance, 117, 311
  current, 46–47
  diffusion constant, 46
  Einstein relationship, 48–50
  lengths, 109

 dopant, 73–75
  diffusivity, 74
  shallow junction, 75
Diode, PN junction, 89
Diode current, 112–115
Diode lasers, 128–133
 application, 132–133
 light amplification, 128–131
 optical feedback, 131–132
Direct recombination, 51, 124
Direct-gap semiconductors, 51, 119–121, 124
Distributed Bragg reflector (DBR), 132
Donors, 5, 10
 in band model, 10–11
 in bond model, 5
 diffusion, 72–74
Doping, 5–6, 70–73
 dopant compensation, 27
 dopant diffusion, 73–75
 gas-source doping, 72
 ion implantation, 70–72
 solid-source diffusion, 72–73
Double-gate MOSFET, 281
Drain-induced barrier lowering (DIBL), 269
 characteristic length, 269
Drain junction, 226, 274, *See* Shallow junctions
Drain saturation voltage, 211
DRAM, *See* Dynamic random access memory
Drift, 38–46, *See also* Carrier scattering
 conductivity, 44–46
 current, 44–46
 drift velocity, 38
 electron mobilities, 38–40
 hole mobilities, 38–40
Drive-in, 75
Dry etching, 68
 material selectivity, 70
Dry lithography, 61–62, 66–68
Dynamic power, 218
Dynamic random access memory (DRAM)
 technology, 238, 240–242
 capacitor, 240–241
 refresh, 241

E

Early, James, 304
Early voltage, 303–304

Ebers–Moll model, BJT, 304–306
  active region, 304
  reverse current gain, 305
  saturation region, 304
Effective channel length, 226–228
Effective density of states, 20
Effective gate capacitance, 178
Effective mass, 12–14
  measurement, 14
Effective oxide thickness, 178
Efficacy, *See* Luminous efficacy
Einstein relationship, 48–50
Electrical equivalent oxide thickness (EOT$_E$), 178, 262
Electromigration, 81
Electron affinity, 159
Electron beam writing, 68
Electron energy barrier, 159
Electron lithography, 68
Electrons, 1–33, 35–58, *See also* Motion; Recombination
  bond model of, 4–8
  concentrations, 19–24
  conduction electrons, 5
  effective mass, 12–14
    measurement, 14–15
Emitter Gummel number, 297
Emitter-to-collector transit time, 308–310
End-point detector, 70
Energy band diagram, 8–10
  in acceptors, 10
  band-gap energies, 9
  of BJT, 292
  conductors, 11–12
  in donors, 10
  insulators, 11–12
  of light-emitting diode, 124
  of MOS system, 158–163
  of PN junction, 90–91
  of quantum-well diode, 127, 130, 131
  of Schottky junction, 134
  semiconductors, 11–12
Energy conversion efficiency, solar cell, 119
Energy diagram and voltage applied, 47–48
Enhancement-mode transistor, 165, 205
Enhancement-type device, 165, 205

Epitaxy, 78–80
  BJT, 307, 314
  LED, 125–126
  MOSFET, strained silicon, 262
  selective, 78–79
Equilibrium carrier concentrations, 28–29, 50
Equivalent oxide thickness (EOT), 271
Erase, nonvolatile memories, 242–244
Etching, 68–70
  anisotropic etching, 69–70
  dry etching, 68
  isotropic etching, 68–69
  plasma etching, 68
  wet etching, 68
Excess carrier concentrations, 50
  BJT, 294
  PN junction, 106, 108–112
Excess minority carrier, 108–112, 294
Extreme ultraviolet lithography (EUVL), 67

F

Fabless design companies, 59
Fermi function, 16–19
Fermi level and carrier concentrations, 21–23
Fermi–Dirac distribution function, 16–19, 329–331
Fermi-level pinning, 136
Fiber-optical communication, 132
Field-effect transistor (FET), 195
  early patents on, 196–197
  HEMT, MODFET, 205–206
  high-mobility, 200–207
  JFET, 206–207
  MESFET, 204–205
  MOSFET, 195–196
Fill factor, solar cell, 123
FinFET, 280–282
Flash annealing, 75
Flash erase, 242
Flash memory, 238–245
Flat-band condition, 158–160
Flicker noise, MOSFET, 235–236
Flip-chip bonding process, 83
Floating-gate memory, 242
Forward bias, carrier injection, 105–107, 113
  in BJT, 292–294
  in PN junction, 109–112

Foundries, 59
Freeze-out, 28
    infrared detector based, 29
Front-end process, 82
Furnace annealing, 75

G

GaAs, III–V compound semiconductors, 6–8
    bond model of, 7
    crystal structure, 7
Gas-source doping, 72
Gate depletion, 174–176
Gate doping, 165–168
Gate-electrode resistance, MOSFET, 231
Gate length, 266–269
Gauss's law, 93
Germanium, 2
Giga-scale integration (GSI), 60
Grain boundary, 76
Gummel plot, 295, 297, 301
Gummel–Poon model, 317

H

Heavy doping effect, 299
Heil, Oskar, 197
Heterojunction, 80
    heterojunction bipolar transistor (HBT), 299–300
High electron-mobility transistor (HEMT), 205–206
High-frequency MOS capacitor C–V, 171
High-frequency MOSFETs, 230–232
High-k dielectric, MOSFET, 262, 270–271
High-level injection effect, 53, 296
High-mobility FETS, 200–207
High-voltage devices, application of, 104
Holes, 1–33, 35–58, *See also* Motion; Recombination
  bond model of, 4–8
  concentrations, 19–24
  current conduction by, 5
  effective mass, 12–14
    measure, 14
  hole energy barrier, 159
Hot carrier injection (HCI), 243–245
Hot electrons, 243
Hot-point probe test, 37–38

I

Imagers, 179–184
  CCD, 179–182
  CMOS, 182–183
  color, 184
Immersion lithography, *See* Wet lithography
Impact ionization, PN junction, 102
  avalanche photodiode based, 133
  electron–hole pair generation by, 104
Impurity ion scattering, 40
In situ doping, 78
Indirect gap semiconductors, 51, 119–121
Inorganic semiconductor materials, 121
Integrated circuits (ICs), MOSFETS in, 259–288
  compact models, 284–285
  concept, 260–261
  device and process simulation, 283–284
  leakage, 259–288
    tunnel gate leakage, 265
  output conductance, 282–283
  scaling, 259–288
  subthreshold current, 263–266
  trade-off between I_{on} and I_{off}, 276–277
  V_t roll-off, 266–269
  W_{dep}, reduction, 272
Interconnect, 80–82
  Cu, 80–82
  low-k dielectric, 82
  multilevel metallization, 81
Interface states, 265–266
International Technology Roadmap for Semiconductors (ITRS), 261
Intrinsic input resistance, MOSFET, 231
Intrinsic voltage gain, 230
Inversion, 164–168
  charge density, 164
  electrons, 164
  gate doping type, 165–168
  layer thickness, MOS, 176–179
  layer, 164
  V_t, choice, 165–168
Inverter, CMOS, 214–216

Index

Ion implantation, 70–72
Ionization energy, 11
Ionized impurity scattering, 40
Isotropic etching, 68–69

J

Junction breakdown, PN junction, 100–104
    avalanche breakdown, 102–104
    high-voltage devices, application of, 104
    peak electric field, 101
    tunneling breakdown, 101–103
Junction depth, 73
Junction field-effect transistor (JFET), 206–207

K

Kilby, Jack, 260
Kinetic energy, 35
    of electrons, 35
        equal-partition principle, 36n1
Kirk effect, 308–310
Knee current, 296

L

Laser annealing, 75
Lasers, *See* Diode lasers
Lattice constant, 2
Leakage power, 218
Leakage, MOSFETS in ICs, 259–288
LEDs, *See* Light-emitting diodes
Legrange, 331
Light amplification, diode lasers, 128–131
Light penetration depth, 119–121
Light-emitting diodes (LEDs), 124–128
    efficiency, *See* Luminous efficacy
    luminous efficacy, 127–128
    materials and structures, 124–127
    quantum efficiency, 125
    quantum well, 126
    solid-state lighting, 127–128
Lilienfeld, J. E., 196
Lithography, 64–68
    electron lithography, 68
    extreme ultraviolet lithography (EUVL), 67
    lithography field, 66
    nanoimprint, 68
    photolithography, 64–68
        immersion, wet, 66, 68
        optical proximity correction (OPC), 66
        phase-shift photomask, 66
        step-and-repeat, 66
        stepper, 66
        steps in, 65
Long-channel IV model, 221, 223–225
Low-frequency MOS capacitor C–V, 171
Low-k dielectric, interconnect, 82
Low-level injection, 53
    PN junction, 111
Low pressure chemical vapor deposition (LPCVD), 78–79
Low voltage-drop rectifier, 141
Luminous efficacy, LEDs, 127
    of lamps in lumen/watt, 128

M

Majority carriers, 24
    distribution, 110–112
Maximum oscillation frequency, BJT, 313
    MOSFET, 231
Mean free path, 40
Mean free time, 36, 40
Memory technologies, 238
Metal migration memory, 245
Metal gate, MOSFET, 262, 271
Metal source/drain MOSFET, 274–275
Metallization, *See* Interconnect
Metal–oxide–semiconductor (MOS), 157–186, *See also* Charge-coupled device (CCD); Complementary MOS; Integrated circuits (ICs); MOS transistor; Multigate MOSFET
    accumulation charge-layer thicknesses, 176–179
    components of charge in, 167
    C–V characteristics, 168–172
    inversion beyond threshold, 164–168
    inversion layer thickness, 176–179
    oxide charge, 172–174
    quantum mechanical effect, 176–179
    surface accumulation, 160–161
    surface depletion, 161–162
    threshold condition, 162–164
    threshold voltage, 162–164
Metal–semiconductor field-effect transistor (MESFET), 204–205
    GaAs MESFET, 204–205

Metal-semiconductor junction
   quantum mechanical tunneling, 141–142
   Schottky barriers, 133–137
   Schottky diodes, 138–140
      applications, 140–141
   thermionic emission theory, 137–138
Miller indices, 2
Miniaturization, 259
Minority carriers, 24
   distribution, 110–112
   injection, *See* Carrier injection
Minority drift current density, 333–335
Modulation doped FET, *See* High electron-mobility transistor (HEMT)
Moore's Law, 260
MOS transistors/MOSFET, 195–247, *See also* Complementary MOS (CMOS) technology
   body effect, 207–209
   drain, 195
   effective channel length, 226–228
   high-frequency performance, 230–232
   high-mobility FETS, 200
   MOSFET IV model, 210–214
      channel voltage profile, 212–214
   N-channel MOSFET, 198–199
   noises, 232–237, *See also* Noises
   output conductance, 229, 282–283
   parasitic source-drain resistance, 225–226
   P-channel MOSFET, 198–199
   properties analog applications, 229–237
   Q_{INV} in, 209–210
   series resistance, 226–228
   shallow-trench-isolation, 195, 198
   source, 195
   steep retrograde doping, 207–209
   structure, 196
   surface mobilities, 200–207
   velocity saturation, 219–220
   voltage gain, 230
   V_t, measurement, 207–209
Motion, of electrons and holes, 35–58, *See also* Diffusion; Drift
   dopant concentration and, 42
   kinetic energy, 35
   temperature dependence of, 42
   thermal, 35–38
Multi-chip modules, 83

Multigate MOSFET, 280–282
   double-gate MOSFET, 281
   triple-gate MOSFET, 282
Multilevel cell technology, 244
Multilevel metallization, 81
Multiple quantum well, LEDs, 126

N

NAND flash memory, 244
Nano-crystal NVM, 244
Nanoimprint, 68
N-channel MOSFET, 198–199
Negative resists, 64
Noises, MOSFET, 232–237
   cross talk, 232
   device noise, 232
   flicker noise, 235–236
   noise factor, 236–237
   noise figure, 236–237
   random telegraph noise, 235
   signal to noise ratio, 236–237
   thermal noise of a resistor, 233
   white noise, 233
Nonideal current, 115
Nonvolatile memory (NVM), 242–245
   charge-trap NVM, 244
   floating-gate NVM, 242–243
   metal-migration NVM, 245
   nano-crystal NVM, 244
   phase-change NVM, 244–245
Noyce, Robert, 260
N-type semiconductors, 5, 25–27, 37–38
   carrier concentration in, 28

O

Off-state current, 263
Ohmic contacts, 133, 142–145
   boundary conditions at, 144–145
Ohmic region, 211
OLEDs, *See* Organic light-emitting diodes
One-sided junction, PN junction, 96
On-state current, 217
OPC, *See* Optical proximity correction
Open-circuit voltage, solar cells, 121–123
Operating life test, 83
Optical fiber, 132
Optical phonon scattering, 43
Optical proximity correction (OPC), 66

Optoelectronic devices applications, 117–133, *See also* Diode lasers; Light emitting diodes; Solar cells
Organic light-emitting diodes (OLEDs), 126
Organic semiconductor materials, 121
Output conductance, MOSFET, 229–230, 282–283
Output power, solar cell, 123–124
Oxidation of silicon, 61–64
   dry oxidation, 61–62, 66–68
   horizontal furnace, 61–62
   vertical furnace, 61–62
   wet oxidation, 61–62, 66–68
Oxide charge, MOS capacitor, 172–174
   fixed oxide charge, 173
   interface traps, states, 173
   mobile oxide charge, 173

P

Parasitic source-drain resistance, 225–226
P-channel MOSFET, 198–199
Peak electric field, PN junction, 101
PECVD, *See* Plasma-enhanced chemical vapor deposition
Penetration depth, light, solar cell, 119
Phase change memory (PCM), 245
Phase-shift photomask, 66
Phonon scattering, 40
Photoconductor as light detector, 9
Photodiodes, 133
Photomask, 64
Photoresist, 64
Photovoltaic cells, 117, *See also* Solar cells
Physical vapor deposition (PVD), 77
Pinch-off concept, 212
   vs. velocity saturation, 225
Planar technology, 61
Planarization, 82
Plasma etching, 68
Plasma-enhanced chemical vapor deposition (PECVD), 78–79
Plasma process-induced damage, 70
PN diode IV characteristics, 112–115
   from depletion region contributions, 114–115
PN junctions
   capacitance-voltage characteristics, 98–100
   carrier injection, 105–107, *See also* Carrier injection
   charge storage, 115–116
   current continuity equation, 107–109
   depletion-layer model, 94–97
      depletion-layer width, 96–97
      in field, 94–96
      in potential, 94–96
   diode lasers, 128–133, *See also* Diode lasers
   excess carriers in forward biased, 109–112
   junction breakdown, 100–104, *See also* Junction breakdown
   LEDs, 124–128
   optoelectronic device applications, 117–133
   photodiodes, 133
   PN diode IV characteristics, 112–115
   PN junction theory, 90–94, *See also* PN junction theory
   reverse-biased PN junction, 97–98
   small-signal diode model, 116–117
   solar cells, *See* Solar cells
PN junction theory, 73, 89–145
   building blocks of, 90–94
      depletion layer, 90–91
      energy band diagram, 90–91
   built-in potential, 91–93
Poisson's equation, 93–94
Polycrystalline solids, 76
Poly-Si gate, 157–160
Poly-Si gate depletion, 174–176
Poly-silicon emitter, 300
Population inversion, diode lasers, 130
Positive resists, 64
Power consumption, CMOS, 218
Predeposition, 75
Primitive cell, 2
P-type semiconductors, 5, 25–27, 37

Q

Quantum efficiency, LEDs, 125
Quantum mechanical effect/tunneling, 141–142, 176–179
   tunneling probability, 142
Quantum well, LEDs, 126
Quasi-equilibrium, 52–54
   boundary condition, *See* Carrier injection
Quasi-Fermi levels, 52–54
   PN junction, 110–112

Quasi-static MOS C–V characteristics, 169–170
  measuring, 171–172
Quaternary semiconductors, LEDs, 126

R

Radiative recombination, 51, 124–125, 128–130
Random dopant fluctuation, 277
Random telegraph noise, 235
Random variation, CMOS technology, 277
Random variation, MOSFET, 277
Rapid thermal annealing (RTA), 75
Reactive sputtering, 77
Reactive-ion etching (RIE), 68–69
Recombination, of electrons and holes, 35–58
  direct, 51, 124–125, 128–130
  radiative, 51, 124–125
  rate of, 51
Rectifier, PN junction, 89
Rectifier, Schottky, 133, 137–141
Reliability, CMOS
  burn-in, 83
  dielectric breakdown, 271–272
  electromigration, 81
  hot carrier injection (HCI), 243
  oxide charge/interface traps, 173, 265–266
  qualification, 83
Renewable energy sources, 117
Resist, photo, 64
Resist strip, 64
Retrograde doping, MOSFET, 207–209, 272–273
Reverse-biased PN junction, 97–98
Reverse saturation current
  PN diode, 112, 114–115
  Schottky diode, 140
Richardson constant, 139
RIE, *See* Reactive-ion etching
Ring oscillators, 218
RTA, *See* Rapid thermal annealing

S

Saturation region
  BJT, 304
  MOSFET, 211–213
Saturation velocity, 219–220

Scaling, MOSFET in ICs, 259–263, 266–274
  channel length, 266–269
  depletion region width, 272–273
  gate oxide thickness, 269–271
  innovations enable scaling, 261–263
  junction depth, 75, 274
Scaling limit, a retrospective, 273
Scattering
  carrier, *See* Carrier scattering
  thermal motion, 36
Schottky barriers, 133–137
  heights, 133–134, 136
Schottky diodes, 133, 138–140
  applications, 140–141
Schottky source/drain MOSFET, *See* Metal source/drain MOSFET
Schrödinger wave equation, 13, 120
SCR, *See* Space-charge region
Selective epitaxy, 78–79, 261–262, 278
Semiconductors, insulators, 11–12
  carrier concentrations, 28–29
Semimetals, 12
Series resistance
  BJT, 312–314
  MOSFET, 226–228, 231
Shallow energy levels, 11
Shallow junctions, 75, 274–275
Shallow-trench-isolation, 195, 198
Shockley, William B., 2
Shockley boundary condition, 106
Short-channel effects, 229–230, 266–269
Short-channel MOSFET, 266–269
  vs. long-channel, 223
Short-circuit current, solar cells, 118, 121–123
SiGe base, BJT, 307–308
SiGe source/drain, *See* Strained silicon technology
Signal to noise ratio, MOSFET, 236–237
Silicon crystal structure, *See* Crystal structure, silicon
Silicon device fabrication process, 60–61, 83–84
  assembly, 82–83
  basic steps in, 60
  CMP, 81–82
  CVD, 77–78
  dopant atoms, introduction, 60
  epitaxial growth, 78–80

interconnect, 80–82
ion implantation, 70–72
lithography, 64–68, See also Lithography
oxidation of silicon, 60–64
plasma etching, 68–70
qualification, 82–83
selective oxide removal, 60
testing, 82–83
Silicon-on-Insulator (SOI), 278–280
Small-signal capacitance, 168
Small-signal diode model, 116–117, 116n5
Small-signal model, BJT, 310–312
Solar cells, 117–124
  direct-gap semiconductor, 119–121
  indirect-gap semiconductor, 119–121
  light penetration depth, 119–121
  open-circuit voltage, 121–123
  output power, 123–124
  short-circuit current, 121–123
  solar cells basics, 117–119
Solder bumps, 83
Solids, 76
  amorphous, 76
  crystalline, 76
  polycrystalline, 76
Source/drain junction, 226, 274, See also Shallow junctions
Source velocity limit, 228–229
Space-charge region (SCR) current, 114
Specific contact resistance, 144
Spontaneous emission, 128
Sputtering, 76–77
  reactive sputtering, 77
Stand-by power, 218
Static power, 218
Static random access memory (SRAM), 238–245
Steep retrograde doping, MOSFET, 207–209, 272–273
Step coverage problem, 77
Step junction, 90
Stepper, 66
Sterling's approximation, 330
Stimulated emission, diode lasers, 128
Stored charge
  BJT, 306–307, 315
  PN junction, 115
Strained silicon technology, 261–263

Stress effect, MOSFET, 261–262, 277
Subthreshold current, 263–268, 277–278
  interface states, effect of, 265–266
Subthreshold swing, 265
Surface accumulation, MOS capacitor, 160–161
Surface depletion, MOS capacitor, 161–162
Surface mobilities, 200–207
Surface roughness scattering, 202
Synchronous rectifier, 141
Systematic variation, CMOS technology, 277

T

Tandem solar cells, 123
Temperature dependence of carrier mobilities, 40–43
Temperature sensor, PN junction, 113–114
Ternary semiconductors, LEDs, 125
Thermal equilibrium, 16–19, 329–331
Thermal generation, 52, 115
Thermal motion of electrons and holes, 35–38
  collisions, 36
  direction change, 37
  mean free time, 36
  scattering, 36
  thermal velocity, 36
Thermal noise, MOSFET, 233–235
Thermal runaway, Schottky diodes, 141
Thermionic emission theory, 137–138
Thermoelectric generator, 37–38
Thin-film deposition, 75–80
  CVD, 77–78
  epitaxy, 78–80
  sputtering, 76–77
Thin-film transistors (TFTs), 76
Threshold condition, MOS, 162–164
Threshold current, diode lasers, 131
Threshold voltage, 162–164, 166–168
Time-dependent dielectric breakdown, 271
Transconductance, 213, 310
Transient enhanced diffusion (TED), 75
Transistor, 2
  BJT, 291–293
  MOSFET, 195–199
Transit time, 306–310

Triple-gate MOSFET, 282
Tunnel leakage, MOSFETS in ICs, 265, 270–272
    reducing gate insulator electrical thickness and, 270–272
Tunneling breakdown, PN junction, 101, 103
Tunneling, metal-semiconductor contact, 141–144

U

Ultra-large-scale integration (ULSI), 60
Ultra-thin-body SOI, 277–282
Unit cell, 1
    of the silicon crystal, 3
Universal effective mobility, 202

V

Vacuum level, 135, 159
Valence band, 8
VCSEL, *See* Vertical-cavity surface-emitting laser
Velocity overshoot, 219–220, 228–229
Velocity saturation, 43–44
    MOSFET, 219–220
        MOSFET IV model with, 220–225
    vs. pinch-off, 225

Velocity saturation region, 225
Vertical-cavity surface-emitting laser (VCSEL), 132
Vertically integrated semiconductor companies, 59
Very large-scale integration (VLSI), 60
Voltage transfer curve (VTC), CMOS inverter, 214–216

W

Wafer charging damage, 70
Wave nature of electrons, 13, 120
Wave vector, 13, 120–121
Wet etching, 68
Wet lithography, 61–62, 66–68
White LEDs, 128
Word line, 238–240
Writable optical storage, 132
Write/erase, nonvolatile memories, 242–244

Z

Zener diode, PN junction, 100
Zener protection circuit, PN junction, 101